NANO
meets
MACRO
Social Perspectives on Nanoscale
Sciences and Technologies

NANO meets MACRO

Social Perspectives on Nanoscale Sciences and Technologies

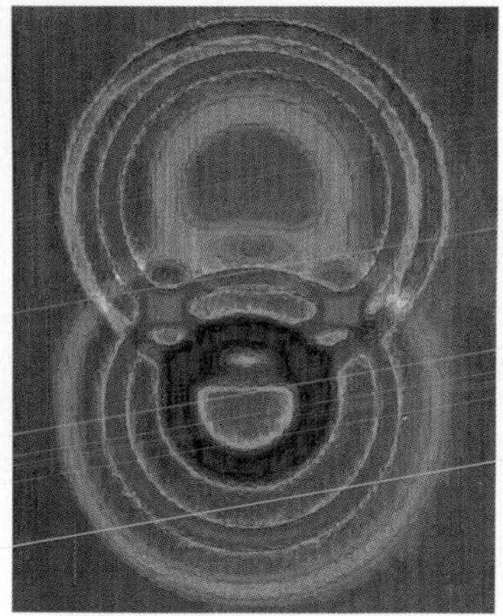

Editors

Kamilla Lein Kjølberg

Fern Wickson

University of Bergen, Norway

Pan Stanford Publishing

Published by

Pan Stanford Publishing Pte. Ltd.
Penthouse Level, Suntec Tower 3
8 Temasek Boulevard
Singapore 038988

Email: editorial@panstanford.com
Web: www.panstanford.com

British Library Cataloguing-in-Publication Data
A catalogue record for this book is available from the British Library.

NANO MEETS MACRO
Social Perspectives on Nanoscale Sciences and Technologies

ISBN-13 978-981-4267-05-2
ISBN-10 981-4267-05-8

Printed in Singapore by Mainland Press Pte Ltd.

Foreword

The Difficult Transition

The new emerging techno-sciences present huge perplexities. Their promise is great, even if it is not always predictable in detail. Perhaps their peril is even greater, but this is much more difficult to define. The idea of a world reduced to *grey goo* by nano-bots now seems far-fetched, but the dangers of nanoparticles passing the blood-brain barrier cannot be dismissed. We are now in a situation where it is impossible to hive off the downside of new developments as something extraneous or accidental, to be managed by some other less interesting sort of practice. All this is very confusing to the researchers participating in the enterprise because their training, equipping them to solve puzzles with very strict boundaries, has left them unprepared for this new sort of science.

For the emerging techno-sciences, it is important to recognize that the context of research impinges on the content in many ways. First there are the priorities that determine which sorts of issues will be investigated and which not. In this way, external forces influence not merely what we will know, but (equally important) that of which we will remain ignorant. Then there are the hazards of the work, which can run through a whole life-cycle, including the research and the materials (affecting research workers and the environment), going on to the manufacturing phase, eventually on to diffusion in the social and environmental contexts, and finally in becoming 'waste'. The sciences of these contextual aspects have traditionally come second, in prestige, rewards and influence, to those which created the problems along with the promise. But now they need high quality and equality with the mainstream sciences, lest we risk quite serious damage from unanticipated consequences. Also, there are now publics and pressure-groups that, with all their appreciation of the benefits of science-based technology, no longer trust researchers and authorities to guarantee their safety. They have power, through protest, politics and the market-place. Scientists who despise or dismiss them are in danger of finding themselves isolated and scorned.

In this context, Post-Normal Science (PNS) can make an important contribution. When scientists try to imagine all this complexity, as a new and urgent part of their job, PNS can be a guide to reflective practice. We say that PNS comes into play when "facts are uncertain, values in dispute, stakes high and decisions urgent". Those scientists who still live in world where they can be secure in their valueless facts do not need PNS – yet. But for the others, it can offer some security in a new and threatening situation. We imagine two axes, one of Systems Uncertainties and the other of Decision Stakes. When both are low, we have traditional Applied Science where safe routine is effective and appropriate. When either is medium, we have Professional Consultancy, as when the surgeon or consulting engineer faces an unpredictable nature and has the client's welfare in her hands. When either is high, as in the case of the new emergent techno-sciences, we are in a different realm. The most important difference is that the peer community who assess problems and solutions must be 'extended' beyond those with formal expertise. This is not only necessary for policies that are good in themselves and for policies that are legitimate in the eyes of the public, but also to ensure the quality of the scientific processes and products.

The transition from Normal to Post-Normal Science now takes place through a set of leading examples. Gene techno-science suddenly discovered that it was no longer normal, and was totally confused. With Nano techno-science, those who manage the development know full well that it's not normal although they are still not fully aware of the significance and the consequences. They are formally determined to avoid the backlash that was so destructive for gene technology but the ideology, attitudes and folkways of science that were so totally dominant for so long cannot be transformed in the space of a few years. Bringing a post-normal perspective to techno-science will require radical changes in the mindset of scientists and of those who sponsor or direct their work. The learning of the necessary lessons may not be totally free of pain.

With Nano techno-science, the severe Systems Uncertainties are known to all and denied by few. The great power of manipulating matter at the nano scale lies in its properties being unpredictably different from those at the meso and macro scales. But, as is universally acknowledged, this unpredictability creates enormous problems for the control of the operations, be they in the lab, in the factory or in the environment. As the pace of technological advance has quickened, it has become increasingly difficult for regulation, on behalf of Safety, Health and the Environment, to keep up or even to catch up. Although there have been some important voluntary initiatives (as between the Environmental Defense Fund and Du Pont[1]) and regulatory agencies all over the world are gearing up for their tasks, the activities of R&D and innovation gallop ahead, into the unknown and the unknowable.

The Decision Stakes are correspondingly high, and ill defined. Certainly, nano methods can transform technologies of all sorts, and can prove to be of immeasurable benefit in a great variety of fields. But the particles, once released into the natural and human environments, will have effects that are indeterminate and possibly undetectable, or at least undetected, in the short run. We do have a few experiments indicating that nanoscale particles can interfere with some physiological processes. While these can serve as general warnings, they give only limited guidance for strategic governance interventions.

We can imagine a 'too late' scenario, when it is retrospectively discovered that some widespread ailments or environmental pathologies can be attributed to particular nano-products. It is inevitable that in retrospect specific early warnings can be discovered, and hence blame can be assigned to scientists, managers and regulators. Depending on how serious the crisis is, there could arise a reaction against the whole

---

[1] See http://www.edf.org/article.cfm?contentID=4821

enterprise. If the social class of 'banksters' were joined by 'sciensters' in public opprobrium, then that core enterprise of our civilisation would enter a new phase in its evolution.

None of that need happen, of course; and it is possible that the huge benefits of nano techno-science will be realised with only minor damages as their price. But we will not know, and cannot know, until such a time as the technology has settled down and the regulators caught up. That moment itself may never arrive, especially if nano techno-science quickly yields its leading position to other, more powerful and exotic technologies, such as synthetic life.

The futurist Ray Kurzweil has a vision of the convergence of all the exciting technologies, when we will have powers exceeding those ever dreamed of by magicians or even science-fiction writers[2]. That vision would seem to be a highly optimistic extrapolation from normal science, when all the puzzles come together in a creative synthesis. We could equally well imagine a convergence of the Systems Uncertainties and Decision Stakes of PNS, when the powers and perils of the sciences that come to constitute our civilisation are inextricably mingled. Citizens will need just to hope for the best, and a constructive opposition to particular innovations may come to be seen by some as fruitless.

Judging by previous history, including nuclear weapons up to now, there is a chance that we will be able to muddle through. As a result of many initiatives like this admirable volume, we may arrive at a situation where scientists, managers and society will have learned to proceed with some genuine caution, and come to understand that it's alright to say no, when there is a danger that the attractions of progress might prove fatal.

Silvio Funtowicz & Jerry Ravetz

---

[2] See, for example Ray Kurzweil (2005) The Singularity Is Near: When Humans Transcend Biology, Viking Penguin

PREFACE

This book is a fortunate spin-off of our own process of trying to come to terms with the rich field of research on social and ethical aspects of nanoscale sciences and technologies. At various times, and in a range of different settings, we have both been asked to teach, present or give an overview of this multifaceted body of work. This book results from our various attempts to find good ways of doing this.

While we come from very different backgrounds (e.g. one of us was raised on a farm in Australia while the other grew up in Norway's largest city), we share a range of academic interests. Both of us took degrees to support an interest in environmental problems and decision-making (Kamilla in Environmental Management and Fern in Political Science and Ecology). Both of us also went on to separately conduct research on genetically modified crops and how they should be governed in the face of uncertainty. We have brought these shared interests in issues of environment, governance and uncertainty, to our research on nanoscale sciences and technologies (nanoST), and they remain important to us.

We were brought together in 2006 at the Centre for the Study of the Sciences and the Humanities (SVT), University of Bergen, to work on a new project entitled "Interdisciplinary Studies of Ethical and Societal Implications of Nanotechnology". This project had a specific emphasis on fostering interaction, dialogue and research across social and natural sciences, and included project partners from a range of different disciplines and countries. Along with most of the project partners, we were new to the field of nanoST. At the initial project meeting, however, we were asked to provide the group with an overview of research conducted on social and ethical aspects of nanoST.

Although we were quite overwhelmed by the richness of the literature in this field, it was still at an early enough stage of development for us to be able to conduct, what was at the time, a relatively comprehensive literature review. This review involved reading all the works that we saw as part of the emerging field of research on social and ethical aspects of nanoST and finding a way to understand and present their diversity to those new to the topic. Through this process, we began to see the works clustering around some common themes and began to categorise them accordingly. This lead us to see value in presenting the field of social and ethical research on nanoST in four categories – Perception, Governance, Philosophy and Science.

After having presented our conceptual map of the field to our project partners, we also presented and discussed it with colleagues and researchers at a small conference and another workshop. We received valuable feedback during these exchanges and were encouraged to publish the work as an academic paper. Our review was published in the journal *NanoEthics* in 2007[1]. In early 2008, we were approached by Pan Stanford Publishing to turn our article into a book. We saw this as a wonderful opportunity to revisit and improve our ideas from the early phase of the project, and took little time to agree to the proposal.

Of course (and fortunately!), after three years of research, the way we understand the field and approach the presentation of it have changed somewhat. During this time, the field itself has also further developed and expanded rather dramatically. The change in our own thinking is perhaps most evident in the way in which we have chosen to rename the different nodes in this book (and indeed in the fact that we now use the more flexible concept of 'nodes', rather than the more rigid notion of 'categories' that we originally used in the article). In this book, the different nodes of social research on nanoST are presented as: *Nano meets Macro*: *In the Making* (formerly Science); *In the Public Eye* (formerly Perception); *In the Big Questions* (formerly Philosophy); and *In the Tough Decisions* (formerly Governance).

---

[1] The article is published as Kjølberg, K. and Wickson, F. (2007) Social and ethical interactions with nano: Mapping the early literature, *NanoEthics*, 1, pp. 89-104.

For *Nano meets Macro*, we have not performed a new review of the field. Instead we have invited authors working on topics across the nodes to contribute with chapters. We specifically aimed for diversity among chapter authors in the sense that they come from different disciplines, different parts of the world and are at different stages of their research careers. In addition to these chapters, the book also contains ten artworks, three short science fiction stories and some poetry, as examples of non-academic commentary and social perspectives on nanoST. Three of the artworks were commissioned, while the others were selected from among the rich flora of existing 'nanoart'. The short stories were successful submissions in a call put out specifically for the book and the poetry was commissioned. While *Nano meets Macro* provides a varied selection of what we believe are important themes and perspectives in the field, it should not be seen as a comprehensive presentation of social aspects of nanoST. Our aim has rather been to show some of the existing diversity in this rich field, and to perhaps provide a useful conceptual apparatus for understanding the field as a whole.

The process of creating this book would not have been possible without help from a number of people and organisations. Firstly, we would like to thank all our authors and artists for agreeing to be involved in this project. We also thank the Norwegian Research Council and the University of Bergen for providing the necessary supporting funding. We thank all our project partners, and the participants at the NANOMAT ELSA conference in 2007 for feedback on our early ideas and the encouragement to develop them further. For detailed commentary on our original article, we thank the anonymous reviewers of *NanoEthics*. To all the staff at SVT, particularly Roger Strand, we extend our gratitude for regular stimulating academic conversation and an enjoyable working environment. We are also of course particularly grateful to Pan Stanford Publishing for their interest in our work and assistance in bringing this book into existence. Finally, we would like to thank our families, friends (and pets) for helping to maintain our sanity when we started to lose ourselves in the sometimes painful, sometimes all absorbing, but always rewarding, process of creating this book.

Kamilla Lein Kjølberg & Fern Wickson

Author Biographies

Lachlan Atcliffe is an Australian-born, currently England-based lawyer. A part-qualified barrister, he recently completed postgraduate study at Bristol University on the legal implications of radio-frequency identification tags. He suspects that no matter how advanced or widespread nanotechnology becomes, it will still have to carry safety labels saying 'Caution: Do not attempt to eat'.

Tore Birkeland is a PhD student in nanoscience at the Department of Mathematics, University of Bergen (Norway). His research interests are numerical methods for atomic physics, and software development issues in computational science. He is the main developer of the atomic physics software package pyprop, which is an attempt to create a common framework for solving many different problems in atomic physics.

Diana Bowman is a Senior Research Fellow in the Monash Centre for Regulatory Studies, Faculty of Law, Monash University (Australia), and a Visiting Fellow in the Institute for Energy and Environmental Law, Faculty of Law, KU Leuven (Belgium). She has undergraduate degrees in both science and law, and has recently completed a PhD. Dr Bowman's research focuses primarily on regulatory and policy issues related to nanotechnology and other new technologies. She was a co-editor of *New Global Frontiers in Regulation: The Age of Nanotechnology*. Recent projects have included assisting the Australian Government with their *Review on the Possible Impacts of Nanotechnology on Australia's Regulatory Frameworks,* and the Food and Agricultural Organisation with their issues paper on *Nanotechnology applications in food and agriculture sectors: Principles and guidance for food safety regulation.*

Sarah Davies is a Research Associate at Durham University. Her research interests focus around 'public engagement' with science. Dr Davies' previous research has examined scientists' talk about publics and the practice of dialogue events. Her current research, on the EU-funded project DEEPEN (Deepening Ethical Engagement and Participation with Emerging Nanotechnologies) involves the analysis of lay ethics of nanotechnologies. She has a BSc in Biochemistry (2001), a MSc in Science Communication (2003), and a PhD (2007) on public dialogue on science, all from Imperial College, London. Dr Davies has also worked in exhibition development at the Science Museum, London, and taught undergraduate science students about communicating science.

Gian Carlo Delgado is a researcher specialized in societal, ethical, legal and environmental aspects of nanotechnology at the Interdisciplinary Research Centre on Sciences and Humanities, National Autonomous University of Mexico (UNAM). Dr Delgado is a member of the National Research System of the Mexican Science and Technology Council (CONACYT), and the editor of the journal "Mundo Nano", published by a group of research entities of the UNAM. He lectures at the Political and Social Sciences Faculty and at the Postgraduate Programme on Latin American Studies; both at the UNAM. Some of his work can be seen at www.giandelgado.net.

Susan Dodds is Professor of Philosophy and Dean of the Faculty of Arts at the University of Tasmania. She is a chief investigator on the Australian Centre for Excellence in Electromaterials Science (ACES), funded by the Australian Research Council. Professor Dodds leads the Ethics program within the ACES.

Robert Doubleday is a Lecturer in Environmental Policy at the Department of Geography, University of Cambridge (UK). His research is concerned with relations between science, citizenship and the governance of emerging technologies. Previously Dr Doubleday was based at the University of Cambridge's Nanoscience Centre, where he worked collaboratively with nanoscientists to explore the social dimensions of their work. He is currently working on a Wellcome Trust funded project that investigates how scientists working in the field of nanobiotechnology make strategic choices about their research careers.

Arianna Ferrari studied Philosophy in Milan (MA 2000) and completed her PhD with a double affiliation at the University of Tübingen (Germany) and the University of Torino (Italy). The topic of her dissertation was ethical and epistemological aspects of genetic modification of animals in biomedical research. Since 2006 Dr Ferrari has been working in the EU project DEEPEN (Deepening Ethical Engagement and Participation in Emerging Nanotechnologies). Her areas of interest include: animal and environmental ethics, bioethics, philosophy of technology, STS, nanoethics.

Arnout R. H. Fischer is Assistant Professor with the Marketing and Consumer Behaviour Group at Wageningen University (the Netherlands). He aims to understand public behaviour through social psychological approaches. Dr Fischer has published on consumer behaviour in relation to food safety and food technologies in academic books and peer reviewed journals. He is involved in several projects and committees in relation to public acceptance of nanotechnology in the Netherlands and internationally.

Hans Fogelberg is a researcher at the STS section of the Department of Sociology at the University of Gothenburg (Sweden). He has a Ms.Sci.Eng. in mechanical engineering and a Tekn.Lic. in History of Technology from Chalmers University of Technology. He completed a PhD in sociology of technology at University of Gothenburg (2000). Dr Fogelberg's main research areas are the sociology of innovation and the history of technology. These research interests include the analysis of innovation in relation to the knowledge economy and risk society. His main empirical fields are sustainable technologies, advanced road transportation technologies, and nanotechnology.

Lynn J. Frewer is Professor of food safety and consumer behaviour in the MCB group in the Social Sciences Department at Wageningen University (the Netherlands). Previously, Professor Frewer was Head of Consumer Science at the IFR in Norwich (UK). She has a background in psychology and has published extensively in the area of risk perception and communication.

Frøydis Gillund holds a Master's degree in Ecology and Natural Resource Management from the University of Life Sciences (Norway). She is currently a PhD student at 'GenØk – Centre for Biosafety'. Her PhD project: 'Genetic Engineering in Aquaculture: Perspectives on Management and Sustainability' focuses on how scientists and policymakers acknowledge and deal with scientific uncertainty related to the introduction of novel technologies in complex ecosystems. Her field of interest is natural resource management, philosophy of science and ethical and social implications of science and technology.

Mercy W. Kamara is a researcher at the Department of Communication, Business, and Information Technology, Roskilde University (Denmark). Her research focuses on 1) the relationship between policy, media representations and public perceptions with regard to controversial technologies; 2) the relationship between science funding policies, scientists' virtues and motivations, and scientific developments; and 3) the nature of public controversies involving sustainable development.

Matthew Kearnes is a RCUK Fellow at the Institute of Hazard and Risk Research (IHRR) and the Department of Geography at Durham University (UK). Since gaining his PhD in Human Geography from the University of Newcastle (Australia), Dr Kearnes has held positions at the Open University and Lancaster University before commencing his current position at Durham in 2006. His research work has focused on understanding the governance of contemporary scientific and ecological practices – particularly the co-production of science and society. Drawing on a background in science and technology studies and social and cultural geography, his recent research has focused particularly on the societal and ethical dimensions of nanotechnology and synthetic biology. He is currently coordinating an ESRC Project entitled: *Strategic Science: Research Intermediaries and the Governance of Innovation*. He is also a co-investigator on the European project DEEPEN (Deepening Ethical Engagement and Participation in Emerging Nanotechnologies) and the ESRC Seminar Series: *Critical Public Engagement*.

Kamilla Lein Kjølberg has a broad research interest in responsible environmental governance. She holds a Master's degree in Natural Resource Management from the University of Life Sciences (Norway), with a double specialisation in Tropical Ecology and Ecological Economics. Her master's thesis was called "When Experts Disagree", and dealt with expert advice on the deliberate release of genetically modified (GM) crops. She has worked with issues related to scientific uncertainty of GM aquaculture as a researcher at 'GenØk - Centre for Biosafety'. For a year she held the position as editor of Gennytt, an online newsletter about GM agriculture. She now works for the Centre for the Study of the Sciences and the Humanities, University of Bergen (Norway), with a PhD dissertation on Nanotechnology and Responsibility.

Renée Kyle is a researcher for the Office of Women (Australian Commonwealth Department of Families, Housing, Community Services and Indigenous Affairs). She was previously a post-doctoral Research Associate at the Australian Centre for Electromaterials Science, funded by the Australian Research Council.

Hildegard Lee is a pseudonym, employed to shield a scientist dabbling in amateur poetry from harsh critique and ridicule. Hildegard finds that expressing herself in poetry poses substantial challenges equal to, and sometimes exceeding, those she faces when expressing herself scientifically. She does, however, enjoy the way this alternative medium opens spaces for her to reflect on the nature of her work. She sometimes questions her choice of pursuing a career in scientific research and occasionally feels that her work is not allowed to benefit enough from her creativity. She would, for example, be very grateful if she could attend conferences and present her work in song, dance or theatre, rather than in powerpoint slides.

Phil Macnaghten is Professor of Geography at Durham University (UK). From a background in social psychology, his research focuses on the cultural dimensions of environmental and innovation policy and their intersection with everyday practice. Professor Macnaghten's current research addresses questions of public participation and the governance of emerging technologies. He is presently leading the European Commission FP6 project DEEPEN (Deepening Ethical Engagement and Participation in Emerging Nanotechnologies).

Colin Milburn is Associate Professor of English and a member of the Science & Technology Studies Program at the University of California, Davis (USA). His research focuses on intersections between science, literature, and media technologies. Associate Professor Milburn is the author of *Nanovision: Engineering the Future* (Durham: Duke University Press, 2008), and he is currently completing a new book about the convergence of video games and the molecular sciences, entitled *Mondo Nano: Fun and Games in the World of Digital Matter*.

Georgia Miller has been the national coordinator of the Friends of the Earth Australia Nanotechnology Project since 2005. She has been engaged in environment and social change campaigns since the mid-1990s, including in relation to GMOs. Georgia is particularly interested in supporting greater public involvement in science policy and decision making, and in making technology development more responsive to social and environmental needs. She has an Honours degree in Environmental Science.

Anne Ingeborg Myhr is employed as a scientist at 'Genøk - Centre for Biosafety' in Tromsø (Norway). She holds a Master's degree in Biotechnology from NTNU (Norway), and a PhD from the University of Tromsø. The title of her PhD thesis is "Precaution, Context and Sustainability. A Study of How Ethical Values may be Involved in Risk Governance of GMOs". Dr Myhr's present research engagements are within the use of genetic engineering and nanotechnology. She is especially interested in governance of risk and uncertainty and how emerging technologies can contribute to sustainability. She is also involved in GenØk's capacity building in risk assessment and management of GMO use and release in the Third World.

Alfred Nordmann received his Ph.D. in Hamburg (1986) and served on the faculty of the Philosophy Department at the University of South Carolina (1988-2002). Following this he became Professor of Philosophy and History of Science at Darmstadt Technical University (Germany) where he founded the nanoOffice. Professor Nordmann's historical interests concern the negotiation of contested fields of scientific knowledge, such as theories of electricity and chemistry in the 18th century, mechanics, evolutionary biology and sociology in the 19th century, nursing science and nanoscale research in the 20th century. Since 2000, Professor Nordmann has been studying philosophical and societal dimensions of nanoscience and converging technologies. His focus, in particular, is on changes in the culture of science and the changing relationship of science, technology, nature and society. From October 2006 to September 2007 he was one of the coordinators of the research group 'Science in the Context of Application' at the ZiF of Bielefeld University (Germany). He remains closely associated as visiting adjunct professor with the Philosophy Department and NanoCenter of the University of South Carolina (USA). In the spring of 2009, he holds the Alcatel-Lucent Fellowship at the IZKT of the University of Stuttgart (Germany).

Rye Senjen has been an active campaigner for social justice for the last 30 years. She has been involved in campaigns on women's issues (including against the trafficking of women) and in the peace movement. Recently she has become very concerned about the potentially devastating effect of nanotechnology on society, the environment and all beings. She has been a member of the Friends of the Earth (FoE) Australia Nanotechnology project, since its inception in early 2005. In late 2006 Dr Senjen was invited to be an expert member on the UNESCO panel for nanotechnology and ethics. In 2008 she co-authored, together with Georgia Miller, the groundbreaking FoE report on nanotechnology and food. She has a life long interest in technology and its effect on society and has worked extensively in the telecommunications industry. She has published numerous scientific articles and a book about the Internet. She holds an Honours degree in Entomology and Horticultural Science, and a PhD in Artificial Intelligence.

Hope Shand is former Research Director of the Canadian-based ETC Group. Hope has conducted research, writing and advocacy work related to agricultural genetic resources and the social and economic impacts of new technologies for the past three decades. With ETC Group (and formerly with RAFI), she has monitored corporate concentration in the life sciences, especially the global seed industry. She lives in Chapel Hill, North Carolina (USA).

Lesley L. Smith lives and writes in Boulder, Colorado (USA). She specialises in hard science fiction and her short fiction has appeared in a variety of venues including 'Analog Science Fiction and Fact'. She is currently hard at work on her third novel, tentatively entitled 'Time Dream'. For more information please visit www.lesleylsmith.com.

Robert Sparrow is a Senior Lecturer in the School of Philosophy and Bioethics at Monash University (Australia). His current research interests include the ethics of nanotechnology, military robotics, human enhancement, and ethical issues surrounding cochlear implants. He has a BA (Hons) from Melbourne University (Australia) and a PhD from the Australian National University.

Roger Strand is Professor and Director of the Centre for the Study of the Sciences and the Humanities, University of Bergen (Norway). He holds a PhD in biochemistry. Professor Strand's research mainly falls within the philosophy of natural science and biomedicine, including research on the ethical and social aspects of bio- and nanotechnology. The focus of his research is on the nature and significance of scientific uncertainty and complexity for environmental and health-related decision-making processes. He is a member of the National Committee of Research Ethics of Natural Science and Technology in Norway.

Geert van Calster co-directs thc Institute of Environmental and Energy Law at KU Leuven (Belgium), where he teaches and researches environmental law, energy law, European economic law, and international trade law. Professor van Calster is a former visiting professor at Erasmus University, Rotterdam (the Netherlands), where he held the chair of international economic law, and in EU law at Monash University (Australia). He was called to the Brussels Bar in February 1999. He is now of counsel (practising) in the DLA Piper Brussels office. He is cited in the Legal 500 as being "well regarded for the interface between economic and regulatory law", and as having an "excellent academic background aligned to pragmatic experience in both public and private sectors." Professor van Calster's current research interests include the regulation of new technologies (nanotechno logies in particular, heading a 5 year research project), and climate change law. His editorial works includes Carbon & Climate Law Review; Nano technology, Law and Business; and Law, Probability and Risk. He directs the Master programme on Energy and Environmental Law at the KU Leuven and is a tenured chair of the Research Fund , KU Leuven, and was a visiting lecturer at Oxford University (September 2006 – September 2008).

Ana Viseu is Assistant Professor in Communications & Culture, and Science and Technology Studies at York University (Canada). Anchoring her work on feminist technoscience and cyborg anthropology, she is interested in ethnographies of the practices of development and use of emergent (and contested) technologies, from both theoretical and material perspectives. Her goal is to understand how particular notions of embodiment, agency, science and technology are reified, created and enacted in emergent technologies. Assisstant Professor Viseu has addressed these issues in two complementary lines of research: wearable computers and nanotechnology. She has published in a number of venues, including the journals of *Ethics & Information Technology* and *Information, Communication and Society.* She is currently engaged in a project that seeks to conduct a discourse analysis of the field of nanomedicine so as to examine the figures of the body that are emerging and being reified within it.

Fern Wickson is a cross-disciplinary scholar with research interests in environmental philosophy and decision-making, the politics of risk and uncertainty, and the governance of emerging technologies. She completed her PhD across the Schools of Biological Sciences and Science, Technology and Society at the University of Wollongong (Australia). Her dissertation was entitled "From Risk to Uncertainty: Australia's Environmental Regulation of Genetically Modified Crops". Prior to this, Dr Wickson undertook a Bachelor of Arts/Bachelor of Science double degree at the Australian National University and an Honours degree in Environmental Politics at the University of Tasmania (Australia). She has lectured across the disciplines of history, politics, philosophy, science and technology studies, biology and engineering.

Lupin Willis is a citizen of the earth's biotic community. When she is not busy growing organic vegetables, campaigning for the protection of oldgrowth forests, and contemplating deep ecology, she likes to write fiction as a way to stimulate reflection on the pace and direction of technological development. She prefers books to television, hand written letters to txts and fruit to fast food.

Contents

Introduction

Nano science and technology is often said to be a 'revolutionary' new field of development with the potential to radically change our lives. But what does this actually mean? What is nano science and technology exactly? Where did it come from? Where is it going? Who's leading this so-called 'revolution'? How could it impact our lives? Do we want it? Do we need it? And; who decides these things? This book is about all these questions, and more. In other words, it is a book about the places and interfaces where *nano* science and technology meets *macro* social phenomena. All the perspectives offered in this book represent different narratives or stories, about what nano is, and what it means in a social context.

Our primary motivation in assembling this collection has been to showcase a diverse range of social views on nano in a way that might help stimulate and facilitate dialogue between those who do nano science and technology and those who study it as a social, political and cultural phenomenon. An additional motivation has been to illustrate how social perspectives on nano science and technology are expressed not just in academia but also in fields such as literature and art, and to try to present all these different perspectives in a way that makes them accessible for a broad audience.

In trying to juggle these multiple motivations, we have structured the book around four general nodes of interest. These are: *Nano meets Macro – In the Making; In the Public Eye; In the Big Questions*; and *In the Tough Decisions*. We believe that this framework helps to organise the wonderful richness and diversity of commentary in this field. Hopefully, the various perspectives offered, can inspire, encourage and assist readers in reflections on the questions posed above, and thereby make the book an interesting, informative and enjoyable read for all!

1. Nanoscale Sciences and Technologies

To attempt a short introduction to 'nano' science and technology is a daunting task. The field in question is incredibly broad and diverse, and both its definition and scope are hotly contested. The breadth of the field begins with the fact that the term 'nano' is used to refer to both 'science' and 'technology'; where 'nanoscience' appears across a wide range of scientific disciplines (physics, chemistry, biology, materials science, computer science, etc.) and 'nanotechnology' has potential application in a range of different sectors (energy, transport, medicine, textiles, communications, etc). To add to this complexity, both present and potential future applications are commonly entangled in the same conversations. Present applications of 'nano' science and technology include things like transparent sunscreens, antibacterial kitchen utensils and stain resistant clothing. More far reaching future visions include the ability to construct absolutely anything through the precise placement of individual atoms; a scenario that thrills the optimists as much as it frightens the pessimists.

In the simplest sense, 'nano' comes from the Greek word for dwarf and refers to the scale length of one billionth (10^{-9}), e.g. one nanometre (nm) equals a billionth of a metre. Common popularisations to help imagine this size include: a nanometre is how much your fingernail grows in a second, or a sheet of paper is around 100 000 nm thick. Contestation begins when one tries to define the boundaries of what constitutes 'the nanoscale'. For many people, the term 'nanoscale' refers to a range between 1 – 100 nm. However, both the beginning and the end of this range remain subject to ongoing debate. Some claim that it should extend as low as 0.1 nm (because atoms and some molecules are smaller than 1 nm) and as high as 300 nm (because the unique properties of the nanoscale can also be observed above 100 nm). The boundaries of 'the nanoscale' are highly significant in both scientific and political terms because they have the possibility to affect everything from funding, to risk assessment and product labelling. The high stakes involved here are an important reason why the definition of the nanoscale continues to be contested. As editors of this book, we use the term 'nanoscale' in a broad and inclusive sense, with a specific aim not to exclude relevant objects of study by referring to a particular range of numbers rather than to pertinent characteristics. Readers are, however, encouraged to keep the ongoing debate about the exact meaning and range of 'the nanoscale' in mind as they go on and read the various contributions in this book.

To capture and acknowledge the diversity that exists in the breadth of work occurring at the 'nanoscale', it is often referred to in the plural as either *nanosciences and nanotechnologies* or *nanoscale sciences and technologies*. In this book, you will find that the authors of different chapters use various terms. While we have encouraged them to explain their choices, as editors we have left this plurality because we feel it accurately represents the diversity that remains in the field. In the title of this book, as well as in the introductions, we have chosen to use the term 'nanoscale sciences and technologies' (nanoST). We do this so as to emphasise that the book deals with a range of different sciences and technologies and to not exclude any of the definitions and foci adopted by our authors.

The fact that 'nano' effectively only refers to a unit of measure is an important reason why a number of extremely different projects and products are collected under the label of nanoST. Just think about the diversity that a concept like 'metre' science and technology would include – all the science and technology that takes place in the size range of 1-100 metres! That would be a very diverse and broad field indeed. Given the incredible diversity in nanoST then, what is it that we are actually talking about? What is it that binds all of these different things together? The reason for talking about 'nano' sciences and technologies as something distinct and unique is that objects at the nanoscale may express different properties from those expressed by larger objects of the same material. Properties, such as colour, conductivity, reactivity and melting point, can all change at the nanoscale. A classic example of this is the way in which nano sized particles of gold appear red and are reactive rather than inert. This expression of novel properties is typically explained by both the presence of quantum effects at this scale (the peculiar phenomena of the subatomic level that are not adequately explained by classical physics), and the increase in surface area to volume ratio that occurs (when an object is divided into smaller pieces, the volume stays the same but more surfaces are created, which often enhances reactivity). Properties of larger objects are also related to nanoscale atomic configuration. For example, both graphite and diamond are made of carbon atoms, but these materials have very different physical properties because of the way in which the atoms are arranged (in sheet form for graphite and in a tetrahedral shape for diamond). One of the early areas of significant development in nanoST was the discovery of, and ability to fabricate, a different atomic structure for carbon, including soccer ball like shapes (fullerenes) and cylindrical tubes (carbon nanotubes). Carbon nanotubes are a 100 times stronger and six times lighter than steel, and can have very high conductivity. They are therefore a good example of the way in which restructuring atoms at the nanoscale can create materials with novel properties. In most definitions, it is the ability to employ, engage, and/or manipulate the novel properties of the nanoscale that is crucial for something to count as nanoST. This means that for many people, nanoST are seen as not just working on the nanoscale, but actively investigating and utilising the novel properties that are in effect there.

This book has a particular focus on *social perspectives* on nanoST. To explain what this means for us and why we think these types of perspectives are important, we will first take a step back and talk more generally about different visions of the relationship between science and society. Here we will indicate some of the ways in which different visions of the relationship between science and society have been dominant in public understandings and political orientations at different points in time, as well as indicate what visions we see rising to prominence around the time that nanoST are emerging. It is important to make clear that although we highlight historical shifts in dominant visions, we certainly do not wish to argue that only one vision has been active at any one time. We believe that different visions of the relationship between science and society very often co-exist (sometimes across different arenas) and tend to compete with each other for prominence and political support. Following our description of different visions of the relationship between science and society, we will discuss the emergence of the concept of ‘ELSA research’ (research on Ethical, Legal and Social Aspects of science and technology) and we present this as another key element in our story of the importance of social perspectives on nanoST. In our description of how this concept became prominent, we also sketch how our notion of 'social perspectives' both incorporates and expands that of ELSA.

While we have chosen to tell this particular story as a way to describe the context and broad framing of this edited collection, this story is of course informed by our own backgrounds and interests. It may therefore not be the same story that others would tell in an introduction to social perspectives on nanoST. As an introduction to this book, however, our version serves to provide an account of the framework within which we as editors have approached the topic.

2. Visions of the Relationship between Science and Society

That science has always interacted with society seems obvious, and yet, the two spheres are often conceptually separated. This separation is supported by powerful ideas of what science is and how society (particularly politics) should relate to it. According to a traditional view, science is seen as an activity that is objective and uninfluenced by beliefs, interests or biases. As a generator and provider of 'objective facts', science is seen as most efficient if largely left alone by society (and particularly politics) to pursue its course towards truth. Many scholars have, however, argued that this conceptual separation between science and society is no longer useful or applicable. Bruno Latour[1] is someone who has been particularly concerned with the conceptual separation between nature (the object of science) on one side and society (the object of politics) on the other. He suggests that this separation (that has been so efficient in bringing about our modern way of living), has now created problems that are not possible to classify as simply one or the other (i.e. as scientific or political). Latour argues that in order to approach these types of problems, we have to admit that 'pure nature' and 'pure society' never really existed, or in his words, that 'we have never been modern'.

---

[1] Latour, B. (1993). *We Have Never Been Modern*. New York: Harvester Wheatsheaf.

We support the position that to overcome many of the social and environmental challenges society is faced with today, it is crucial to recognise the ways that science and society interact and co-create each other. To point to some of the historical shifts and theoretical work informing this position, we will now present a brief (and therefore inevitably rather superficial) overview of varying visions on the science/society relationship. In presenting this overview, we try to show why there is now a trend moving away from the dominance of a story that sees science and society as separate, independent spheres, to one where they are seen as entangled in a mutual process of co-production. We encourage those interested in more detailed descriptions of these visions to refer to the cited literature, as well as to the host of other works that can be found commenting on this topic.

2.1 Science: From a gentleman's activity to the way to win a war

Science,[2] in its early history (particularly in 18th and 19th century England), has been characterized as an activity predominantly carried out by 'gentlemen' – independent, wealthy men (or men supported by wealthy benefactors), who had the time and money available to engage in research activities. Science was therefore seen as a predominantly curiosity driven quest for knowledge undertaken by privileged individuals. Although some individuals engaged in early scientific research already saw its potential for application and social benefit,[3] they usually did not have the required resources and/or extensive theoretical grounding to see this social benefit realized (e.g. in the way that industry was later able to do with such incredible success). 'Technologies'

---

[2] We will not provide a precise definition of science here, as the definition itself will vary in the visions we are about to describe. In a very general sense, however, it can be thought of as referring to systematic processes of rational inquiry into the nature and behaviour of the physical world that result in knowledge capable of predictive explanation.

[3] For example Francis Bacon (1561-1626) was a particularly early example of someone who intellectually anticipated the application of science for social benefit while Isaac Newton (1643-1727) was a rare example of someone able to apply their scientific knowledge to technological development (e.g. in his telescope).

(interpreted broadly as tools) were certainly in use during this period. They were, however, primarily products of craftsmanship and tended to be used as instruments to help create scientific knowledge rather than being products created by it.

It is common in stories about the relationship between science and society to point to the importance of World War II (WWII) for creating a significant shift in the previously dominant vision. While the development of technologies based on scientific knowledge arguably occurred in fields such as chemistry and physics from at least the mid 19th century (e.g. in the development of dyes and electric power), this shift was given an enormous boost after WWII. During the war, science was actively mobilised by the nations involved to solve technical problems paramount in war (such as the development of superior weaponry, communications, medical treatment etc.) The active use of science and the development of science-based technologies in WWII are seen to have created a significant shift in the understanding of the relationship between science and society. This shift is often specifically linked to a series of letters exchanged between the US President Roosevelt, and the director of the US Office of Science and Development, Dr Vannevar Bush.

The key question posed by President Roosevelt to Dr Bush was: After science had been so successfully employed in advancing the war effort, how might it contribute to the betterment of society in times of peace? In other words, how could the government assist the advancement of science so that it might continue to offer benefits to society? Vannevar Bush's response was documented in a report entitled "Science – The Endless Frontier".[4] In this report, Bush recommended that the President invest in "basic scientific research", arguing that the generation of new knowledge was the crucial first step towards science-based technologies, and the subsequent social benefits that were assumed to follow. Basic research was described by Bush in this report as essential "scientific capital".

---

[4] Bush, V. (1945). Science: The Endless Frontier http://www.nsf.gov/od/lpa/nsf50/vbush1945.htm (last accessed 26.05.09).

The post WWII conception of the relationship between science and society is often referred to as "the linear model"[5] or as the "traditional social contract of science", where social contract refers to an unwritten set of mutual expectations. In this model, science and society (often understood as politics) are still perceived as distinct and separate spheres. There is, however, a unidirectional or linear exchange between these spheres: society puts money into science and science gives society the resulting knowledge and innovation. Between these stages of exchange, society allows science to conduct unhampered research, with the idea that the serendipitous nature of scientific investigation requires freedom from social/political control and constraints. It is important to emphasise that according to this model, science is seen as existing in two distinct forms: 'Academic Science', where public money is invested in basic research, and 'Industrial Science' where the 'scientific capital' of basic research is combined with the financial capital of industry and business to generate products and applications. According to this model, the primary role of politicians and the state is not to directly invest in the development of technological applications and innovations (seen as the preserve of industry and the market), but in the development of the necessary scientific capital for this, i.e. basic research. Importantly, within this concept of the science/society relationship, the scientific community is seen to have the responsibility, and ability, for self-regulation. This means that quality assurance and control over both the direction of investigation and the application of new knowledge can be largely left to the scientific community.

2.2 The paradoxes of modern science and technology

The dominance of this understanding of the science/society relationship in the post WWII years created a high level of public investment in basic scientific research and fed a hugely successful industrial application of science in new technologies. As human health, safety and personal

---

[5] Guston, D. (2000). *Between Politics and Science: Assuring the Integrity and Productivity of Research*, Cambridge University Press.

comfort have improved through this scientific and technological progress, however, it has been increasingly common to point out that as a society we now find ourselves facing a paradoxical situation: the same progress that we have benefitted so enormously from has also created new threats to both human health and safety, and to the environment. This paradox is closely linked with another: as the application of science-based technologies has generated new risks (e.g. those from pesticides, cars, heavy metal contamination etc.), humans remain reliant on what Ulrich Beck[6] calls the "sensory organs of science" to see, understand and control these risks. Within this situation, science becomes an important player within the political sphere - its expertise is sought to help inform decision-making, very often around remediation of the very problems that it has in part created!

The majority of the negative impacts generated by science-based technologies have of course been unexpected. Spectacular examples of this (such as ozone holes and climate change) have led to not only an increasing awareness of the potential for new technologies to have unintended negative consequences, but also of the limitations of science's ability to predict all potential impacts. In other words, there is now a greater sensitivity to the limitations of scientific knowledge and the way in which these limitations constrain the ability to understand and predict what will happen when new technologies are introduced into complex, interacting, and evolving social and biological systems.

Due to increasing awareness of these paradoxes, it is now becoming common to suggest that while science remains an important informant for political decision-making, the level of confidence in scientific expertise has been eroding among the public (particularly in certain sectors such as food and nations such as the UK). One therefore sees the story of the relationship between science and society shifting. This is particularly occurring through the development of an enhanced academic critique of the position and power of science in political decision-making.

---

[6] Beck, U. (1992). *Risk Society: Towards a New Modernity*, London: SAGE.

In the early 1990s, Silvio Funtowicz and Jerome Ravetz[7] put forward the argument that different circumstances require different types of relationships between science and policy. In what is often referred to as the theory "post-normal science"[8] (PNS), they differentiated three types of science /society relationships based on the 'level of uncertainty' and the 'decision stakes' involved. Through this, they created a useful way of thinking about different roles that science can play in political decision-making, and particularly, the circumstances that may warrant broader involvement in judgments of its quality. Their approach specifically challenges the view of the traditional social contract, where quality control of science was always seen as best managed within the scientific community without involvement from society in the form of either politicians or citizens.

Funtowicz and Ravetz argue that when uncertainty and decision stakes are low,[9] the form of science Rayner[10] terms 'consensual science' is appropriate. The quality of this science is seen as aptly managed by traditional peer review processes. When decision stakes and uncertainties increase to a medium level,[11] however, the mode of scientific problem solving becomes more like professional consultancy. In this situation, different individuals can make different judgements on the appropriate scientific methods and the recipient of the scientific advice therefore becomes an important contributor in the evaluation of quality. Finally, when decision stakes and uncertainties are high,[12] Funtowicz and Ravetz

---

[7] Funtowicz, S. O. and Ravetz, J. R. (1993). Science for the post-normal age, *Futures* 25, pp. 739-755. Funtowicz, S. and Ravetz, J. R. (1994). Uncertainty, complexity and post-normal science, *Env. Tox. and Chem.* 13, pp. 1881-1885.

[8] The term post-normal draws on the term 'normal science' famously coined by the philosopher of science Thomas Kuhn, who described science as a type of puzzle solving that occurred through steady advance but which was punctuated by conceptual revolutions or paradigm shifts.

[9] *Low* here refers to a situation where uncertainty is largely technical and the research is 'mission oriented' with a straightforward end use.

[10] Ravner, S. (1992). Cultural Theory and Risk Analysis. *Social Theories of Risk*. S. Krimsky and D. Golding (eds). Westport, Praeger: 83-115.

[11] *Medium* refers to a situation where uncertainty is methodological and includes questions about the reliability of particular theories/research approaches, and research has become 'client serving' rather than mission oriented.

[12] *High* refers to situations where uncertainties are of an epistemological or an ethical variety and the research is 'issue driven' and involves conflicting values and purposes.

argued that a new type of science for policy is required, a post-normal science. In this situation, quality assurance requires extension to a broader community (involving a wide range of stakeholders, including interested citizens) into what the two scholars call "extended peer review". Given the enormous potential and impact envisaged for nanoST, this field can certainly be seen to have high decision stakes. NanoST also arguably involve a high level of uncertainty relating not only to how to characterise, detect and measure nanoscale particles, their toxicological behaviour, and the levels and routes of potential exposures, but also related to social and ethical questions around the manipulation of matter on this scale. For nanoST, it can therefore be argued that it is meaningful to apply the model of PNS, which implies that any interaction between science and policy should be exposed to a process of extended peer review.

Other scholars have also advocated a shift away from the linear model as a way to understand and structure the science/society relationship, but have done so by arguing that the process or 'mode' of knowledge production itself has shifted. Where before it may be argued that there was a clear sense of distinction between 'Academic Science' and 'Industrial Science' (or between basic and applied science, science and technology), there is now said to be a range of more hybrid forms. Now, for example, governments fund applied research, university researchers seek industry funding, industry conducts basic research, basic knowledge is generated through technology creation, technology creation proceeds without the fundamental science being fully understood, and so on. Given the description of nanoST we have provided, one can certainly imagine how drawing a clear line between academic and industrial research, or basic science and applied technology is difficult in this case. The shift into more hybrid forms of knowledge production has been referred to by John Ziman as 'post-academic science'[13] and by Michael Gibbons *et al.*[14] as 'mode 2' science. Whether this represents a shift in

---

[13] Ziman, J. (1998). Why must scientists become more ethically sensitive than they used to be? *Science*, 282: pp.1813-1814.

[14] Gibbons, M., Limoges, C., Nowotny, H., Schwartzman, S., Scott, P. and Trow, M. (1994). The New Production of Knowledge: *The Dynamics of Science and Research in Contemporary Societies*. London: SAGE.

actual relations and practices or just a shift in dominant stories about relations and practices remains open to debate, as does the extent to which the hybridity of a 'mode 2' style science is a normative proposal or a descriptive account.

2.3 Towards scientific quality as social robustness

For Gibbons and colleagues, one important element in their vision of the relationship between science and society is the way in which public funding agencies have shifted from institutions primarily responsible for maintaining basic research at universities, to instruments for attaining national social and economic goals. The linear model suggests that funding basic research will always, in the end, lead to societal benefits. In a mode 2 model, however, researchers are asked to explicitly justify their projects in terms of their usefulness for society to achieve funding. This situation creates pressure for research to orient itself towards 'realworld' practical problems and economic growth creation, rather than to curiosity driven conundrums as in the previously dominant vision (even if this may sometimes only be manifest in applications for funding rather than in actual research work). From the point of view of public funding, this shift could be seen as driven by a realization of just how large an impact science and technology have on the social and biological world. As the mentioned paradoxes of science and technology become more apparent, society has arguably begun to increasingly demand opportunities to 'speak back' to science, to request specific types of knowledge and applications, and particularly, to request help in solving pressing practical problems.

In this situation, where science and society are seen as closely intertwined, ethics has received more attention in the narratives. While the vision of industrial science, with its intention to generate and release products into society, could be seen as always having had a direct connection to ethical questions (e.g. is this right/good/desirable?), basic academic research was traditionally considered largely free from entanglement in social and ethical dilemmas (except perhaps those around how research is carried out). As long as basic research was seen as a process of uncovering truth, it did not have to deal with social and ethical questions to the same extent, because truth was seen by definition as desirable and good. According to the understanding of 'mode 2' or 'post-academic' styles of knowledge production, however, where boundaries between basic and applied research are blurred and science is seen as a tool to be directed towards social priorities, the position of a science free from social, ethical and political scrutiny, critique and curtailment, disappears.

What appears then is a corresponding shift in the notion of what constitutes 'good' science. While 'good' science was seen as that which revealed truth; this has gradually shifted to that of 'good' science as reliable knowledge – knowledge that works and can be used and applied with success. Now, according to Gibbons and colleagues, however, one starts to see 'good' science as 'socially robust' knowledge - that which society wants, accepts, and deems valuable. This is not to suggest that truth becomes what society wants, accepts and deems valuable, but rather that social criteria for deciding the types of questions, problems and products publicly funded science should be occupied with, is given enhanced importance.

The concept of 'socially robust' science essentially implies a new social contract between science and society where the expectations on science take a new form. According to this model, knowledge generation should be more transparent and participatory. The public needs to be more informed about how science works and what research is under development, as well as have a say in how public money is spent. Furthermore, science should be more ethically sensitive and reflective about both its role in society and its potential implications, recognizing that some of these will be unanticipated and possibly negative. The new social contract also entails that one should be prepared to accept the inclusion of voices other than those of scientific experts in political decision-making, specifically acknowledging that the more experiential knowledge of other stakeholders and laypeople can also count as a form of expertise in certain circumstances. In many ways, the 'post-normal science' concept advocates similar changes to that of 'socially robust science'. Both concepts make the case for a new form of science that is more sensitive to its role in society and its ethical dimensions, more reflexive about its potential impacts and limitations, more prepared to engage in direct interactions with members of society and more open to broader notions of what constitutes quality.

Currently the situation is that a range of these different visions of the relationship between science and society can be seen to co-exist and compete in different spheres. However, over the last decade or so (and certainly since controversies such as those around genetically modified organisms and mad cow disease), there has arguably been a general tendency to request a more socially embedded and ethically sensitive science. It is within this context, in which a new social contract vision of the relationship between science and society has gained increasing currency and political support, that nanoST have entered the scene. From the very beginning of their becoming a political priority in most countries, nanoST have had to relate to visions of a new social contract for science. In many ways, nanoST are presented as one of the first 'new' fields of science that is being actively advanced through a 'mode 2' style of knowledge production, being asked to develop in a socially robust way, and being recognized at an early stage as involving both high levels of uncertainty and high decision stakes. This context is seeing social perspectives on nanoST become not only a widespread subject of research, but also a factor of increasing political importance. The relevance of research on social dimensions of nanoST is in line with, and supported by, a commitment to fund ELSA (Ethical, Legal and Social Aspects) research on emerging fields of science and technology.

3. ELSA: Research on Ethical, Legal and Social Aspects

The concept of ELSA[15] research as a field in its own right was brought into being with the human genome project (HGP). When the HGP was launched in 1990, a dedicated program specifically for research on ethical, legal and social aspects (ELSA) was created as an integral part of the project. Thereafter, ELSA research received 3-5% of the annual HGP budget. Since the development of this program, the inclusion of funding for ELSA research has developed into something of a norm for large-

---

[15] While the term ELSA has been that widely adopted in Europe, in the USA this field of research is often referred to as ELSI (Ethical, Legal, and Social Implications/Issues).

scale science and technology research projects, particularly those that are given national priority status and those considered to be pushing boundaries at the 'frontier' of knowledge.

ELSA forms of research became particularly prominent around the development of biotechnology in both medicine and agriculture; a field that has been both highly promising and subject to widespread social critique and concern. As politicians struggled with decision-making on biotechnology development, they began to increasingly seek out, fund and employ research looking into social, ethical and legal aspects. For cynics, this type of research might be seen as being supported simply to facilitate the smooth introduction of products into the consumer market. For others, however, this support represents a genuine recognition that broader issues and concerns (often initially raised in debates in the public sphere) are important and require careful research and political attention. There is now also an increasing tendency to see support for ELSA research as based on a belief in its ability to help steer new fields of science and technology in a socially robust direction; both directly through altering technology trajectories and more indirectly through working to enhance the reflexivity and ethical sensitivity of the scientists involved.

While ELSA research rose to prominence around biotechnology, it was argued that the research in this case often came too late and with too little influence on political and scientific processes. In other words, that it came when most of the important decisions were already taken and scientific and technological trajectories had already been set (for example through extensive financial investment and product development). When nanoST emerged as a political and economic priority at the beginning of the new millennium, there was therefore a sense in which this represented an opportunity for ELSA research to 'get it right'. It now had the chance to be initiated early in innovation processes and incorporated into both scientific processes and political institutions 'upstream' before important decisions had been taken. For nanoST, there has been a remarkable emphasis on the importance of early attention to ethical and social aspects (e.g. in important policy documents such as the US NSF (2001) and in the UK RS/RAE (2004)).[16] Substantial funding has therefore been allocated to ELSA research on nanoST and ELSA researchers have become increasingly involved in important political initiatives as well as in a range of 'experimental' collaborative efforts with nanoST researchers

When we reviewed the field of ELSA and related research on nanoST in 2006,[17] we found a collection of highly self-reflective scholars, struggling to take advantage of their new opportunities and the heavy weight of the chance to 'get it right' this time. As described above, with nanoST, ELSA has in some ways gone from being a strand of research that has been in opposition to/functioned as a critical voice within the research establishment, to one that is now invited and included (and to some extent institutionalised) in various scientific and political arenas. This new situation is also something that ELSA researchers are struggling to come to terms with: How to seize the opportunities that are being offered, without losing their critical distance and potential?

In this book we document some of these struggles through showcasing some of the diverse research taking place on ethical, legal and social aspects of nanoST. However, we also try and stretch the concept of ELSA so as to capture broader social commentary taking place on nanoST. ELSA research has often been dominated by social science approaches and we wish to extend the notion of social perspectives to include more humanities based perspectives but also to specifically include perspectives being expressed on social, ethical and legal aspects from outside academia. With this in mind, we would now like to outline the book's overall structure, style, motivation and goals.

---

[16] Roco, M. and Bainbridge, W. (eds). (2001). *Societal Implications of Nanoscience and Nanotechnology*. Final Report from the Workshop held at the National Science Foundation in Sept. 28-29, 2000. Arlington, VA: NSF. .http://www.wtec.org/loyola/nano/NSET.Societal.Implications/nanosi.pdf (last accessed 26.05.09). Royal Society & The Royal Academy of Engineering (RS/RAE). (2004). *Nanoscience and nanotechnologies: Opportunities and uncertainties*, London: Royal Society http://www.nanotec.org.uk/finalReport.htm (last accessed 26.05.09).

[17] Kjølberg, K. and Wickson, F. (2007). Social and Ethical Interactions with Nanotechnology: Mapping the early literature, *NanoEthics* 1, pp. 89-104.

4. Nano meets Macro: Social Perspectives on Nanoscale Sciences and Technologies

4.1 What do we mean by 'social perspectives'?

In ELSA research, 'the social' appears as just one of three perspectives, in addition to 'the ethical' and 'the legal'. It may therefore be natural to assume that this book is limited to a certain selection of what takes place under the label of ELSA research. We, however, think exactly the opposite and use the term 'social perspectives' in a broad way to include all research offering commentary on interactions between science and society. For us, 'social perspectives on nanoST' *includes* ethical and legal perspectives, as well as historical, anthropological, sociological, philosophical, etc - in other words, work from any discipline that relates to and explores interactions between nanoST and society.

Interesting and relevant commentary on nanoST in a social context does, however, also take place outside scientific institutions, research organisations and universities. In our broad understanding of 'social perspectives', we also include non-academic perspectives. This is demonstrated through the inclusion in this book of science fiction stories, poetry and visual artworks. These contributions should be seen as perspectives that engage many of the same questions and issues that are raised through the academic chapters. However, perhaps some readers find that they can engage with these contributions in a different way, or that these contributions provoke alternative responses and reflections. We think that this is what makes them particularly valuable and relevant for inclusion in a book dealing with nanoST in a social context. Other forms of social commentary on nanoST (e.g. those expressed in the media, policy and industry) as well as other forms of artistic contributions (such as movies and music), would also potentially represent relevant examples of social perspectives on nanoST for us, although they are not as easy to include in an anthology.

4.2 Outline of the book

The rich diversity of social perspectives is partly brought about by the way in which traditional disciplinary boundaries are often crossed in research on nanoST in a social context. To show the value of the way the field asks cross-disciplinary questions, we found it best to avoid using terms traditionally associated with particular disciplines when describing its overall shape. In our own efforts to understand and conceptually organise the richness and diversity in social perspectives on nanoST, we therefore began to view the different individual efforts as relating to a select number of overarching themes and interests.[18] In this book we call these themes or attractors of interest: 'nodes', in the sense that they form centering points of component parts, and link together otherwise quite different perspectives and commentaries. While it has been fruitful for the purpose of this book to not think within disciplinary boundaries, we should be clear that we are not at all questioning the value of traditional disciplinary research. Rather, we are seeking to show how people from different disciplines can meaningfully approach a shared topic of interest from very different perspectives. In this book we conceptualize and present social perspectives on nanoST as clustered around four different nodes of interest in the interface between science and society. These nodes are described as the places where *Nano meets Macro*: *In the Making; In the Public Eye; In the Big Questions;* and *In the Tough Decisions*. The idea and content of these nodes developed in our thinking through a process of mapping the themes, issues and questions people are interested in when they study nanoST in a social context. The table below shows the type of questions and topics we have identified, and how they have crystallized into our four nodes.

This book is organized into four sections, each presenting a range of academic and artistic works that can be seen as relating to a particular node. We have not aimed to give a comprehensive account of the various perspectives possible on each of the nodes. Our aim has rather been to showcase some of the existing diversity, and to hopefully provide a

---

[18] Kjølberg, K. and Wickson, F. (2007). Social and ethical interactions with nanotechnology: Mapping the early literature, *NanoEthics* 1, pp. 89-104.

useful conceptual structure for understanding the shape and dimensions of the field as a whole.

As indicated in the table below, the first node and section links together interests somehow related to social perspectives on nanoST *In the Making*. These perspectives are interested in the process of creating nanoST and/or the history of their development; in other words, concerned with how nanoST are made or brought into being. The second node links together different perspectives and research interests related to nanoST *In the Public Eye*. Here we find a particular interest in how nanoST is represented, understood and perceived in the public sphere. Thirdly, there is the node that attracts those interested *In the Big Questions*. These are those deep questions, about how we do/should relate to technology, nature and each other and specifically how nanoST developments, concepts and visions engage with these (or in Douglas Adams' words, the big questions about 'life, the universe and everything'). The last node is the one that deals with nanoST *In the Tough Decisions*. Here we find perspectives on the politics and governance of nanoST and debate about how these processes of decision-making and control may best be approached.

CENTRAL CONCEPTS:	INTERESTED IN nanoST AS:	FIELD OF CONCERN:	NANO MEETS MACRO:
Laboratory practice, historical context, driving forces, instrumentation, images, cross-disciplinarity, scientific education…	A (new) field of research and development that spans various scientific disciplines	How nanoST are constructed and developed through scientific practices and social policies	***In the Making***
Perceptions and opinions, rhetoric, metaphors, hype, fear, visions, science fiction, media representations…	A development represented in a range of mediums and understood differently by various members of society	How nanoST are understood, presented, talked about and represented in the public sphere	***In the Public Eye***
Personhood, human/machine relations, human/nature relations, the natural vs the artificial, morality, ethical norms…	A complex phenomenon raising or engaging fundamental philosophical questions	How nanoST interact with questions of metaphysics, epistemology and ethics	***In the Big Questions***
Institutions, expertise, risk, uncertainty, public participation, engagement, responsibility, power, values, decision-making, trust, law…	A political/decision-making problem requiring new approaches to governance, regulation and decision-making	How nanoST are approached in the processes and institutions for decision-making, regulation and governance	***In the Tough Decisions***

To complicate this lovely clean picture just a little, it needs to be said that not all social perspectives on nanoST deal with issues and questions from just one node. Although we find it conceptually useful to separate the work into these four nodes, there are certainly instances where perspectives are given on topics of interest across them. We view this cross-nodal work as extremely valuable. For example, in addressing the question of how to achieve socially robust nanoST, it may be necessary to think about how nanoST is represented 'in the Public Eye', how it relates to 'the Big Questions' as well as how to approach 'the Tough Decisions'. For us, this does not diminish the value of the concept of the different nodes, it simply highlights the way in which all models are necessary simplifications of a complex and intertwined reality. The value of the concept of different nodes around which social perspectives on nanoST cluster, is that it helps us see higher level themes and issues of interest and how these can be approached by people from different backgrounds and from different perspectives. The contributions in this book are organised according to the node that we see them as primarily concerned with. However, some of the occasions where we see engagement with questions of interest across nodes will be highlighted in the section introductions. The introductions to each of the four sections will also provide more detail on both the node of attraction and on the content of the different academic and non-academic contributions presented within them.

4.3 Book style

The style of this book reflects our motivation to try to provide an approach to understanding and structuring the richness and diversity seen in social perspectives on nanoST. We have specifically attempted to design the book so that it may be used as a tool for interdisciplinary discussion and specifically asked all our authors to keep this in mind when writing their contributions. In our own experiences with trying to teach social perspectives on nanoST in a number of different settings, we have struggled to find texts that could be used as a starting point for discussion, primarily because they are so often written in dense disciplinary language. While the chapters in this book could be seen to have overcome the many challenges in addressing a multidisciplinary audience with varying degrees of success, we hope that at least the range of different perspectives offered and the way they have been positioned around common themes or nodes may help with the task of cross-disciplinary communication. Certainly our use of the concept of different nodes was an attempt to conceptually organise the field in a way that makes its richness apparent and accessible to people working outside it. Additionally, we have included suggested "Questions for Reflection" at the end of each contribution, in the hope that this may also facilitate interdisciplinary teaching and learning. In the situation that nanoST finds itself in, where actors from all sides are calling for enhanced interdisciplinarity (both between the natural sciences and between the natural and social sciences), there is a growing need to develop tools and skills for interdisciplinary communication and understanding. Our hope is that this book may help stimulate reflection and communication between people from different backgrounds and with varying entry points to their interest in nanoST. We therefore hope that our approach to this book has resulted in an interesting and enjoyable read for people interested in nanoST from the social sciences, the natural sciences and beyond!

Nano Meets Macro:
In the Making

This section presents perspectives on the emergence of nanoscale sciences and technologies. The contributors in this section are interested in how nanoST are created or 'made', both in laboratories and through policy documents. The field of concern forming this node of interest can therefore be described as the construction and development of nanoST through scientific practices and social policies. Its central concepts include: historical context, driving forces, laboratory practice, instrumentation, images, crossdisciplinarity, and scientific education. Those working with these issues are all interested in how the social and the scientific meet and combine in the making of nanoST. They have a particular interest in how and why nanoST have recently emerged as a 'new' field of inquiry, or they focus their attention on how the research within this field takes place in practice. This means that this node contains perspectives not only on how the social (such as policy documents, funding programmes and priorities etc.) affect the development of nanoST, but also on how the creation of nanoST in scientific laboratories incorporates social processes, concerns and influences.

This section consists of three academic chapters, two short pieces of poetry and three artworks, that all relate to nanoST 'In the Making'. In chapter 1, 'Historical Context of the US National Nanotechnology Initiative', Hans Fogelberg takes an historical perspective and focuses on the creation of nanoST as a political priority. He gives particular attention to the National Nanotechnology Initiative (NNI) of the United States of America, which he uses to explore the relationship between nanoST and earlier political support for the field of materials research. In doing this, Fogelberg presents some of the historical context framing the NNI and argues that this policy document fails to engage important questions around how to move from knowledge generation to innovation.

In chapter 2, 'Questioning Interdisciplinarity: What roles for laboratory based social science?', Robert Doubleday and Ana Viseu also begin by highlighting the political context within which nanoST have emerged. Their particular emphasis is, however, on the calls for the science and technology to develop in a more 'socially robust' way. One of the approaches that is often highlighted as promising for incorporating societal considerations into research and development processes is the integration of social scientists into nanoST laboratories. Doubleday and Viseu use their chapter to describe their experiences as 'in house' social scientists of two nanoST laboratories, one in the United Kingdom and the other in the United States. In comparing their experiences, they find similarities in a mismatch between the expectations held by the natural scientists they were involved with and their own hopes to conduct more traditional laboratory based ethnographies. In this way, they present both the way in which they would have liked to study nanoST 'in the making' and some of the challenges facing the integration of social scientists in the construction and development of nanoST.

Chapter 3, 'The Science and Politics of Nano Images', is a collaborative effort between a natural and a social scientist examining the importance of imagery in the making of nanoST. Roger Strand and Tore Birkeland explore both the construction of scientific images of the nanoscale, as well as the use of nanoST visualisations in documents promoting and communicating policy. In their examination of the construction of scientific images, Strand and Birkeland focus on images from probing style microscopes such as the STM and AFM, while in their exploration of nano imagery in science policy, they use the example of a particular European Commission brochure. In their analysis of nanoST imagery, the authors highlight differences between the views and modes of analysis of social and natural scientists and develop a method for organising the work of interpreting nano images that can be used across these different views.

In chapter 4, Hildegard Lee gives us two examples of 'Poetry from the Laboratory'. In the first example, Lee employs the haiku form of poetic expression and seeks to stimulate reflection on the role that funding plays in the making of nanoST. In the second example, she uses the lighter tone of a limerick to draw attention to the challenge of integrating social and ethical concerns into nanoST during the laboratory stage of their creation.

In the artwork 'Eigler's Eyes 2', Chris Robinson parodies an iconic image within nanoST, namely the STM image 'Quantum Corral' created by IBM scientist Don Eigler. In his artwork, Robinson questions what we really see in STM images by playing with the absurd through combining objects from various scales in the same image. Crossing the cultures of art and science, Robinson specifically wants to challenge us to reflect on how we perceive and represent truth.

Kristoffer Kjølberg uses his artwork 'Triangular Masterpiece no. 5' to suggest that working with abstract art can mirror the construction of nanotechnology. He does this through presenting an image that is the product of a unique constellation of standard component parts. He then encourages us to consider the way in which these parts are like particles and how artists are like scientists, assembling and reassembling them in a variety of ways. This piece persuades us to reflect on the process of nanoST in the making and how this process relates to art in the making.

The final contribution in this section, the artwork 'NanoFireball' by Cris Orfescu also encourages us to reflect on how the making of nanoST relates to the making of nanoart. This artwork is described by Orfescu as involving the creation of a 'nanosculpture' that is then imaged through scientific instrumentation and artistic process. In his written presentation, Orfescu describes a particular concept of 'NanoArt' and relates this to the use of nanoscale particles in artworks throughout history. In this piece, Orfescu is challenging us to consider what constitutes the boundaries between science and art and the dizzying potential we find in both.

While this section contains a rich variety of perspectives on nanoST 'In the Making', it is worth mentioning some of the issues and aspects that we consider to also belong to this node of work, but which are not included as chapters or contributions in this book. One of the important themes of this node, that is not given attention here, is that relating to the development and use of instrumentation in nanoST in the making. Another missing field is that referring to educational practices and processes, and particularly to the unique challenges of crossdisciplinarity in education for nanoST. Finally, although Doubleday and Viseu present their work as social scientists embedded in nanoST laboratories, anthropological style ethnographic investigation into nanoST research cultures is also a theme of interest in this node that we have not been able to capture in this book.

NanoST is being made not only in a range of laboratories around the world, but also in socio-political instruments such as national policy initiatives, funding programmes and educational directives. The way in which the social and the scientific meet and become entangled in this process of creation is the subject of particular attention in this section of the book. Some of the contributions presented here do, however, also reach out to engage questions and themes in the nodes that are the focus of the latter sections of the book. Examples of this kind of cross-nodal work include questions such as: how are images of the nanoscale important not only for the development of nanoST but also for the public understanding of the issue? What do we really 'see' in images of the nanoscale and how do particular styles of image presentation shape our understanding of the way the world is and what is possible in our interactions with it? How does the challenge of responsible governance extend into laboratory practice and the daily life of scientists? and How does the political construction of nanoST create implications for the governance of innovation?

Chapter 1

Historical Context of the US National Nanotechnology Initiative

Hans Fogelberg

By analysing materials research and its historical links to industry and national policy, this chapter provides an understanding of the circumstances that framed the creation of 'nanotechnology' as top priority in the US national science policy and innovation policy. The chapter uses this historical description to explain why nanotechnology initiatives bring high expectations for innovation outcomes, while they do not provide a plan for how this development should be organised.

Nano Meets Macro - Social Perspectives on Nanoscale Sciences and Technologies
by K L Kjølberg & F Wickson

www.panstanford.com
978-981-4267-05-2

1. Introduction

This chapter traces the historical context of nanotechnology, using the lens of an innovation studies perspective. Two themes run in parallel: (1) the historical role of materials science for nanotechnology research topics and organisation, and (2) the role of the understanding of innovation in science policy for the design of nanotechnology programmes. These two themes are brought together to explain the framing of the US national initiative on nanotechnology development. A critique is developed on what can be considered a lack of an innovation perspective in this type of nanotechnology initiative. The chapter focuses on US developments, but the argument may have broader relevance. National initiatives on nanotechnology now appear in other nations (Barben *et al.* 2008), and while there are important differences, there are also striking similarities with respect to the role of materials research for how nanotechnology is organised (Fogelberg 2008).

The US has been a forerunner in organising materials research and creating a historically specific context for this research: first during a formative phase and an expansion through links to rationales of national security (military frame); and later, through rationales of national economy (civil frame). The US was not the first nation to (re)formulate materials research into the framework of a national nanotechnology programme, but it was first to design a programme for science-based economic growth in this technology area that was so large that it affected the national research portfolio. This initiative both inspired and forced other nations to launch similar programmes. These programmes, or national nanotechnology initiatives (NNIs), that now spread across industrialised nations can be regarded as science policy plans for the content, form, and process of nano-innovation.

This chapter suggests that such plans should be understood against the background and paths of earlier materials research. This includes a longer history of national policies on science, technology development, and innovation in important high-technology areas. The more normative dimension has to do with future innovation management in nanotechnology. Our capability to harvest the possible benefits of nanotechnology while avoiding negative impacts may in fact stand and fall with our capacity to understand these longer historical threads.

2. 'Innovation' as an Entry-Point to the History of Nanotechnology

2.1 Theoretical conceptualisations

The innovation perspective emphasises that research is embedded with and linked to society. This means paying attention to the various 'contexts of application' that surround scientific practice and that embed this practice in a societal frame (Gibbons *et al.* 1994). The innovation approach assumes that such links and embedding provide channels for exchange of resources and knowledge between groups of actors, and that this explains how and why innovation occurs.

Links between science and social, economic, and technological contexts have probably always existed, even though analysts usually agree that such links have become more pronounced in later years. The phenomena of integration between science and its contexts started taking shape during the second half of the 19th century. It became more clearly expressed from at least WW II and the second half of the 20th century, when many nations began to develop a more explicit policy for science. Materials research was in the midst of this change. It emerged as a field under this era and under a policy paradigm of mutual trust termed the "social contract for science" (Guston 2000, p. 5). That model states that if politics provides the resources for scientific work, with few or no strings attached, innovations will flow back from these investments automatically. It thus allows a relative autonomy for academic research under the assumption that there will in the end be beneficial results from this research (Guston 2000).

This social contract model was operational during a specific historical period after the WW II until the 1980s.[1] Part and parcel of the above social contract is also the so-called 'linear model' of innovation. This overly idealistic model states that innovation is a unidirectional

[1] Historians vary in the time span of this period, from 1970s to 1990s.

sequence, that runs through the steps of: 'basic research', 'applied research', 'development', followed by 'production' and 'diffusion' (Godin 2006). The model implies that interference with, or limitations on the first step may prevent value creation in later steps. The notion of 'strategic science' is a more recent policy concept (Stokes 1997; Rip 2002; 2004), aimed at solving the problem of intervention in the first step without interfering with its positive function as outlined in the linear model. Strategic science is perceived as research somewhere in-between basic and applied research. This and other new conceptualisations are overlays that are added to the earlier conceptualisations of research. The earlier social contract for science produced a range of new technologies and it provided a platform for many of the basic technologies of today's high-technology nations.

A substantial part of this production, however, was probably linked to the fact that military funding was substantial and allowed this system to operate in a mode of growth. Science developed as a growth-prone social system, but one that was not immune to limits to growth (Price 1975). Research funding as a fraction of GDP is one limiting factor, and changes in rationale for funding from military to civil, another. In combination these factors create a need for more sophisticated mechanisms for the distribution, management and evaluation of the resources that societies use for research. The transition from growth to a 'steady-state' situation may have occurred in Europe in the 1970s and in the US in 1990s (Ziman 2000). This led to an increased effort of managing resources, particularly by developing more science policy, which in parallel with other changes resulted in a very different environment for knowledge production (Ziman 2000).

This transition has resulted in a distinct stream of science policy analysis aimed at understanding these transformations. This has in turn led to rapid conceptual development, with the notion of the 'knowledge economy' being the most obvious. Within innovation and science policy studies, we meet terms such as the 'triple helix' model of innovation, describing innovation as a three-party interaction and communication act that includes universities, funding agencies and industries. The academic milieu, industrial interests, and policy actors are being brought closer together by this 'triple helix' process, which not only produces new knowledge, technology, and markets, but also reshapes the institutions that participate in the innovation process (Etskowitz & Leydesdorff 1995; 2000).

The notion of 'Mode 2' research (Gibbons *et al.* 1994), for example, points to a shift in the way in which university research is conducted. According to this notion, the context of academic knowledge production has changed. The transition is from relative autonomy with respect to society and from a strong disciplinary structuring of the inner life of the university (Mode 1), to a more blurred (Mode 2) situation. Here, knowledge is more socially distributed, and external criteria and direct value creation become important parts of what constitutes research (Gibbons *et al.* 1994). 'Academic capitalism' denotes the business-like behaviour among research groups that is driven by changes in the funding context to which groups need to adapt (Slaughter & Leslie 1997). The notion of the 'entrepreneurial university' points to the more general cultural and organisational changes that the development of a knowledge economy has led to inside universities (Clark 1998). And finally, the concept of 'post-academic science' represents a similar argument. 'Post-academic science' refers to "radical, irreversible, world-wide transformation in the way that science is organized, managed and performed" (Ziman 2000, p. 67). As social phenomena, at least since the 1970s, it is said to affect all knowledge institutions of the modern society. Ziman points to the following underlying factors for this process (Ziman 2000, p. 67-82): the advancement of knowledge has required the organizing of larger and transdisciplinary collectives; the knowledge production system has reached a financial limit to growth, and new institutions and professions are invented in order to manage the resources that are available; the impact of earlier science-based technologies on society has spurred the expectation that direct value can be created from research, and that this value creation is manageable by policy measures. This led to a changing social contract for science and the advent of a 'science policy'. This policy is today increasingly collapsed with the term 'innovation policy'. The discourses of science and economy are no longer separate.

2.2 The understanding of 'basic research' in relation to 'innovation'

All the above concepts represent attempts to theorise around the same phenomena, and all stress the importance of placing science in a social and historical context, tracing what can be regarded as the new and emerging 'reflexive systems of innovation' (Fogelberg & Sandén 2008; Fogelberg 2008). Such innovation perspectives and contextual approaches try to trace a joint or coupled development that links scientific progress, industrialisation and social change. Or put differently, academic research and technology application are studied as one social (or socio-technical) system, rather than as two de-coupled systems. This implies a refutation of the linear model of innovation, and has led analysts to a partial rethinking of entrenched dichotomies such as basic/applied and science/technology.

Historians and sociologists that study research and knowledge-based innovation using contextual approaches have shown that actual research involves a constant blurring and renegotiation of such dichotomies and boundaries (see early formulation in Barnes & Edge 1982). 'Basic research', then, is something we need to analyse empirically rather than to prescribe what it is in advance. Historical studies have in this vein shown how the research that actors consider as 'basic' has been more linked to societal and historical contexts than previously understood (see e.g. Rosenberg & Nelson 1994; Latour 1987).

Advanced (basic) materials research is a case in point. Scientific knowledge, laboratory instruments and the fabrication techniques of advanced materials research have long been involved in a process of 'transfer' from the context of academic research to contexts of application. The success of this scientific and technological heritage is significant, but perhaps most visible in the longer development of microelectronics, as well as in more recent developments in biomaterials. But the successes of this research and development also raises questions on how such innovation processes occur, and what role policy can play in the management of future innovation. This is the obvious question for proponents of nano-innovation: how to create a nano-policy that produces developments in desirable areas without taking away the variety and open-ended character of research.

The innovation lens applied in this chapter thus situates basic research 'in context' and asks the presumingly simple question: What is the historical context of the formulation of a nanotechnology initiative? It assumes that a central characteristic of basic research is that it is an activity conducted in a social and historical context. It also assumes that this context explains why a particular research field and domain exist, and why it can be transformed into something we regard as new.

3. The Standard History of Nanotechnology Innovation

The field of nanotechnology has, despite its relative immaturity, already developed a shared view on its own origin and evolution. The existence and details of this 'standard history' of nanotechnology have been described and analysed by others (Baird & Shew 2004; Mody 2004; Shew 2008). It typically contains the following elements:

(1) Richard Feynman, a famous American physicist, held a speech at Caltech in the late 1950s about the in-principle (theoretical) possibility of designing completely new properties and technical functions in materials by controlling matter at the atomic scale. The level of control of matter was, however, at this point in time not sufficient to allow the applications that Feynman suggested. (2) In the early 1980s Heinrich Rohrer and Gerd Binnig developed the scanning tunnelling microscope (STM) and this instrument became the first in a series of probe microscopes with atomic resolution that was considered to open a window to more precise control. (3) In 1990 D.M. Eigler and E.K. Schweizer used probe technology to manipulate individual atoms and managed to 'write' their company logo IBM with a few atoms. This visual achievement and pervasive representation of control suggested that the level of control Feynman had asked for was now in place. (4) In parallel, Eric Drexler developed a vision of a molecular engineering era based on a new production paradigm for our material world. He envisioned molecular 'machines' and 'assemblers' that could be designed to utilise minimal resources to build and produce useful real-world materials, bottom-up. He imagined both radical potential and radical risk (Drexler 1986; Drexler *et al.* 1991). (5) Finally, the appropriate societal response to

these possibilities now at hand was the action taken by forward-looking policy actors. They started to formulate a programme for a new technology and future economy based on the mastering of nanoscale objects and systems. This vision is most forcefully embodied by the US National Nanotechnology Initiative (NNI).

Such a shared standard history usually exists in other technology fields as well, and while they probably indicate something about the historical development of technology, they are usually too simple and linear to be useful as historical descriptions of science, technology or innovation. Standard stories are historical reconstructions of a development, and as such tend to overemphasise the links between often separate processes and instances in the larger sea of history. However, the particular elements of the 'Feynman-STM-IBM-Drexler-NNI' story can be viewed as representing important entry to certain aspects of the history of nanotechnology. The historical and contextual approach thus encourages us to ask questions about what the elements in the standard history represent. The following can be suggested as starting point for such analysis.

'Feynman' indicates that materials physics has a longer tradition of engaging in questions about application and use, e.g. through the advanced materials research described below. Particular scientific communities have been funded for historical reasons, creating particular institutional development paths and links to industry and society.

'STM' represents research as a highly advanced material culture, where particular instruments and laboratory capacities represents windows to new developments in science and technology. The STM technology has produced images that have been used as pervasive representations of nanotechnology providing a 'nanopresence' (Mody 2004, p. 120) for a larger public. But the actual role of STM for research and industry is probably exaggerated (Baird & Shew 2004; Mody 2006a), and this might also be true for the role of STM for nanotechnology research.

That the text that gained attention was 'IBM', indicates that there are in place particular industrial sectors that have been permeable with academic research, and created or shared the instruments with university researchers. That research technology can bridge academic and commercial worlds is well known from historical studies (see e.g. Joerges & Shinn 2001).

'Drexler' can be seen as a representation of the fact that scientific ideas and advanced speculation on technological developments can spread beyond the boundaries of academic knowledge communities and research organisations. In this case it opened up more public discourses on nanotechnology, which in turn initiated repercussions on both policy and science.

The processes that provide us with new and advanced technologies are no longer crafted inside discursively and technologically 'closed worlds' (Edwards 1996). Technological change and innovation have become more 'public' for a simple reason: since the end of the Cold War legitimacy for allocating resources to research has increasingly been sought using reference to civil and economic rationales. Funding of research on nanotechnology is subject to an open competition with other societal areas and goals. Hence public contexts have become increasingly important for both patrons of science funding and for researchers.

The 'NNI' is in a sense the natural response by publicly funded systems for research and innovation to this new situation. The implementation of a science policy for nanotechnology did not come silently, as many of its predecessors in materials science did. Nanotechnology is surely 'scientific' and 'technological', but also 'political' and 'public' (Glimell 2004).

The use of an historical innovation perspective on 'the standard story' thus encourages reflection on the idea that there are social and cultural contexts in which scientists, engineers and their institutions are embedded that need to be analysed better. The next section describes a history of nanotechnology that is more centred on the research on new advanced materials and on the national contexts that produced such competences and capabilities.

4. Materials Research Origins of Nanotechnology

4.1 A contextual and historical approach

What are the research trajectories that allow high-tech nations to craft and implement a science policy that turns the fabrication and analysis of nanoscale objects into a technology-based programme for economic growth? Where do such research capacities come from, and what do those historical threads and paths mean for the prospects of both current and future nanotechnology developments?

This section pays attention to the fact that there are funding institutions at national levels, research collectives within universities and inside government laboratories, that with little effort and with quite substantial credibility have relabelled and changed their work and organisation to something 'nano'. This is not to be mistaken for the claim that 'nano', at the end of the day, is a just an empty word, and that it is not really representing something new or a real change. The claim developed here is in fact the opposite. The ethos, organisation and natural contexts of 'use', that have been so familiar with materials science and engineering, are now *expanded*. Through nanotechnology policy programmes it starts to become a role model for a much larger portion of modern sciences than ever before.

4.2 Evidence from the history of science

Historians of science have shown that few research areas and historical contexts are more important for explaining the rapid adoption of 'nano' than the field of materials science and engineering (Bensaude-Vincent 2001; Mody 2004; 2006a; 2006b; Choi & Mody 2009). This is especially true if we include in materials research areas such as surface science and microelectronics, as well as their instrumental communities. Materials science is contextual by definition. Bensaude-Vincent captures this natural embeddedness between materials science and the usefulness of that research in the following way:

> *The notion of materials combines natural science and the humanities; it combines physical and chemical properties with social needs, industrial or military needs. From this coupling of natural and human aspects embedded in the definition of materials follows a basic feature of materials science: Knowing and producing are never separated. Materials science couples scientific research with engineering application of the end-product* (Bensaude-Vincent 2001, p. 223).

This section draws on the above mentioned historians work and views on the particular historical importance of materials research. However, it also attempts situate their results from science studies more closely to the area of innovation. The section argues that the particular historical specificities surrounding materials research later framed how proponents designed national programmes and strategies for nanotechnology.

4.3 The long search for the 'scale' that governs material properties

An early trend in materials science, that served as an intellectual fundament for the later establishment of interdisciplinary laboratories, was the ambition of scientists to move materials research away from its earlier focus on practice-, use- or material-specific knowledge. The trend was towards more general and fundamental scientific knowledge about the scale that governs the properties of materials (Leslie 1993), and of the "processing of materials to their properties and uses" (Committee on Material Science and Engineering 1974, p.1). In a more recent statement from a major architect of the US NNI, nanotechnology is defined as "the lowest scale where we can transform matter under control for practical purposes" (Roco 2004, p. 9).

Even though important domains of materials research, such as 'microelectronics' and 'solid state physics', do not signal explicit association with 'nano', one should be clear that researchers in the field worked with relevant theories, such as quantum theory, and were well aware that the nanoscale was a crucial dimension defining material properties. This scale is not something new with nanotechnology but can rather be considered the longer-term intellectual ambition of advanced materials research, as it has developed from the late 1950s. Also scientific instruments have evolved since then, however, contemporary scientists were not without any resources to investigate matter in earlier years. Instruments were available already in the late 1950s and early 1960s, for e.g. crystal growth and thin-film production and analysis. Their toolbox of what we today would call 'nano-instruments' also included electron microscopes for visualizing nanoscale structures (Leslie 1993).

The goal of 'control' of matter and the basic 'epistemic culture' (Knorr-Cetina 1999) of the research communities in question - i.e. the way knowledge workers go about tackling their task including their material resources for doing so - is part of a continuum and an evolutionary development. The three important types of research communities of laboratory milieus are: the producers of new scientific instrumentation and equipment, theory groups, and experimentalists. These represent the different communities of research that 'trade' and develop 'language' that allows them to communicate and advance their research field (Galison 1997). Theory in physics predicted the in-principle possibility of a 'precise control' of technology very early, but instrumentation and capacity to do experimental work lagged behind. The development of the latter however improved significantly within the specific historical context of advanced materials research.

4.4 National contexts of materials science

The importance of early military and space contexts for materials research should not be underestimated. Leslie (1993) provides a research and university-centred account of the historical development of materials science and engineering, with a particular focus on microelectronics. Lécuyer (2006) follows the same process, but from the observation point of industrial development, using instead a firm-centred history of the development of the microelectronics industry of Silicon Valley. Both identify the important driver for the establishment of materials research as a military financed expansion. Together these authors show the intrinsic and close links that existed in the US between universities, military funding, and industrial development. The scope of materials research activity was huge, and it allowed advanced academic science to coexist with industrial development. The 'productivity of research' in the sense of Guston (2000) became a built-in feature of the evolution of this innovation system for materials development, especially in the area of microelectronics. It provided basic research with important links to specific contexts of application.

The role of military and space rationales and contexts for materials research started to decline at the end of the Cold War. New contexts gained increasing attention, and roles, as the legitimising source for science policy. It is not very surprising that the civil rationale that comes closest to mobilising large funding and concerted policies is the 'national economy' and the threat of nations falling behind in a global (now post Cold War and civil) technological competition (see e.g. McCray 2005). This transition from military to civil framing of materials science in relation to 'the national' is a familiar development in most high-technology nations. But it does not occur at the same point in time, and is not equally strong among nations. The beginning of this change was felt in the US in the mid 1970s, and concerned "changing patterns in spending on basic and applied research and between civilian-oriented and defense- or space-oriented research and development; and the growing federal awareness of the importance of materials" (Committee on Material Science and Engineering 1974, p. 8, 12).

The description that historians provide on the situation for US materials science in the late 1950s is strikingly similar (at least in general) with the framing of nanotechnology today. There was a view of a perceived 'lag' in technology advancement that required national mobilisation and new efforts in materials research. The national significance of this research, which was something the major actors agreed on, required that the nation embark on coordinated and very large-scale research programmes. The type of research recommended was multi- or interdisciplinary in nature, yet it was not merely 'applied' research but in a significant way still assumed to be 'basic'.

4.5 *Development of the interdisciplinary research paradigm*

This new paradigm for contextual research was implemented, first, through so-called Interdisciplinary Laboratories (IDLs), and later, through Materials Research Laboratories (MRLs). Funding was for several years mainly military based, where "[t]he defence establishment virtually created materials science as an academic discipline, funding all but a tiny fraction of American materials research during the Cold War years." (Leslie 1993, p. 213). This interdisciplinary research resulted among other things in microwave components, semiconductors, laser technology, thin-film technology, and infrared optics (Leslie 1993).

The outcome of the work inside such milieus was not only new technologies. From the view-point of military research funding patrons, it turned out to have developed too protected spaces for academic research. It provided scientists space for doing what we in many cases would call basic or fundamental science.

Following the first era of MRLs in 1980s were two larger and long-term funding mechanisms. First, the National Science Foundation funded 'centers' programme: the Science and Technology Centers (STCs) and second, the Engineering Research Centres (ERCs). The latter was more aimed at industry links and bringing together the "discovery-driven culture of science and the innovation-driven culture of engineering" (Parker 1997, p. 1). The MRLs became the precursor of the more recent Materials Research Science and Engineering Centers (MRSECs) (National Research Council 2007).

The MRSEC programme is represented by more than 100 interdisciplinary research groups and topics that are grouped under 8 research themes (see details in National Research Council 2007, p. 185-190). A comparison of the focus of these groups with the formulation of the research topics that are considered relevant for nanotechnology gives interesting results. A close reading of key preparatory reports of the US nanotechnology initiative (Roco *et al.* 1999; Siegel *et al.* 1999; NSTC 2000) shows striking similarities.

The differences between nanotechnology research and materials research start to fade when comparing the detailed descriptions of the relevant science and technology areas. This is true also for the description of the existing and future potential applications, and the ways this development should be organised. The following match the common description of nanotechnology. Materials research "is carried out by scientists and engineers with training and background that includes physics; chemistry; materials science and engineering...//...and, increasingly, the biological sciences". Materials research is "interdisciplinary by definition and by evidence of the diverse backgrounds of its practitioners." And "the most exciting and important advances occur at the interfaces between traditional disciplines, forever altering the scope and boundaries of those disciplines" (National Research Council 2007, p. 34).

The Materials Research Science and Engineering Research Centers are now defined as a crucial part of the US NNI. Thus the merging is now almost complete. The description of the type of researchers, and the type of research conducted under material research, is almost exactly the same as that which occurs in the descriptions of nanotechnology research.

4.6 Interlude

Before entering the next section it is enlightening to share the thoughts of two well known innovation analysts that reflect on the situation for US research just a few years before nanotechnology proponents started to formulate and mobilise researchers and policymakers:

> *[t]he end of the Cold War has eroded the rationale that has served over the past 40 years to provide the justification for government support of university research in a number of fields of vital importance to American industry. The first order of business, in our view, is to assure that government support of university research in the engineering disciplines and applied sciences, such as materials and computer science, not be orphaned by sharp cutbacks in military R&D that are almost certain to occur over the coming years. One element that is essential is to articulate clearly that a major purpose of*

government funding of university research in these fields is to assist American industry. (Rosenberg & Nelson 1994, p. 345).

One response to their request would be to implement a new 'science as endless frontier' programme in the Vannever Bush tradition (Bush 1945). Such a programme both legitimises basic research in the eyes of the public, while it also promise resources for basic research and an attractive portion of integrity for researchers doing fundamental research. The National Nanotechnology Initiative (NNI) does all this.

5. The US National Nanotechnology Initiative (NNI)

5.1 Analysing documents as 'script' for nano-innovation

This section describes how materials research and science policy structure the design of the national initiative for nanotechnology development. It analyses several of the major documents under the Interagency Working Group on Nanoscience, Engineering and Technology (IWGN). These documents, including the subsequent nanotechnology initiative, are studied as plans, or 'script', for how new knowledge is to be produced and transformed into societal value.

With the release of the NNI in 2000, 'nanotechnology' was added to and became a major part of the domains that funding agencies and governments have to consider. Nanotechnology became a major 'techno-socio-political innovation strategy' of many nations (Wullweber 2008, p. 28). Nanotechnology raced past other science areas in recognition. Now at the end of the first decade of the 21st century, a nation that does not have strengths in nanotechnology research will be considered a weak technology nation. This section describes how this situation came into place. It also shows that documents lack, both in general and in detail, explicit descriptions of the process of innovation. The major argument of this section is thus that knowledge transfer is in most cases taken for granted, and as something that is not in need of explanation or explication.

5.2 Pre NNI documents

A series of preparatory studies were conducted in the U.S. in the mid and late 1990s, which prepared the way for the U.S. national initiative on nanotechnology. One of the preparatory studies was a benchmarking study of the international developments in research in nanotechnology areas, *Nanostructure Science and Technology* (Siegel *et al.* 1999). This report demonstrates the importance of advanced and physics-centred materials research for nanotechnology. The affiliation of panel members and their backgrounds are linked to materials science organisations. Several of the panel members hold a PhD in physics and site visits were made at advanced materials research departments at various locations in Europe and Asia.

The conclusions of the report are, first, that nanotechnology is already fact and reality. Because of advances in materials science "it is abundantly clear that we are now able to nanostructure materials for novel performance" (Siegel *et al.* 1999, p. xiv). This ability has consequences for the development of new materials. It exceeds earlier capabilities of the field to such an extent that it "represents the beginning of a revolutionary new age in our ability to manipulate materials for the good of humanity" (Siegel *et al.* 1999, p. xiv).

This conclusion must be understood against the background and heritage of earlier described materials science, especially microelectronics. That area had already produced a wide range of useful materials and applications, some of which the above report and similar studies display as 'nanotechnology'. That work began already in the 1920s and an atomic scale conception of the 'performance' of materials. The level of control of matter of that scale was known to determine the level of performance of real-world materials. Nanotechnology was perceived to *enhance* the level of control in the field of materials research, representing a quantitative change rather than a qualitative change. What nanotechnology seemed to offer was an unprecedented ability to control structures that, in effect, would dramatically change and revitalise materials science. For a field already accustomed to interdisciplinary work, and already convinced of the importance of cross-boundary work for the development of real-world applications, more interdisciplinarity meant larger possibilities for success (Siegel *et al.* 1999; See similar view in Hu & Shaw 1999).

The topic of nanotechnology is positioned closer to the research policy and political context by the *Nanotechnology Research Directions* report of the National Science and Technology Council's (NSTC) Interagency Working Group on Nanoscience, Engineering, and Technology (IWGN) (Roco *et al.* 1999). The White House statement accompanying the report expresses a clear expectation of the outcomes of an innovation process. Assistant to the President for Science and Technology, Neal Lane, describes that the report "outlines the necessary steps on how advances made in nano-science, engineering, and technology can help to boost our nation's economy, ensure better health care, and enhance national security in the coming decade" (in the preface page of Roco *et al.* 1999).

The actual steps needed for science to produce this societal value, however, remain unclear. The report refers only in general terms to a grand coalition of actors working together for a common goal. Five major types of such actors are mentioned: academe, private sector, government R&D laboratories, government funding agencies, and professional societies. But the urgency of a national plan is now voiced. The President's Council of Advisors on Science and Technology point out that "[n]anotechnology is the first economically important revolution in science and technology (S&T) since World War II that the United States has not entered with a commanding lead...//...Now is the time to act." (PCAST 1999).

5.3 The US NNI

President Clinton included the *National Nanotechnology Initiative: Leading to the Next Industrial Revolution* (IWGN 2000) as part of the budget for fiscal year 2001. This provided the area with unprecedented funding and, above all, it created recognition of nanotechnology as a new policy area for industrial competition. It was expected to lead to a revolution "likely to profoundly affect existing and emerging technologies in almost all industry sectors and application areas" (IWGN 2000, p. 21).

Yet these projections did not come with a detailed plan for the actual process of innovation. The framing of the initiative was an 'investment strategy' rather than an innovation strategy, yet innovation was implied. Nanotechnology research funding was for example increased from $270 million in FY2000 to $495 million in FY2001. The components identified for FY2001 were: fundamental research ($195 million), grand challenges ($110 million), centres and networks of excellence ($77 million), research infrastructure ($87 million), and finally, ethical, legal, and social implications and workforce training ($28 million) (IWGN 2000, p. 15).

5.4 Assessments of the NNI

The first assessment of the NNI was performed in 2005, by the National Nanotechnology Advisory Panel (NNAP 2005). It was now increasingly clear that nano-innovation would not happen automatically. Reflecting on past achievements, it was realised that the technology procurement of semiconductor technology during the Cold War was a primary reason for the diffusion of microelectronics to consumer products. While military technology procurement may still play a role, few expect this to be the main driver for the development of today's nanotechnology. The assessment instead promoted the use of public funding for basic research at universities, and to rely on the transfer of technology through other mechanisms. The panel signalled that it would start to raise more questions about the apparent lack of 'technology transfer', 'innovation' and 'economic growth' in the years to come (NNAP 2005).

The second assessment of the NNI, from 2008, was one of the first documents to provide explicit information on the organising of innovation. Nanotechnology innovation was no longer expected to occur automatically. The diffusion of nanotechnology turned out to have been slow or occurred in modest ways: "Transfer of nanotechnology know-how and ideas from university research labs to industry occurs primarily when students are hired by existing companies or start new ones" (NNAP 2008, p. 24). Expectations were much higher. There were obvious obstacles for innovation that the national programme needed to address, where government "plays a central role in overcoming the barriers in the process of nanotechnology innovation and commercialisation." (NNAP 2008, p. 24).

The conclusion was that the future NNI must continue to support the long path of multidisciplinary centres in materials research, but *increase* the presence and partnership with industry in such milieus. A new programme was also needed to address learning of how nanotechnology innovation actually works, through monitoring and analysis (NNAP 2008). Not only have actors started to question that innovation is automatic, they also realise that we actually know very little about how nano-innovation occurs, and how it may be stimulated and managed. In a way, nanotechnology innovation has just begun to be a real concern of NNI promotors.

6. Conclusion

The road not taken in current US national nanotechnology initiative is to take responsibility for the details of nano-innovation in a way that matches the promises of benefits and role for the national economy. The involved actors, including political and policy actors, have displayed confidence in the capacity of science to deliver social value, by itself, and automatically.

The current framing of nanotechnology research and innovation resembles that of the earlier activities in materials research in several ways. It builds on and promotes multi- or interdisciplinary research and institutional mechanisms in the form of 'centres'. However, it also remains within the old 'social contract for science' model, in the sense that actors have expected outcomes to be an automatic result of organising research in such centres and multidisciplinary groups.

That expectation neglects the fact that those earlier centres for materials research were naturally embedded within a larger system of innovation for military funded microelectronics technology. That system was later transferred to a civil framing, but kept its close epistemic and innovation links between academic research and corporate milieus. With the more recent decline of fundamental research inside corporations, there is no longer this natural interface for the transfer and communication of highly advanced materials knowledge. Part of the earlier (seemingly automatic) mechanisms for knowledge transfer has eroded. Several important aspects of the contexts of research that provided its workability have changed.

A partial response from nanotechnology proponents has been to link the question of technology transfer and value creation in nanotechnology to the expected role of the entrepreneurial university. Yet, this is perhaps a futile attempt in relation to the loss of the earlier function of materials research in relation to industrial development. To overcome this obstacle is a true challenge for nanotechnology promoters, and indeed, maybe also for our capacity to develop a society that relies on sustainable (nano)technologies. In conclusion, nanotechnology initiatives bring a discrepancy between expected benefit and a plan for how it will realise its goals. And the road *not* taken in current nanotechnology initiatives is, paradoxically, the road towards innovation.

Questions for Reflection:

1. What is an historical/contextual explanation of the origin and current state of nanotechnology?
2. How would you describe the relationship between science, technology and the market in the case of nanotechnology?
3. You have been asked to design a long-term strategy for 'sustainable nanotechnology development' in your country. How would you approach this task?

Bibliography

Baird, D. and Shew, A. (2004). Probing the history of scanning tunneling microscopy. In Baird, D., Nordmann, A. and Schummer, J. (eds) *Discovering the Nanoscale*, Amsterdam: IOS Press, pp. 145-156.

Barben, D., Fisher, E., Selin, C. and Guston, D. H. (2008). Anticipatory governance of nanotechnology: Foresight, engagement and integration. In Hacket, E. J., Amsterdamska, O., Lynch, M. and Wajcman, J. (eds) *The Handbook of Science and Technology Studies,* Cambridge, MA: The MIT Press, pp. 979-1000.

Barnes, B. and Edge, D. (eds) (1982). *Science in Context: Readings in the Sociology of Science,* Milton Keynes, England: The Open University Press.

Bensaude-Vincent, B. (2001). The construction of a discipline: Materials science in the United States, *Historical Studies in the Physical and Biological Sciences*, Vol 31, Part 2, pp. 223-248.

Bush, V. (1998 [1945]). *Science – The Endless Frontier. Reprint Edition*, North Stratford: AYER Company Publishers.

Choi, H. and Mody, C. (2009). The long history of molecular electronics: Microelectronics origins of nanotechnology. Forthcoming in *Social Studies of Science*, Vol. 39, Issue 1.

Clark, B. R. (1998). *Creating Entrepreneurial Universities: Organizational Pathways of Transformation*, Oxford: Pergamon-Elsevier Science.

Committee on Material Science and Engineering (1974). *Materials and Man's Needs. Materials Science and Engineering. Summary Report of the Committee on the Survey of Materials Science and Engineering*, Washington, D.C.: National Academy of Sciences.

Drexler, K. E. (1986). *Engines of Creation. The Coming Era of Nanotechnology*, New York: Anchor Books.

Drexler, K. E., Petersen, C. and Pergamit, G. (1991). *Unbounding the Future: Nanotechnology Revolution*, New York: Morrow.

Edwards, P. (1996). *The Closed World: Computers and the Politics of Discourse in Cold War America*, Cambridge, MA: The MIT Press.

Etzkowitz, H. and Leydesdorff, L. (1995). The triple helix university-industry-government relations: A laboratory for knowledge-based economic development, *EASST Review*, Vol. 14, No. 1, pp. 14-19.

Etzkowitz, H. and Leydesdorff, L. (2000). The dynamics of innovation: From national systems and "Mode 2" to a triple helix of university-indiustry-government relations, *Research Policy*, Vol. 29, pp. 109-123.

Fogelberg, H. (2008). Den svenska modellen för nanoteknik – mer effektiv än reflexiv? *Nordic Journal of Applied Ethics*, Vol. 2, No. 2, pp. 53-72.

Fogelberg, H. and Sandén, B. (2008). Understanding reflexive systems of innovation: An analysis of Swedish nanotechnology discourse and organization, *Technology Analysis & Strategic Management*, Vol. 20, No. 1, pp. 65-81.

Galison, P. (1997). *Image and Logic: A Material Culture of Microphysics,* Chicago: University of Chicago Press.

Gibbons, M., Limoges, C., Nowotny, H., Schwartzman, S., Scott, P. and Trow, M. (1994). *The New Production of Knowledge. The Dynamics of Science and Research in Contemporary Societies*, London: SAGE Publications.

Glimell, H. (2004). Grand visions and Lilliput politics: Staging the exploration of 'The Endless Frontier'. In Baird D., Nordmann, A., Schummer, J.. (eds) *Discovering the Nanoscale*, Amsterdam: IOS Press, pp. 231-246.

Godin, B. (2006). The linear model of innovation: The historical construction of an analytical framework, *Science, Technology & Human Values,* Vol. 31, pp. 639-667.

Guston, D. H. (2000). *Between Politics and Science: Assuring the Integrity and Productivity of Research,* Cambridge, UK: Cambridge University Press.

Hu, E. L. and Shaw, D. T. (1999). Synthesis and assembly. In Siegel, R. W., Hu, E. L., Roco, M. C. (eds) *Nanostructure Science and Technology: A Worldwide Study,* prepared under the guidance of the IWGN and NSTC, World Technology (WTEC) Division at Layola College, Maryland.

IWGN (2000). The *National Nanotechnology Initiative: Leading to the Next Industrial Revolution.* Report by the Interagency Working Group on Nanoscience, Engineering and Technology of the National Science and Technology Council Committee on Technology, February 2000.

Joerges, B. and Shinn, T. (eds) (2001). *Instrumentation Between Science, State and Industry,* Dordrecht: Kluwer Academic Publishers.

Knorr-Cetina, K. (1999). *Epistemic Cultures: How the Sciences Make Knowledge*, Cambridge, MA: Harvard University Press.

Latour, B. (1987). *Science in Action. How to Follow Scientists and Engineers Through Society,* Cambridge, MA: Harvard University Press.

Lécuyer, C. (2006). *Making Silicon Valley: Innovation and the Growth of High Tech, 1930-1970*, Cambridge, MA: The MIT Press.

Leslie, S. W. (1993). *The Cold War and American Science: The Military-Industrial-Academic Complex at MIT and Stanford,* New York: Columbia University Press.

McCray, W. P. (2005). Will small be beautiful? Making policies for our nanotech future, *History and Technology,* Vol. 21, No. 2, pp. 177-203.

Mody, C. (2004). How probe microscopists became nanotechnologists. In Baird, D., Nordmann,A. and Schummer, J. (eds) *Discovering the Nanoscale*, Amsterdam: IOS Press.

Mody, C. (2006a). Corporations, universities, and instrumental communities. commercializing probe microscopy, 1981-1996, *Technology and Culture*, Vol. 47, No. 1, pp. 56-80.

Mody, C. (2006b). Nanotechnology and the modern university, *Practicing Anthropology*, Vol 28, No 2, pp. 23-27.

National Research Council (2007). *The National Science Foundation's Materials Research Science and Engineering Centers Program: Looking Back, Moving Forward.* Report by the MRSEC Impact Assessment Committee, Solid State Sciences Committee, National Research Council.

NNAP (2005). *The National Nanotechnology Initiative at Five Years: Assessment and Recommendations of the National Nanotechnology Advisory Panel,* the President's Council of Advisors on Science and Technology (PCAST), May 2005.

NNAP (2008). *The National Nanotechnology Initiative: Second Assessment and Recommendations of the National Nanotechnology Advisory Panel*, the President's Council of Advisors on Science and Technology (PCAST), April 2008.

NSTC (2000). *National Nanotechnology Initiative: The Initiative and Its Implementation Plan.* Report by the National Science and Technology Council. Committee on Technology. Subcommittee on Nanoscale Science, Engineering and Technology, July 2000.

Parker, L. (1997) *The Engineering Research Centers (ERC) Program: An Assessment of Benefits and Outcomes.* Report from Engineering Education and Centers Division, Directorate for Engineering, National Science Foundation. Arlington, Virginia, December 1997.

PCAST (1999). *PCAST Letter to the President Endorsing a National Nanotechnology Initiative*. www.ostp.gov, reports archive (accessed December 11, 2008).

Price, D. S. (1975). *Science Since Babylon*, New Haven & London: Yale University Press.

Rip, A. (2002). Regional innovation systems and the advent of strategic science, *Journal of Technology Transfer*, Vol 27, pp. 123-131.

Rip, A. (2004). Strategic research, post-modern universities and research training, *Higher Education Policy*, Vol 17, pp. 153-166.

Roco, M. C. (1999). Research programs on nanotechnology in the world. In Siegel, R. W., Hu, E., Roco, M. C. (eds) *Nanostructure Science and Technology: A Worldwide Study,* Prepared under the guidance of the IWGN and NSTC. International Technology Research Institute, World Technology (WTEC) Division at Layola College, Maryland.

Roco, M. C. (2004). The US National Nanotechnology Initiative after 3 years (2001-2003), *Journal of Nanoparticle Research*, Vol. 6, pp. 1-10.

Roco, M. C., Williams, R. S., Alivisatos, P. (1999). *Nanotechnology Research Directions: IWGN Workshop Report: Vision for Nanotechnology R&D in the Next Decade*, International Technology Research Institute, World Technology (WTEC) Division at Layola College, Maryland.

Rosenberg, N. and Nelson, R. (1994). American universities and technical advance in industry, *Research Policy,* 23, pp. 323-348.

Shew, A. (2008). Nanotech's history: An interesting, interdisciplinary, ideological split, *Bulletin of Science, Technology & Society*, Vol 28, No 5, pp. 390-399.

Siegel, R. W., Hu, E. and Roco, M. C. (eds) (1999). *Nanostructure Science and Technology: A Worldwide Study,* prepared under the guidance of the IWGN and NSTC. World Technology (WTEC) Division at Layola College, Maryland.

Slaughter, S. and Leslie, L. L. (1997). *Academic Capitalism. Politics, Policies, and the Entrepreneurial University*, Baltimore & London: The John Hopkins University Press.

Stokes, D. E. (1997). *Pasteur's Quadrant: Basic Science and Technological Innovation*, Washington , D.C.: Brookings Institution Press.

Wullweber, J. (2008). Nanotechnology – An empty signifier à venir? A delineation of a techno-socio-economical innovation strategy, *Science, Technology & Innovation Studies*, Vol 4, No 1, pp. 27-45.

Ziman, J. (2000). *Real Science. What It Is, and What It Means,* Cambridge, U.K.: Cambridge University Press.

Chapter 2

Questioning Interdisciplinarity: What Roles for Laboratory Based Social Science?

Robert Doubleday & Ana Viseu

During the last decade it has become commonplace for governments and funding agencies across the world to call for the integration of societal issues and concerns into scientific practice. One path often proposed to achieve this goal is the extension of interdisciplinarity to include a role for the social sciences in technological and scientific endeavours so as to contribute to the development of 'socially robust' technologies. Nanotechnology is often showcased as a primary site for such collaborations. However, to date there has not been significant discussion of how the inclusion of social science as part of the interdisciplinary scope of nanotechnology has worked in practice. Such an examination is all the more important as policy documents suggest this integration is a relatively straightforward process, while in practice it demands the alignment of the agendas of numerous stakeholders. This chapter draws on ethnographic accounts of two social scientists working in two nanotechnology laboratories located in the US and UK over the course of three years. It analyses our experiences by focusing on the institutional context for the two projects, and by describing the ways in which these two collaborations worked in practice. Our goal is to provide a critically constructive understanding of the potential of such interdisciplinary collaborations.

Nano Meets Macro - Social Perspectives on Nanoscale Sciences and Technologies
by K L Kjølberg & F Wickson

www.panstanford.com
978-981-4267-05-2

1. Introduction

Since the year 2000, when the National Nanotechnology Initiative (NNI) was created in the United States of America (US), nanotechnology has become a priority for many national science and technology policies. Its arrival as a strategically important area of research policy has been accompanied by discussion of how best to ensure its 'responsible development'. Although responsible development means different things to different people, policy makers have proposed a raft of measures designed to ensure public confidence in the development and regulation of nanotechnologies. These measures rely heavily on notions of interdisciplinarity as bringing together the sensibilities and agendas of different stakeholders and disciplines, so as to foster accountability and innovation. Two areas of concern stand out, the first relates to the anticipation and regulation of environmental, health and safety risks posed by nano engineered materials. A second area of importance has been education and the 'social and ethical' aspects of nanotechnologies. In addressing these questions, particularly the latter, policy makers have turned to social science as a source of insight and practical input to an expanded interdisciplinary field of nanotechnology.

Within this framework the social sciences assume the important role of facilitating the development of more 'socially robust' technologies. The assumption is that the social sciences will work as mediators, facilitators and representatives for social and ethical issues, helping to 'create' technologies that are more closely aligned with wider society's concerns and interests, thus avoiding the political and economic costs of public controversy (see for instance, Kearnes & Wynne 2007). Importantly, it is also implicitly assumed that consideration of the 'social' aspects of technologies will automatically include the interests and concerns of the public at large. In other words, the social sciences and the public become if not synonymous, then at least largely overlapping realms, and inclusion of one means inclusion of both. It is therefore not surprising that one suggested component of such policies is to include social science research in interdisciplinary collaborations located in publicly funded nanotechnology facilities. The authors of this chapter were involved in two early examples of such projects conducted in the US and United Kingdom (UK), that saw their integration in technical facilities as researchers trained in the study of the sociocultural dimensions of science, or 'science and technology studies' (STS).

In this chapter we draw upon our experiences to reflect on the ways in which this policy mechanism of institutionalised interdisciplinarity is put into practice. To do so we focus on a few of the successes and failures, enchantments and disenchantments that together constituted our three-year experience. Our purpose is to issue some words of caution regarding how these interdisciplinary collaborations should be set-up and what they can achieve. We start by introducing the science policy context, and proceed to discuss some of the theoretical and methodological resources from STS that we drew on in conceiving our role as social science researchers in the laboratory. We then discuss our ethnographic experience, and argue that in practice the scope for our research was constrained by strong underpinning assumptions about the role of social science and particular power relations that this policy mechanism facilitates. Finally, we conclude by suggesting ways of framing future research projects to make the most of collaboration between social and physical sciences on issues emerging during the laboratory phase of fabricating the facts and artefacts of nanotechnology.

2. Nano Policy Context – The Promises of Interdisciplinarity

While the term 'nanotechnology' remains technically ambiguous (Wood *et al.* 2007), one thing is clear: it has become an important term in science policy. According to industry estimates, global funding for nanotechnology R&D has grown dramatically in recent years to reach $13.5 billion in 2007 (Lux Research 2008). In this section we discuss the context for this rapid rise in nanotechnology funding and chart distinctive features of its emergence over the past decade. In particular, we focus on policies to support the 'responsible development' of nanotechnologies, and the potential contribution of the social sciences.

The first major nanotechnology research programme was the US National Nanotechnology Initiative (NNI), launched by President Clinton in January 2000. Its budget grew rapidly and planned expenditure in

2009 is over $1.5 billion. As Johnson (2004) argues, the timing of nanotechnology's ascendancy as a category of public research funding is not coincidental. The 1990s saw a re-ordering of research priorities in the US as the Cold War came to an end. At the same time there were also greater expectations on research as a driver of economic competitiveness, as well as increased demands for political oversight of public R&D spending. It is in this new context that nanotechnology emerges as a strategic priority for public investment. The central place of the NNI in US science policy was formalised by the passage of the 21st Century Nanotechnology Research and Development Act in 2003 (P.L. 108-153). As Fisher and Mahajan (2006) have argued, the Act puts forward contradictory goals, namely those of fostering "rapid technological implementation" (Fisher & Mahajan 2006, p. 5) and "conduct[ing] technology development with more effective regard for societal considerations" (Fisher & Mahajan 2006, p. 5; see also Guston 2008; Bennett & Sarewitz 2006).

The US model was quickly followed by a race to nano (Fisher & Mahajan 2006) that translated into the establishment of coordinated nanotechnology research programmes in other parts of the world; notably with Japan and South Korea launching national programmes in 2001, and in 2002 the European Union announcing a prominent place for nanotechnology in its sixth framework programme (Roco 2003).[1] In all these jurisdictions the course of nanotechnology policy has been shaped by more general realignments of relations between science, government and wider society.

By the 1990s it was clear that a number of trends were combining to challenge the established way of publicly funding science. In addition to political demands for a greater economic return on public 'investment' in research, there was also increasingly visible public questioning of the direction of scientific and technological development. In Europe, mishandling of the 'Mad Cow Disease[2] crisis and public controversy

---

[1]It is worth noting that the UK Government did not immediately follow with a nationally coordinated nano research programme, a decision that was roundly criticised by a prominent parliamentary committee (House of Commons 2004).
[2]Or Bovine Spongiform Encephalopathy (BSE) and its link to the human disease variant Creutzfeldt-Jakob Disease (vCJD), see Jasanoff 1997.

over the introduction of genetically modified foods, crystallised a demand for greater public participation in the formation of science and technology policy. While in the US questions about scientific fraud undermined public confidence in the self-regulation of science. Collectively these events led to a renegotiation of relations between science, the state and citizens (Gibbons *et al.* 1994; Guston 2000; Jasanoff 2005).

It is against this backdrop of demands for simultaneously demonstrating more directly the social utility of research, and opening up the criteria of what would count as socially desirable technologies, that nanotechnology policy took shape. This historical context helps account for the emphasis placed by policy makers on the 'responsible development' of nanotechnology. For example, as a concrete measure to promote responsible development the 21st Century Nanotechnology R&D Act required a one-time study to be carried out on how to manage the impacts of nanotechnology in areas such as environmental risks, privacy, military technology, and human enhancement.[3]

Similarly, when the European Commission set out its nanotechnology strategy in 2004 it committed to 'the responsible development of nanotechnology' as part of a process that would "fully integrate societal-considerations into the R&D process," enabling the development of nanotechnology "according to democratic principles" (EC 2004, p.18-19). That same year the UK Royal Society and Royal Academy of Engineering (RS & RAE) published an influential report recommending policies to support the responsible development of nanotechnologies. The report emphasised the importance of anticipating potential barriers to the widespread use of an emerging technology at early stages of its development. This early identification would allow for the adaption of social institutions, such as regulatory frameworks, but also, crucially, for altering the trajectory of technological development in response to social aspirations and concerns. This RS & RAE report set an agenda for the

---

[3]Similar efforts were conducted in parallel with the Human Genome Project, which included a program dedicated to the study of its 'Ethical, Legal, and Social Implications' (ELSI).

anticipatory steering of research and development policy, and highlighted a role for public engagement in this process.

In sum, both in the US and the UK 'responsible development' entails integrating the social sciences in the R&D process as means to anticipate the impacts of nanotechnology—including attempting to pre-emptively address and mitigate any possible negative public reactions. Such integration has been proposed as resulting in more 'socially robust' technologies (Wilsdon & Willis 2004; Nowotny *et al.* 2001; Gibbons 1999). For instance, Macnaghten *et al.* (2005) set out an agenda for social science to study the production of scientific and technological knowledge and, in particular, identify assumptions about social worlds that are intimately bound up with the research and development of nanotechnologies. These social assumptions can then be tested through public dialogue projects. The authors argue that the responses generated through such public engagements when presented back to scientists and policy makers become a means of checking and enriching otherwise implicit social models of innovation. However, Joly and Kaufman (2008) have argued that in processes of public engagement little attention has been paid to establishing the mechanisms through which the responses can be incorporated into larger technical research and development agendas. Joly and Kaufman recognise this as a deep power asymmetry that they argue leads the social to be 'lost in translation'. In other words, the findings of such activities are too often not followed up, while the activities themselves become exemplary of integration and interdisciplinarity. It follows then that this approach does not attempt to blur the distinctions between science and society, nor to highlight the ways in which they are mutually constitutive as advocated by STS scholars (see next section). Instead it preserves a distinction between science and society, giving primacy to the technical. One means put in place by funding agencies to foster the interdisciplinary development of nanotechnology is the incorporation of social scientists and humanities scholars in nanotechnology facilities. We now turn to the field of STS and the contribution it can make to understanding the every-day complexities of laboratory science.

3. Lab Studies – Cultural Studies of Science

While from a policy perspective it seems self-evident that the goal of creating socially robust nano sciences and technologies is facilitated through interdisciplinarity, and particularly, the integration of social scientists in technical endeavours, its practice is more complex (see Barry *et al.* 2008). To study this complexity we rely on the field of STS, an umbrella term for different theories, methodologies and approaches that explore the "origins, dynamics, and consequences of science and technology" (Hackett *et al.* 2008, p. 1), while putting forward that science, technology and society are co-constructed rather than being related in a cause and effect relationship (Hackett *et al.* 2008; Law 1991; Jasanoff *et al.* 1995). Work in STS that focuses on mundane aspects of laboratory science helps us understand the complexities of extending the interdisciplinary scope of nanotechnology to include social sciences, and it is to this issue that we now turn our attention.

Until the 1970s social studies of science were concerned with the social instituitions of science, such as career structures, and avoided consideration of scientific method or the content of scientific knowledge (Knorr Cetina 1995). Within this approach, 'the social' remained outside the realm of scientific knowledge, only becoming relevant as an explanation of scientific failure or fraud (Bloor 1976). However, as the field of STS developed, the production of scientific knowledge and the every-day practices of science became an increasingly active area of research, with 'laboratory studies' emerging as one method to conduct this research.

In this context laboratories emerged as privileged sites for the study of scientific production. To study the development of scientific knowledge STS scholars use the same method that anthropologists have long used to study 'foreign tribes': ethnography (see Geertz 1973; Marcus & Fisher 1986). Ethnographic studies rely on immersion in particular settings (participant observation) as well as interviews, to observe and interpret what the 'tribe members' say and do (Geertz 1973). The analogy here is a powerful one: that scientists have their own cultural norms and conventions and are as esoteric a group as those remote tribes, albeit a much more powerful one in contemporary societies (Knorr Cetina 1995). Social scientists and humanities scholars have used these ethnographic methods to examine the "cultural apparatus of knowledge production" (Knorr Cetina 1995, p. 144; see also Doing 2008). By placing science within culture, and thus within society, these studies have shown that scientific objects are not "natural givens" but rather "'technically' manufactured in laboratories ... [and] also inextricably *symbolically* and *politically* construed" (Knorr Cetina 1995, p. 143). In other words, "facts" are not simply discovered, observed, or described, instead they are produced in particular and local cultural settings: these dictate how problems are defined, what instruments are used to characterise them, the rhetoric and language used to argue them, the venues where results are published, and the kinds of funding to be drawn upon (see for instance, Latour & Woolgar 1989; Knorr Cetina 1999; Traweek 1988; Fujimura 1987).

What the aforementioned lab studies did was to draw attention to the fact that these issues are not extra or un-scientific, instead they are the stuff of science, part of the scientific process. By the same token, as science and technology develop the social and natural worlds are re-configured. The argument then is twofold: not only should we recognise that scientific practice is itself a technical, cultural, social and political activity; we must also recognize that its products are deeply implicated in the definitions, character and boundaries of the social (Latour 1987; 1993; Jasanoff 2004). The example of nanoscale research is a case in point: what STS puts in evidence is that scientists and engineers bring their values, beliefs, norms and assumptions to their research. These are not a sign of bias or lack of objectivity, but are instead part of the culture of different research laboratories, which are themselves part of a larger society that prioritises and funds research at the nano scale. Importantly, as the science develops and produces further knowledge the society changes. For instance, if the application of nanotechnology to medicine is successful we are likely to see not only changes in medical practice but also in personal self-care. For STS'ers the goal is not to deny the existence of reality, but rather to show the many ways in which science is deeply implicated in the construction of natural and social realities (see for instance, Stengers 2008). In this sense, the goal of interdisciplinary lab studies is not only to examine the "cultural apparatus" of scientific activity so as to understand how knowledge is produced, but also, to perhaps 'intervene' (Hacking 1983) in this process, or to put in evidence of how things could have been otherwise (Haraway 1991; 1994).

In most cases, lab studies have come about as informal collaborations, whereby an interested social scientist contacts a counterpart in the natural sciences and negotiates access to study the work of scientists in the lab. These studies are marked by a high degree of independence on the part of the ethnographer which facilitates the movement of the ethnographer between belonging to, and being removed from, the research site; a movement that some have argued is key to ethnography (Strathern 1999). Other more formal arrangements have also been put in practice, and these become more common as science policy mandates interdisciplinarity – as illustrated by our own examples. For instance, Jane Maienschein, a historian of science who worked as science policy adviser to a US Congressman, argues that the social sciences and humanities have a role to play in science, namely those of being "partners with different perspectives", and helping scientists communicate with the public (Maienschein 2002, p. 142; see also Maienschein *et al.* 2008). In the nano realm we find two examples that are somewhat similar, those of Gorman *et al.* (2004) and Fisher *et al.* (2006). In both these cases the goal of the ethnographic engagement was less to examine the practices of the production of nano, but to examine and enhance interdisciplinary collaboration. To that effect Gorman *et al.* (2004, p.69-71) put forward the concept of "trading zones" that is, zones of shared expertise, and Fisher *et al.* (2006, p.485) aim to foster reflexivity on the part of the scientists so as to create "governance from within".

In their analysis of the role of the social sciences in nano, Macnaghten, Kearnes and Wynne (2005) argue for a "radically different" approach whereby social sciences: "[r]ender scientific cultures more self-aware of their own taken-for-granted expectations, visions, and imaginations of the ultimate ends of knowledge, and [render] these more articulated, and thus more socially accountable and resilient" (Macnaughten *et al.* 2008, p.278). The issue, however, is how we can accomplish this goal, and the literature on ethnography shows that this is not an easy task. In fact, a common difficulty for the ethnographer lies in the kinds of relationships to subjects and the study site. As mentioned above, the status, role and independence of the ethnographer, and thus her possibilities for action, are largely dependent on the relations established (Brettell 1993; Van Maanen 1988).

In what follows we use our own examples as the 'in-house' social scientists in two nanotechnology centres, located in the United States and the United Kingdom. We will start by examining the institutional contexts that led to our hiring and framed our job descriptions, and will then provide a few examples of how these jobs were put into practice.

4. Cultural Studies of Nanoscience: Two Laboratory Studies in the US and the UK

The science policy landscape that we have outlined has encouraged the creation of posts for social scientists and humanities scholars to be integrated within nanotechnology facilities. Such integration is expected to enhance interdisciplinarity and help create sciences and technologies that are more socially robust. However, in the policy arena interdisciplinarity remains conceived much in the same way as any other scientific grant proposal that demands that scientists include discussion of 'wider societal benefits' and commitments to 'outreach activities'[4] as part of their research proposal. Such activities are then too often delegated to a third party with a tangential role within the actual research agenda. These are the questions that concern us: How is interdisciplinary integration being put into practice? What is it expected to accomplish? To answer these questions we rely on ethnographic data generated in the course of our 3 year experience in two nano facilities.

Our stories are remarkably similar: we both started our positions in 2004 and had similar goals in mind, namely to conduct ethnographic laboratory studies. Our expectation was that the proximity to scientists and engineers provided by our formal integration in these facilities would provide us with first hand access and experience of their cultural apparatuses of knowledge production, namely the tacit understandings,

---

[4]The term 'outreach' here refers to activities aimed at disseminating the research and its findings to non-scientific audiences.

practices and processes through which nano was brought to life. In other words, we thought that by being close to our subjects we would be able to obtain a rich, nuanced and detailed description of how facts about nano are produced and become "visibly-rational-and-reportable-for-all-practical purposes" (Lynch 1993, p. 14, citing Garfinkel 1967, p. vii). We also thought that such closeness would make it easier for us to intervene in the knowledge production of nano by highlighting the overlooked or implicit values, decisions, beliefs and cultural expectations that underlie the making of nano. We naively pictured this as a 'win-win' situation: we would learn about the practices of doing nano, and they would learn to recognise the social dimensions of their work. Instead, we were asked to focus our gaze outwards, where the public is assumed to be located, often antagonistically so. Our ethnographic activities were looked upon with disinterest or suspicion which, as detailed below, eventually led to their falling apart. Our goal is not to conduct a cross-cultural analysis of the US and UK cases, although the similarity of our experiences should be noted. Instead, we want to flesh out some of the challenges and strengths of the current model, so as to illuminate other ways to extend interdisciplinarity to the study and production of science.

4.1 Viseu at the Cornell NanoScale Facility (CNF) and the National Nanotechnology Infrastructure Network (NNIN)

In 2003 the United States' National Science Foundation (NSF) put out a programme solicitation for the creation of a National Nanotechnology Infrastructure Network (NNIN). Envisioned as an "integrated national network of user facilities that will support the future infrastructure needs for research and education in the burgeoning nanoscale science and engineering field" (NSF 2003, p. 2), the NNIN allows individual researchers and small companies to have access to expensive labs and instrumentation and as such it is a key component of the United States' aspiration to leadership in the nano arena. NSF's solicitation specified the numerous requirements that applicants would need to fulfil in order to be considered for the $14 million dollar per year award.[5] Among

[5] For an initial period of five years.

these, NSF prioritised research on the "social and ethical implications" of nanotechnology (NSF 2003, p. 9), specifying that the "quality and appropriateness" (NSF 2003, p. 12) of these plans would be used as an additional review criteria. The 13 node network led by the Cornell NanoScale Facility (Cornell University) recruited Cornell's Science and Technology Studies department (STS) to help outline and coordinate these studies, and later won the competition to run the NNIN.

In the spring of 2004 I was called for an interview for the position of Research Associate in social and ethical issues at the NNIN, and more specifically the Cornell NanoScale Facility (CNF). The interview had two parts, the first at the STS department, the second at the CNF. The STS component focused on my research abilities, particularly my ethnographic research experience, as we discussed the possibilities of studying the culture of scientific and technological production at the CNF. That the 'social' would emerge from research was our common ground. At the CNF however there was an expectation that these issues could be defined *a priori*, as I was specifically asked to name nano's social and ethical issues. The idea that one can detail what is socially problematic or unethical in nano before conducting research is still a commonly held view, one that is closely associated with the belief that the role of the social sciences is to 'predict' the dangers and possibilities that lie ahead.[6] More importantly as Lewenstein (2005) has argued, attempts to define "what counts as a 'Social and Ethical Issue' in nanotechnology" are exercises of power that "preclude our ability to understand the principles inherent in issues that make them social or ethical" (Lewenstein 2005, p. 206), thus also promoting certain sites of inquiry while hindering others. In other words, what is being defined is what constitutes a 'legitimate object of study' and how it should be best approached. Placing the social and ethical dimensions of nanotechnology outside the practices of its production, summons up images of social scientists as mediators between the social and technical worlds, educating the general public about science, and scientists about 'the

---

[6] Which then prompts scientists and engineers to decry the need for early studies in the social and ethical dimensions of science and technology by citing the past failures of the powers of prediction (Nardi 2001).

public's' concerns. In my case this task of mediation had, no doubt, some very pragmatic successes but it also complicated the equally important task of conducting an ethnographic laboratory study. In what follows I use the examples of three projects—web portal, training video, and an ethnography—that I conducted while at the CNF to discuss these issues.

Much of my work at the CNF revolved around two projects—a training video and a web portal—that aimed to inform and educate the users of the NNIN and the general public on the societal and ethical issues of nano. The web portal (SEI n.d) constitutes the public face of NNIN's 'societal and ethical' activities. It provides an overview of the network's activities as well as a database of resources that goes well beyond the NNIN's own efforts. The web portal was the NNIN's preferred medium to engage the public-at-large, and aimed to inform and educate it. In this respect, my experience stands in contrast to that of Doubleday where face-to-face interactions were privileged. The website provided opportunities for learning but little space for exchange or discussion. It is therefore an important part of public engagement but by the same token an incomplete one.

Another project that I engaged in entailed the production of a video on the social and ethical implications of nanotechnology. The audience for this video are new users to the NNIN's 13 facilities: the video is incorporated in the new users' orientation session and is followed by a discussion and/or a short quiz. A pilot version was produced by an engineering undergraduate student, who spent the summer of 2005 doing research at the CNF under my and Lewenstein's supervision. Although at first Priscilla's colleagues wondered aloud about what she could learn from someone who is not a "real scientist" and therefore doesn't know "anything about all of this," our video generated a long discussion when it was presented to all 81 undergraduate students. The video was based on interviews conducted at the CNF's cleanroom and thus provided an empirically oriented discussion of the social and ethical dimensions of nanotechnology. It gave students the opportunity to reflect on issues that are commonly understood to be part of the ethics repertoire of the sciences (Foukal *et al.* 2005), namely the role of students in research, lab safety, and the ethics of publishing, to name a few, but also on issues that are no less important such as the need to communicate and engage the public or the role played by hype in science. The video's most important message was that all these dimensions are part of the scientific process, rather than being extra-scientific. This message is still at the core of the current version of this video (Kysar *et al.* 2007). However this latest version is more abstract, offering a general framework to the practical integration of these concerns within the ordinary practices of science.

Both the video and the web portal are examples of the successful integration of the social sciences in the NNIN. In fact both were prioritised and privileged as means for the evaluation of the overall 'societal and ethical' effort, both internally and externally, that is, not only for those within NNIN but also for the NSF, the funding agency. And this is what makes them problematic, in at least three ways: (1) the issue of who is competent to evaluate interdisciplinary efforts; (2) what criteria should be used to conduct such evaluation; and finally, (3) the assumptions about the role of the social sciences that are being reified through it. The first issue seems to have been recognized as problematic by the US House of Representatives, since it is requesting, in its amendments to the 21st Century Research and Development Act, that the National Nanotechnology Initiative, which as stated above, oversees the nano R&D research efforts in the US, establish a "subpanel [within its Advisory Panel] to enable it to assess whether societal, ethical, legal, environmental, and workforce concerns are adequately addressed by the [NNI]" (H.R.554 2009). The second is equally important, by prioritising methods of evaluation that rely on the principle that scientific findings are those that can be empirically verified, the funding agencies are thereby facilitating and rewarding quantitative and 'short term' research and results. These have value but they must be complemented with more complex kinds of research—such as ethnographies or case studies—that are common in the social sciences and humanities and that generally take longer to yield results. The third issue comes as an effect of the previous two and it speaks to the ways in which such modes of evaluation reinforce dominant assumptions of what social scientists can do, namely to communicate the science to the public at large, and to represent the public in the lab. The science here is less a research site, and more a launching pad; the social sciences more of a service entity that fills in for an absence in the natural sciences (Barry *et al.* 2008). The last example that I provide illustrates how these three can become problematic.

As mentioned above my goal upon arrival at the CNF was to conduct ethnographic research. Because this involves being proficient in the 'language' and 'skills' of nanofabrication, I asked a colleague to take me into the cleanroom and teach me about the techniques of nanofabrication. The experiment was successful and I learned much, however, it also underscored the disciplinary boundaries that must still be overcome, since I was later told that my training was going to be included as 'outreach' in the next NSF review report. I started conducting my ethnography by participating and taking notes of the day-to-day activities. A few months afterwards I was asked to speak at the "Short Course on Technology & Characterisation at the Nanoscale" that the CNF organises. I decided that rather than focus exclusively on general matters that are important to discussions of the social and ethical issues in nanotechnology—such as public acceptance, regulation, intellectual property and health and safety—I would also discuss some of the issues that I had been uncovering in my ethnographic research. I discussed the shifts in professional identities experienced by CNF personnel who work both as researchers and lab technicians, the negotiations surrounding definitions of 'contamination' and finally the relationship between the machines that populate the CNF's cleanroom and their 'owners.'

The talk was well received by the audience, so I was surprised when a few days later during the weekly staff meeting my colleagues (and I) were publicly warned that everyone had to be careful about what they said around me. This indicated that my ethnographic study of the CNF, until then largely ignored, was understood as a sort of antagonistic activity. This view of the social sciences as adversaries rather than partners is not uncommon and may be fuelled by the fact that it is imposed upon scientists by the funding agencies, but in practice it made it difficult for me to continue my ethnographic research. Another reason that contributed to the falling apart of my ethnography was the difficulty that I faced in separating my role as staff from that of ethnographer. Being a resident staff member made me privy to a number of things that I wouldn't otherwise be, both at the level of personal exchanges and institutional activities. The ethics of separating the 'personal' from 'work' were compounded by the difficulties of studying those who hired me and paid my salary and the ensuing power asymmetries. These asymmetries did not manifest themselves in direct orders or confrontations but in mundane transactions that are laden with power, such as the need to demonstrate my 'usefulness' to the facility, and being evaluated accordingly. Power asymmetries have been identified as one of the problems with formal participatory methods to engage the public in science (Joly & Kaufmann 2008), my experience, and that of Doubleday, show that more work must be done examining their role in calls for institutional interdisciplinary integration.

4.2 Doubleday at the University of Cambridge Nanoscience Centre

The first case of 'embedding' social science in a UK nanotechnology laboratory was proposed in 2003 by Professor Welland, director of the Nanoscience Centre at the University of Cambridge. UK policy proposals for promoting social science collaborations with nanoscientists were implemented in a more *ad hoc* fashion than those in the US discussed above. Unlike the US, the UK did not have a coordinated national research policy for nanotechnology, and experimental inclusion of social science in interdisciplinary research was left as a matter for individual researchers.

The Cambridge Nanoscience Centre project called for a social science post-doctoral research associate to work with Welland on the 'social implications of nanotechnology'. The broad job description included supporting interactions among nano and social scientists, facilitating public and stakeholder dialogue, and time to carry out independent research. I applied for this position as someone with prior research experience in the study of public controversy over emerging technologies, and more importantly with a desire to complement my knowledge about nanotechnology by conducting ethnographic research at the Centre. When I took up the post in May 2004 Welland acknowledged that the initial framing of the project was broad and told me that I should define the work myself. I hoped to use the position to put into practice suggestions that ethnographic research carried out collaboratively with nanoscientists could contribute to greater reflection among laboratory scientists on the social dimensions of their work. This reflection would in turn lead to the development of more 'socially robust' nanotechnologies. However in practice, as with the Cornell example, practical constraints quickly emerged which shaped the focus of the project.

As in the case of Viseu's work at the Cornell NanoScale Facility, my research project was in some ways successful, and was certainly regarded as such by Welland and the funders of the Nanoscience Centre. However, my ethnographic project can be said to have failed in two ways that are important for the argument developed in this chapter. First, as the lone lab-based social scientist reporting to the director of the Nanoscience Centre there were significant practical constraints to carrying out an ethnographic laboratory study. Second, focusing on the engagement of junior laboratory scientists with social science overestimates the degree of influence individual scientists have over technological trajectories. The examples below illustrate why this is so.

Soon after I arrived at the Nanoscience Centre, Welland suggested that I help coordinate a public engagement project. Since Welland was also my direct boss, his suggestions took on an added strength. The 2004 RS & RAE report had argued that public engagement was a necessary element of the responsible development of nanotechnologies. Welland had discussed ways to put this recommendation into practice with Dr Parr, Greenpeace UK's Chief Scientific Adviser. Through these initial discussions with Parr a consortium of four partners was formed to run a citizens' jury on nanotechnology during the summer of 2005.[7] The aim of the *NanoJury* project was to adapt standard citizens' jury methods to facilitate deliberation among a small group of citizens about developments in nanotechnology. The group of citizens heard evidence from six witnesses who represented a range of perspectives from academia, government, industry, and environmental and development non-governmental organisations. The *NanoJury* then developed a series of recommendations for nanotechnology policy.

During the *NanoJury* project my task was to mediate between scientists at the Nanoscience Centre and the citizens' jury project. To

---

[7] See Gavelin *et al.* 2007 for a brief description of the NanoJury and its findings.

Welland my role made perfect sense because I was someone whose prior research experience meant that I had some familiarity with the process and aims of citizens' juries and my day-to-day knowledge of the Nanoscience Centre meant that I was able to relay relevant information about nanoscience research to the *NanoJury*. My direct involvement as a 'go-between' passing from the lab to the public engagement project also meant that the scientists themselves could remain back at the laboratory, away from the process of engagement and able to continue with their research. This division of labour justified my presence at the Centre, but it also worked to limit scientists' direct involvement in the public engagement, and also served to shift my research focus away from the laboratory itself, reducing my opportunities to carry out ethnographic research at the Nanoscience Centre.

The final report of the *NanoJury* raised a series of quite broad questions about how science policy priorities are set. The jury findings called for greater public debate about the aims of innovation policy and more explicit effort to develop technologies to meet democratically defined needs. My task was then to communicate the results back to scientists. However the topics covered by the jury mostly concerned national research policy, and were not of immediate relevance to the graduate students and early career scientists I presented to. This mismatch between the register of the public engagement project and the concerns of laboratory scientists could not be overcome by the social scientist acting as translator between the wider public and scientists. As Joly and Kaufman (2008) suggest, greater attention needs to be paid the instituitional mechanisms by which wider public engagement can contribute to shaping technological development. It is exactly to this question of how and at what point research agendas are established in the laboratory that laboratory ethnography can contribute through its in-depth study of the practice rather than the formal accounts of how science works.

The practical constraints represented by my attention to the citizens' jury and away from the laboratory illustrate a strong initial assumption held by many scientists at the Nanoscience Centre that social aspects of nanotechnology were only relevant when considering public attitudes to the technology; and additionally, that public attitudes were only significant inasmuch as they were either positive or negative towards nanotechnology. This view is illustrated by a conversation I had with one research student after a seminar I gave. I was asked: "So, what will you tell the public about nanotechnology?" In response I attempted to turn the question around to see how he would talk about nanotechnology. He said "But I'd just tell them it's wonderful!" When I pressed him, he admitted that he did not think nanotechnologies were necessarily wonderful. He went on to say: "Well, I think there are some things that worry me - like privacy – and sensors that could be used for surveillance - and military uses." In the ensuing discussion about challenges to privacy from increasingly sophisticated and miniaturised devices, I asked if he thought that these topics were something he would like to reflect on more. He said that he would not as it would not make any difference what he thought or did, because the research would carry on in that direction anyway.

This exchange illustrates the limits of policies that seek to promote the responsible development of nanotechnologies through the interaction of lab scientists and social scientists. On one hand this conversation, and many other similar conversations that I engaged in, showed that the laboratory scientists were aware of and interested in social and ethical dilemmas raised by nanoscience research. On the other hand reluctance of the researcher to discuss the ultimate objectives of nanotechnology research spoke to his sense of lack of influence over the direction of development, and his dependence on being able to contribute towards the wider set of technological goals in order to progress in his career.

This latter point about the importance of junior researchers understanding and conforming to underlying assumptions about the appropriate goals for nanotechnology research was supported a year later when I was canvassing interest in a digest of information on nanotechnology policy developments. I had hoped that a regular summary of policy debates might spark some discussion among scientists at the lab about the wider political and social context of their research. However, the only response my email elicited was from the same graduate student, who was then approaching the end of his PhD. He said it would be most helpful if I could suggest which areas of nanoscience research would attract the most funding as he was in the process of working out what direction to take his research in next. This was not the response I had expected, but illustrated the significant pressures on junior researchers as they seek a career in science. It also illustrated the limits of the model of providing opportunities for these scientists to reflect on the social and ethical aspects of their research through engagement with social scientists as a significant element of the responsible development of nanotechnologies.

These two episodes, my work on the *NanoJury* and conversations with graduate students, illustrate two constraints of the wider 'embedded' social science model. The first demonstrates the assumption that the role of social science is to engage with the potentially problematic social aspects of developments in nanotechnology, and that these are defined in terms of their potential for negative public reactions. Therefore the social scientists is seen as a mediator between the laboratory and an external public. The second episode points to limits in the assumption that focusing on working with individual researchers can alter the trajectories of nanotechnologies.

My initial intention on taking up my role as an 'embedded' social scientists was to use ethnographic methods to study how commitments to long-running visions about the purposes of particular technological developments were entered into in the everyday work of laboratory science. My hope was to use this study to develop a better understanding of the relationship between the wider public and political context for research and the strategic decisions that academic scientists make about their research. However, as I have shown, a variety of pressures to turn my attention away from the laboratory towards external publics limited my capacity to carry out research on laboratory practice. This dynamic was, in some ways, similar to that experienced by Viseu at Cornell. We now turn to a discussion of the similarities of our experiences to draw out lessons for policies that support the responsible development of nanotechnology through expanding the scope of interdisciplinarity to include the social sciences.

5. Discussion

One of the striking conclusions about our experiences as lab-based social scientists in the US and UK is the similarity of our experiences. This similarity is important in the face of the distinct national policy contexts and local institutional settings under which we were operating. In the US, nanotechnology policy is nationally co-ordinated, with layers of political oversight and expectations that social science research will be conducted alongside nanoscience and technology. Within this framework the Cornell lab was funded not as a site to develop a research agenda but one to provide fabrication facilities for visitors—academic and non-academic. By contrast the Cambridge lab was more concerned with 'basic' academic research and the training of graduate students. Moreover, in the UK forms of direct public participation are more readily recognised (and prioritised) within the policy culture as an appropriate response to questions of the social dimensions of nanotechnology. These framing differences make the similarities between the two experiences all the more important because they reveal common unspoken assumptions about the role of the social sciences in their integration with natural sciences. It is on these commonalities that we focus here.

We both found that our projects—and to a large degree our ability to contribute to the 'opening up' of nanotechnology research and development—suffered under the weight of expectation placed on the post of lab-based social scientists as contributing directly to enhancing the social responsibility of nanotechnology research. This expectation sustains four significant misconceptions about the role of the lab-based social scientist. It assumes:

- That there is one 'public' with homogeneous and predictable views.
- That the social scientist can mediate between science (or the lab) and 'the public'.
- That 'the public' in question is potentially antagonistic towards the laboratory (and conversely, that being better informed will result in greater acceptance).
- That science is only social to the extent that it has to accommodate or take this 'public' into account.

These assumptions underlie the emphasis on communication with an imagined public through web-sites and public engagement projects, with the unstated goal of educating this public. The positioning of social science as mediating between the public and the lab had the associated effect of identifying the social scientist with a 'public' perspective.

Also damaging to the project of carrying out an ethnography of laboratory practice was the duty placed on us to contribute tangibly and immediately to the 'responsible development' of nanotechnology. At fault is the model of collaboration put forward in institutional reports and documents, such as the early NSF account of the social dimensions of nanotechnology (Roco & Bainbridge 2001). Here the editors argue that locating social science research in the nanotechnology centres will, by itself, lead to "openness, disclosure, and public participation" (p. iv). In this logic a nanotechnology laboratory that contains social science will be more responsive to public concerns and aspirations, and thus more responsible, and less likely to generate future public resistance to commercialisation of the technology. It is in the framework of this concern with a resistant public that turning of the social scientist's gaze towards the inside of the laboratory is seen as potentially dangerous, as illustrated by the falling apart of our ethnographic projects.

This vision of collaboration is misguided in its objectives—that social sciences can predict and appease (Ebbensen 2008) public attitudes towards emergent technologies; it is also blind to the enduring power asymmetries (Joly & Kaufmann 2007) that are endemic to these initiatives and stem, not only from deep-seated assumptions relating to labour divisions but also from budgetary differences which dictate objectives and research agendas.[8] The issue of funding should not be underestimated because as the Director of the CNF told Viseu, the reason why there was only one social scientist on site (hardly a critical mass) is because the facility's mandate is to provide technical facilities. If the goal is to challenge traditional and 'irresponsible' modes of scientific production, then we must also challenge these asymmetries.

---

[8]See Marcus (2002) for a discussion of these power asymmetries and how they are particular harmful for the social sciences.

Which leads us to another problem with the current vision of interdisciplinary collaboration. It rests on a flawed model of science, one that ignores the local, contingent and collective character of scientific production, that is, its social character. Past laboratory studies in the field of STS have underscored this by describing science as an irreducibly collective activity involving numerous scientists, engineers, students, technologies, institutions, funding agencies, and scientific journals, among others. Science and innovation are collective activities that take time and follow unpredictable paths of development. This, of course, is not to say that they cannot be steered in the public interest, nor that social science cannot play a constructive role in this process. It does however suggest that placing the emphasis on individual laboratory scientists as the locus of responsibility is mistaken.

What then do our observations and experience suggest for taking forward the opportunities for collaborative laboratory studies presented by developments in nanotechnology? We suggest recasting the role of the social sciences. Not as a 'service' discipline that is brought in to fill a vacuum in the sciences and is placed in a relation of subordination (Barry *et al.* 2008). Not as an 'applied' science that strives to find answers for previously known questions, or solutions for existing problems, in this case applied to policy making or generating public acceptance. But, perhaps as a 'basic' science, one that is exploratory and, one that seeks to expand knowledge. In this framework, interdisciplinary collaborations, and particularly ethnographies, are conducted to describe the tacit understandings, choices, and conceptual frameworks that are later naturalized as part of the scientific process. We would not want to suggest that there is no room for the social sciences to intervene directly in the conduct of natural sciences. On the contrary, we believe that this intervention is crucial and caution that the lingering questions of autonomy, authority and legitimacy posed by the current model must be studied further.

6. Conclusion

We started this chapter describing the lure of interdisciplinarity as a solution to the problem of developing 'socially responsible' nano sciences and technologies. We argued that while collaborative engagement of social science with nanotechnology research was understood as unproblematic from a policy stance, little had been written on its actual practice. Moreover, we examined and critiqued the assumptions behind such calls for interdisciplinary integration as being based on the assumption that the social sciences can (and should) represent the public, and thus that their integration will decrease the possibility of resistance by the public to nanotechnology. Arguing that this model is flawed we turned to the field of science and technology studies in search for other models for the study of scientific development, namely that of the ethnographic study of scientific knowledge production.

We then described our work as the 'in-house' social scientists in two nano facilities located in US and the UK, and the difficulties we experienced. It is important to emphasise that both the Cornell University and University of Cambridge laboratories where we worked were receptive and open to much of our work, and our colleagues among the nanoscientists were supportive and curious about our social science projects. The limits to interdisicplinarity we have discussed are not in any way criticisms of the specific arrangements of our projects, but rather speak to much more general concerns about the role of social science in nanotechnology.

In short, these difficulties put in evidence the following: (1) the deep power asymmetries between the social sciences and the natural sciences that are largely ignored in the current science policy. These asymmetries, we suggest, pertain not only to funding but also to the power to define research problems, questions, agendas, and importantly, to evaluate the research activities; (2) the ways in which the current models of institutional integration do not seek to reconfigure the relationship between the social and natural sciences, or between the natural sciences and society, but instead reify deep seated assumptions as to their nature and character; and finally, (3) the ways in which the formal integration of social scientists poses challenges to the work of those scientists namely in terms of the always complicated issue of managing the relationship between researcher and subjects. This formal integration creates a situation where the researcher must be 'useful' and the subjects are the 'bosses.' Our stories are then a cautionary tale against, not the practice of interdisciplinary collaboration, but rather its formalization into science policy. One solution is to return to the kinds of collaborations promoted by STS, whereby the social scientist possesses the power and autonomy to frame his/her own research.[9]

Questions for Reflection:

1. What sort of interesting and/or useful information do you think an ethnographic study of (your own) scientific laboratory could provide?
2. How do you think nanoscale science and technology changes when natural scientists collaborate with social scientists?
3. How are the goals for research in nanoscale science and technology decided? Can you suggest ways to encourage dialogue between groups that might have different goals (researchers, funding bodies, citizens etc.)?

[9] Doubleday would like to acknowledge financial support from the Wellcome Trust (Society Award WT080193MA), which made further reflection on his role a lab-based social scientist possible. Viseu's work was supported by the National Science Foundation under Grant no. ECS-0335765. Support came also from the Centro de Estudos Sociais, Universidade de Coimbra, Portugal, through a grant from the Fundação para a Ciência e Tecnologia.

Bibliography

Barry, A., Born, G. and Weszkalnys, G. (2008). Logics of interdisciplinarity, *Economy and Society*, 31, pp. 20-29.

Bennett, I. and Sarewitz, D. (2006). Too little, too late? Research policies on the societal implications of nanotechnology in the United States, *Science as Culture*, 15, pp. 309-325.

Bloor, D. (1991 [1976]). *Knowledge and Social Imagery* (second edition with a new foreword), Chicago: University of Chicago Press.

Brettell, C. B. (ed) (1993). *When They Read What We Write: The Politics of Ethnography*, Westport, Connecticut & London: Bergin & Garvey.

Doing, P. (2008). Give me a laboratory and I will raise a discipline: The past, present, and future politics of laboratory studies in STS. In Hackett, E. J., Amsterdamska, O., Lynch, M. and Wajcman, J. (eds) *The Handbook of Science and Technology Studies*, Cambridge, MA: MIT Press, pp. 279-296.

Department for Trade and Industry (2005). *Response to the Royal Society and Royal Academy of Engineering Report: "Nanoscience and Nanotechnologies: Opportunities and Uncertainties"*, HM Government, London.

Ebbesen, M. (2008). The role of the humanities and social sciences in nanotechnology research and development, *NanoEthics*, 2, pp. 1-13.

European Commission (2004). *Towards a European Strategy on Nanotechnology,* Brussels: European Commission. ftp://ftp.cordis.europa.eu/pub/nanotechnology/docs/nano_com_en.pdf (last accessed Mar 30, 2009).

Fisher, E. and Mahajan, R. L. (2006). Contradictory intent? US federal legislation on integrating societal concerns into nanotechnology research and development, *Science and Public Policy*, 33, 1, pp. 5-16.

Fisher, E., Mahajan, R. and Mitcham, C. (2006). Midstream modulation of technology: Governance from within, *Bulletin of Science, Technology & Society*, 26(6), pp. 485-496.

Fujimura, J. H. (1987). Constructing 'Do-Able' problems in cancer research: Articulating alignment, *Social Studies of Science*, 17, pp. 257-293.

Garfinkel, H. (1967). *Studies in Ethnomethodology*, Englewood Cliffs, N.J.: Prentice-Hall.

Gavelin, K., Wilson, R. and Doubleday, R. (2007). *Democratic Technologies? The Final Report of the Nanotechnology Engagement Group.* London: Involve. http://www.involve.org.uk/assets/Publications/Democratic-Technologies.pdf (last accessed Mar 30, 2009).

Geertz, C. (1973). *The Interpretation of Cultures: Selected Essays*, New York: Basic Books.

Gibbons, M. (1999). Science's new social contract with society, *Nature*, 402, pp. C81-C84.

Gibbons, M., Limoges, C., Nowotny, H., Schwartzman, S., Scott, P. and Trow, M. (1994). *The New Production of Knowledge: The Dynamics of Science and Research in Contemporary Societies*, London: Sage.

Gorman, M. E., Groves, J. F. and Shrager, J. (2004). Societal dimensions of nanotechnology as a trading zone: Results from a pilot project. In Baird, D., Nordmann, A. and Schummer, J. (eds) *Discovering the Nanoscale,* Amsterdam: IOS Press, pp. 63-73.

Guston, D. H. (2000). *Between Politics and Science: Assuring the Integrity and Productivity of Research*, Cambridge: Cambridge University Press.

Guston, D. H. (2008). Innovation policy: Not just a jumbo shrimp, *Nature*, 454, pp. 940-941.

Guston, D. H. and Sarewitz, D. (2002). Real-time technology assessment, *Technology in Society*, 24, pp. 93-109.

Hackett, E. J., Amsterdamska, O., Lynch, M. and Wajcman, J. (2008). Introduction. In Jasanoff, S., Markle, G. E., Petersen, J. C. and Pinch, T. (eds) *Handbook of Science and Technology Studies*, Thousand Oaks, CA: SAGE Publication, pp. 1-8.

Hacking, I. (1983). *Representing and Intervening: Introductory Topics in the Philosophy of Natural Science*, Cambridge: Cambridge University Press.

Haraway, D. (1991). *Simians, Cyborgs and Women: The Reinvention of Nature*, New York: Routledge.

Haraway, D. (1994). A game of cat's craddle: Science studies, feminist theory, cultural studies, *Configurations*, 2, pp. 59-71.

House of Commons Science and Technology Committee (2004). *Too Little Too Late? Government Investment in Nanotechnology*, Fifth report of session 2003-2004, London: House of Commons.

H.R. 554 – 111th Congress (2009). National Nanotechnology Initiative Amendments Act of 2009, *GovTrack.us (database of federal legislation).* http://www.govtrack.us /congress/bill.xpd?bill=h111-554&tab=summary&page-command=print (last accessed Mar 28, 2009).

Jasanoff, S. (1997). Civilization and madness: The great BSE scare of 1996, *Public Understanding of Science,* 6, pp. 221-232.

Jasanoff, S. (2004). The idiom of co-production. In Jasanoff, S. (ed) *States of Knowledge: The Co-production of Science and The Social Order*, London: Routledge.

Jasanoff, S. (2005) *Designs on Nature: Science and Democracy in Europe and the United States,* Princeton: Princeton University Press, Chapter 9.

Jasanoff, S., Markle, G. E., Petersen, J. C. and Pinch, T. (eds) (1995). *Handbook of Science and Technology Studies*, Thousand Oaks, CA: SAGE Publication.

Johnson, A. (2004). The end of pure science: Science policy from Bayh-Dole to the NNI. In Baird, D., Nordmann, A. and Schummer, J. (eds) *Discovering the Nanoscale*, Amsterdam: IOS Press, pp. 217-230.

Joly, P. B. and Kaufmann, A. (2008). Lost in translation? The need for 'upstream engagement' with nanotechnology on trial, *Science as Culture*, 17, pp. 1-23.

Kearnes, M. P. and Wynne, B. (2007). On nanotechnology and ambivalence: The politics of enthusiasm. *NanoEthics*, 1, pp. 131-142.

Knorr Cetina, K. (1995). Laboratory studies: The cultural approach to the study of science. In Jasanoff, S., Markle, G., Petersen, J. C. and Pinch, T. (eds) *Handbook of Science and Technology Studies*, Thousand Oaks, California: SAGE Publication, pp. 140-166.

Knorr Cetina, K. (1999). *Epistemic Cultures: How the Sciences Make Knowledge*, Cambridge, MA: Harvard University Press.

Kysar, D., Viseu, A. and Guston, D. (2007). Societal and ethical implications of nanoscale science and engineering (video), societal and ethical issues in nanotechnology — National Nanotechnology Infrastructure Network (NNIN). http://nnin.org/nnin_ethicstraining.html (last accessed Mar 13, 2009).

Latour, B. (1987). *Science in Action: How to Follow Scientists and Engineers Through Society*, Milton Keynes, Philadelphia: Open University Press.

Latour, B. (1993). *We Have Never Been Modern* (translated by Catherine Porter), Cambridge, MA: Harvard University Press.

Latour, B. and Woolgar, S. (1979). *Laboratory Life: The Social Construction of Scientific Facts*, Beverly Hills & London: Sage Publications.

Law, J. (1991). Introduction: Monsters, machines and sociotechnical relations. In Law, J. (ed) *A Sociology of Monsters: Essays on Power, Technology and Domination,* London and New York: Routledge, pp. 1-23.

Lewenstein, B. (2005). What counts as a 'social and ethical issue' in nanotechnology? *HYLE-International Journal for Philosophy of Chemistry*, 11, pp. 5-18.

Lux Research (2008). *Nanotechnology Corporate Strategies*, New York: Lux Research Inc.

Lynch, M. (1993). *Scientific Practice and Ordinary Action*, Cambridge: Cambridge University Press.

Macnaghten, P. M., Kearnes, M. P. and Wynne, B. (2005). Nanotechnology, governance, and public deliberation: What role for the social sciences? *Science Communication*, 27, pp. 1-24.

Maienschein, J. (2002). Innocent reflections on science and technology policy, *Technology in Society*, 24, pp. 133-143.

Maienschein, J., Laubichler, M. and Loettgers, A. (2008). How can history of science matter to scientists? *Isis*, 99, pp. 341-349.

Marcus, G. E. (2002). Intimate strangers: The dynamics of (non) relationship between the natural and human sciences in the contemporary U.S. University, *Anthropological Quarterly*, 75, pp. 519-526.

Marcus, G. E. and Fischer, M. (1986). *Anthropology as Cultural Critique: An Experimental Moment in the Human Sciences*, Chicago: University of Chicago Press.

Mody, C. (2004). How probe microscopists became nanotechnologists. In Baird, D., Nordmann, A. and Schummer, J. (eds) *Discovering the Nanoscale,* Amsterdam: IOS Press, pp. 119-133.

Nowotny, H., Scott, P. and Gibbons, M. (2001). *Re-thinking Science: Knowledge and the Public in an Age of Uncertainty*, Cambridge: Polity Press.

National Science Foundation (NSF) (2003). *National Nanotechnology Infrastructure Network (NNIN) Program Solicitation* (NSF 03-519). http://www.nsf.gov /pubs/2003/nsf03519/nsf03519.pdf (last accessed June 18, 2006).

Nardi, B. (2001). A cultural ecology of nanotechnology. In Roco, M. and Bainbridge, W. (eds) *Societal Implications of Nanoscience and Nanotechnology*, final report from the workshop held at the National Science Foundation in Sept. 28-29, 2000, Arlington, VA: NSF, pp. 246-251. http://www.wtec.org/loyola/nano /NSET.Societal.Implications/nanosi.pdf (last accessed Feb 7, 2004).

Foukal, P., Melsheimer, F., Albrecht, G., Marque, J., Thomsen, M., West, J. O., Kirby, K., Houle, F., Finegold, L. and Whitbeck, C. (2005). Letters: Ethics concern draw many questions, some answers, *Physics Today*, 58, pp. 12-16.

Public Law 108-153 – 108th Congress (2003). *21st Century Nanotechnology Research and Development Act.* http://frwebgate.access.gpo.gov/cgi-bin/getdoc.cgi?dbname =108_cong_public_laws&docid=f:publ153.108.pdf (last accessed June 12, 2008).

Roco, M. C. (2003). Broader societal issues of nanotechnology, *Journal of Nanoparticle Research*, 5, pp. 181-189.

Roco, M. C. and Bainbridge, W. (eds) (2001). *Societal Implications of Nanoscience and Nanotechnology*, final report from the workshop held at the National Science Foundation in Sept. 28-29, 2000. Arlington, VA: NSF. http://www.wtec.org/loyola/nano/NSET.Societal.Implications/nanosi.pdf (last accessed Feb 7, 2004).

Royal Society & The Royal Academy of Engineering (2004). *Nanoscience and Nanotechnologies: Opportunities and Uncertainties*, London: Royal Society http://www.nanotec.org.uk/finalReport.htm (last accessed Aug 8, 2008).

Societal and Ethical Issues in Nanotechnology (SEI) (n.d.) *Societal and Ethical Issues in Nanotechnology*, USA: National Nanotechnology Infrastructure Network http://www.sei.nnin.org/ (last accessed Jan 18, 2008).

Shapin, S. and Schaffer, S. (1985). *Leviathan and the Air Pump: Hobbes, Boyle and the Experimental Life*, Princeton, New Jersey: Princeton University Press.

Stengers, I. (2008). A constructivist reading of process and reality, *Theory, Culture & Society*, 25, pp. 91-110.

Strathern, M. (1999). *Property, Substance and Effect. Anthropological Essays on Persons and Things*, London: Athlone Press.

Traweek, S. (1988). *Beamtimes and Lifetimes*, Cambridge: Harvard University Press.

Van Maanen, J. (1988). *Tales of the Field: On Writing Ethnography*, Chicago: Chicago University Press.

Wilsdon, J. and Willis. R. (2004). *See-through Science: Why Public Engagement Needs to Move Upstream*, London: Demos.

Wood, S., Jones, R. and Geldart, A. (2003). *The Social and Economic Challenges of Nanotechnology*, Swindon: Economic and Social Research Council.

Wood, S., Jones, R. and Geldart, A. (2007). *Nanotechnology: From the Science to the Social*, Swindon: Economic and Social Research Council.

US National Science and Technology Council (US NSTC) (2004). *The National Nanotechnology Initiative: Strategic Plan*, Washington D.C.: National Science and Technology Council.

Chapter 3

The Science and Politics of Nano Images

Roger Strand & Tore Birkeland

This chapter provides an introduction to two perspectives on nano images: that of nanoscience and that of research on the ethical, legal and social aspects (ELSA) of nanoscience and nanotechnology. Whereas the nanoscientist may ask what object or state of affairs on the nanoscale is evidenced by the image, the ELSA researcher might be more interested in questions such as: 'What can I learn about the research and the researchers making this image, in terms of their objectives, world views, cultural conventions, interests or values?' Discussing a few examples, we argue that ELSA research can be enriched by an understanding of the scientific reading of nano images. Conversely, nanoscientists may learn from ELSA research in order to arrive at ethically and politically sound use of images.

Nano Meets Macro - Social Perspectives on Nanoscale Sciences and Technologies
by K L Kjølberg & F Wickson

www.panstanford.com
978-981-4267-05-2

1. Introduction

Nano images—images of objects or states of affair at the nanoscale—are increasingly found not only in the scientific literature, but also in documents on nanoscience and nanotechnology policies; in education and popularisations; and to an increasing degree in popular culture and art. For instance, at the time of writing (April 2009), 'nanoart' had its own page in the English and Italian versions of the online encyclopaedia Wikipedia.

In this chapter we will discuss the use of nano images. In doing so, we will try to combine two rather different perspectives that in our opinion ought to be more exposed to each other. One perspective is that of research on the ethical, legal and social aspects (ELSA) of nanoscience and nanotechnology. The other perspective is that of nanoscience and nanotechnology itself. Discussing a handful of examples, we wish to show what kind of questions, answers and reflections can be produced when an ELSA-of-nano researcher meets a nano researcher. Indeed, this is how the chapter came about: as an effort of collaboration and co-production between a natural scientist (Birkeland) and an ELSA researcher (Strand). Rather than excelling in one or the other of the two directions, the chapter tries to provide an introduction to the two perspectives and produce a common ground from which interdisciplinary exchange of ideas may take place. The common ground will have the shape of a simple 'method' that does little more than organising the work of interpreting nano images to avoid possible misunderstandings in interdisciplinary debate.

For a number of closely interrelated reasons, the distance between nano and ELSA of nano should not be underestimated. The differences can be seen as a typical instance of what C. P. Snow (1961) once called the "two cultures" of natural science on the one hand and the humanities (and social sciences) on the other. With regard to scientific images, such as nano images, the differences will, in part, relate to emphasis and interest. When confronted with, say, a scanning tunnelling microscopy (STM) image, the nanoscientist might be more interested in what he or she may learn from the image about the natural world: 'What structure in the nano world do I see evidence of in this image? Is it a valid and reliable representation of the nano structure?' The ELSA researcher might – although not always and not necessarily – be more interested in questions such as: 'What can I learn about the research and the researchers making this image, in terms of their objectives, world views, cultural conventions, interests or values? How will the image be read and interpreted by different audiences?' While one perspective places emphasis on the 'natural', the other may place its emphasis on the 'social' (or cultural and political).

We shall see that the differences can go deeper than a question of emphasis. Although philosophical nuances exist within the scientific communities, natural science is often taught and performed under the (possibly implicit) assumption that there exists an objective, uniquely correct description of the physical world, and that science is the pursuit of that description. A slightly less subtle way of saying this is that (natural) science is the pursuit of the truth about the world out there. Within the social sciences and the humanities, however, it is frequently assumed (and argued) that there can be no unique objective truth, and that also the practice of natural science should be described and understood partly in social, cultural and political terms. This latent disagreement is sometimes exposed in full in ELSA research in which social scientists study what natural scientists do. For instance, nanoscientists may think that they are merely engaged in the innocent description of nature as it really is, while the ELSA researchers may see the work of the nanoscientists as—in part—historically, culturally and politically contingent, and rightly the subject of ethical or political controversies.

Such differences in perspectives need not be severe. Indeed, the underlying motivation behind this chapter is that interdisciplinary discussion is both possible and valuable. For the sake of clarity, however, it may be useful to sometimes accentuate the difference. In this respect, the interpretation of images is actually an important instance of such disagreements across the 'two cultures', down to the definition of what an image is, or if nano images really should be called images (Pitt 2004; 2005). There is no consensus in the literature on how to define terms such as 'image', 'picture' and 'visualisation', and one should always be aware of the risk of definitional and conceptual confusion. In this chapter, we will use the term 'picture' synonymous with 'illustration', that is, as a general word for graphic material defined as "something that visually explains or decorates a text".[1] We will reserve the word 'image' for pictures that are intended to represent, or interpreted to represent, objects or states of affair. That is, we shall discuss images also in terms of their having a *reference* in the real world. This choice is not at all trivial since it takes us out of most social science literature on how to understand images. As will be clear below, however, it is a central element in our effort to try to develop a common ground for interdisciplinary debate. The argument would be that even if one does not believe that nano images really can represent nano objects, it is still highly relevant to understand if, when and how such images are understood as representational by nanoscientists. ELSA researchers, being social scientists, have knowledge and skills in how to interpret images; but so have nanoscientists. What we aim at, is an exchange of some of the knowledge and skills involved on both sides. This implies that most readers will find some part of the chapter quite introductory, but hopefully not trivial.

Our approach is as follows: We will begin with a discussion of political aspects of images and pictures in general, exemplified by an important brochure about nanotechnology intended for the general public. This will serve as an introduction to how one may interpret the use of images in terms of their choice of motive and composition. We will then gradually move to nano images proper, that is, images that are intended to represent objects or states of affair at the nanoscale. Our main argument in that part is that a proper understanding of how nano images are interpreted by nanoscientists resolves (or rather renders irrelevant) the philosophical quarrels about "images not really being images". The main message throughout the chapter is that one should always remain critical when interpreting images, and that there are many different intellectual resources to draw upon when doing so. Finally, the main points are summarised in a simple 'method' for an interdisciplinary analysis of images.

---

[1] Definition of "illustration" by the Merriam-Webster online dictionary, http://www.merriam-webster.com/

2. "Nanotechnology—Innovation for Tomorrow's World"

The title of this chapter, "The Science and Politics of Nano Images", might be surprising to some. However, we shall see that images – also nano images – have political aspects. By 'politics', we do not merely refer to institutions (such as elections and governments), but more generally to processes in society involving or relating to power or authority, for instance with respect to decisions on rights and duties, wealth and welfare. In this sense, images can be made, used, interpreted or passively received in a political manner and with a political influence.

Our main example here is the 60-page brochure *Nanotechnology – Innovation for tomorrow's world* (European Commission 2004). This brochure was originally published by the German Ministry of Education and Research. It was later adopted by the European Commission (EC) and translated into 18 languages. Before proceeding with this chapter, downloading a copy of the brochure is recommended[2]. Page numbers in this chapter refer to the English version, but we suggest that you first leaf or scroll through the brochure in a *language you do not know*, in order to get a first impression of the use of images.

According to the European Commission, the purpose of the brochure is "to illustrate to the public what nanotechnology is" (EC 2004, p.3). 'Illustrate' is the appropriate word. The brochure combines text and bright, colourful pictures in a powerful graphical design. Among the many pictures, you will find:

- nano images (images of objects and states of affair at the nanoscale)
- images of nano laboratory equipment and scientists
- images of nanotechnology as well as potential fields of application for nanotechnology
- images of macroscopic objects (plants, animals, micro-organisms; the Earth, the sky, stars)
- imaginative and artistic pictures and drawings

---

[2] At the time of writing the brochures were available at URL http://cordis.europa.eu/nanotechnology/src/pe_leaflets_brochures.htm. In case of problems with finding the document, please do not hesitate to contact the first author of this chapter via URL http://www.uib.no/svt.

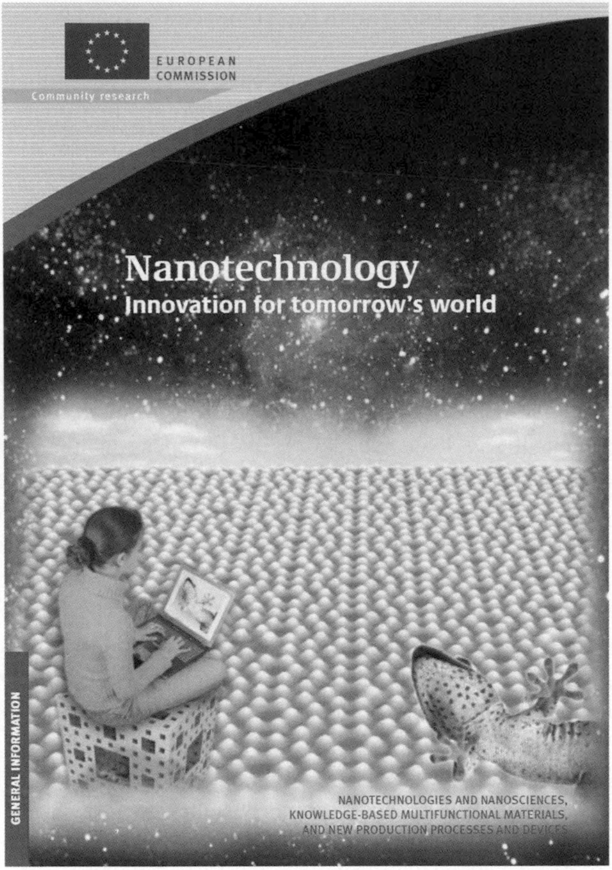

Figure 1. Facsimile of cover page of the brochure *Nanotechnology—Innovation for tomorrow's world* (EC 2004). Reproduced courtesy of the European Commission. For colour reference turn to page 559.

In many cases, pictures are of the collage type, mixing two or more of the types mentioned above. We propose that the use of pictures in this brochure as a whole makes a number of implicit messages:

- The world of nanotechnology is colourful and beautiful.
- Nanotechnology enables us to see on the nanoscale. The pictures of nanoparticles and atoms are images showing that the particles exist and what they really look like.
- Nanotechnology is advanced technology. Nanoscientists use hi-tech equipment, and most of them are men. (Women are depicted on the front cover and on page 35; however, it is unclear at best whether they are presented as nanoscientists).
- There is continuity between phenomena in nature and man-made nanotechnology. Nanotechnology is essentially natural.
- There is a continuity between structures at the nanoscale, via macromolecules and organisms and all the way up to astronomic scale (the Earth, stars and galaxies). Similarly, there is a continuity between nanotechnology and space technology.
- There will be many useful applications of nanotechnology in the future, in fields such as ICT, transport, health, food, sport and leisure.

Messages conveyed may be more or less intentional and more or less explicit from the perspective of the authors. For instance, when this brochure was published, the European Commission at the same time was spending money and effort to recruit women to science through their science-and-society policies. The fact that almost all nanoscientists pictured in the brochure appear to be male is unlikely to be interpreted as an explicit statement about the gender of these scientists or as an internal protest against the policies to recruit women. It is important to understand, however, that the 'incidental' or unconscious character of the choice to depict the scientists as male, does not make it less powerful or more innocent. The message may still have an impact on the views and opinions of the receiver. Indeed, when neither the sender nor the receiver is aware of the content of the message, it can be difficult for the receiver to think critically about it.

3. Combinations of Nano Images with Other Images

As an example of non-scientific use of nano images, the EC brochure is a good case to study because it contains so many pictures of different types, and it mixes them in creative ways. In this section we shall focus on how nano images are mixed with other images in pictures of a collage type. We will highlight three types of mixtures or juxtapositions of nano images with other images:

- nano image and image of scientific equipment and/or personnel
- nano image and image of plant, animal or micro-organism
- nano image and image of the Earth, stars and galaxies

Examples of all three types can be found in the EC brochure, and they can all convey interesting explicit and implicit messages. First, the juxtaposition between the nano image and the image of a scientist and/or scientific equipment may simply say: 'The data that were used in the making of this nano image were collected by this scientist who used this laboratory equipment/this measuring device.' More subtly, one may identify implicit messages in such mixtures: That the nano image is *what you would see* if you used that measuring device (which often is not true, because the nano image can be the result of later, sophisticated processing of the raw data by software); that the acquisition of such nano images *requires* sophisticated equipment and the expert skills and knowledge (perhaps indicated by a white lab coat and a stern-looking middle-aged male scientist); in sum, that there are some experts who can see directly into the nanoscale, to whom the reader presumably does not belong and accordingly in whom he or she will have to *trust*.

The second type of mixture is that between the nano image (or a textual description of nanotechnology) and a living organism. Typical examples of organisms would be the lotus flower or the gecko. The message here is one of relating nanotechnology to nature (see Wickson 2008 for an overview of such narratives). The organisms have surprising and impressive capacities: The lotus flower is self-cleaning and the gecko can walk on the ceiling. These capacities can be understood at a molecular level, and the explanation is given by nanoscience.

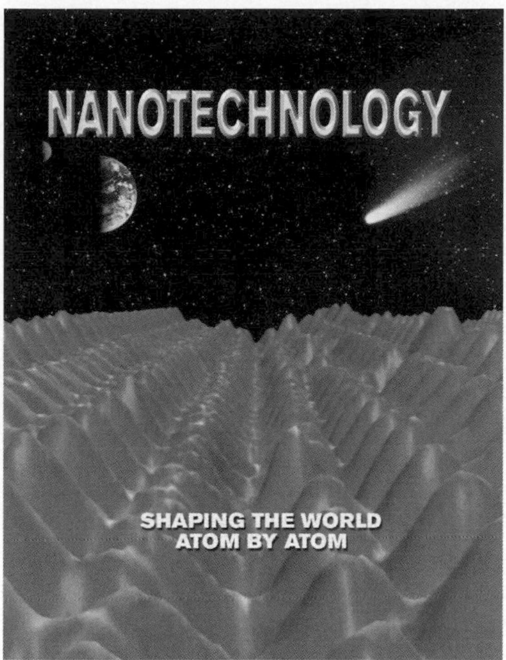

Figure 2. Facsimile of cover page of the brochure *Nanotechnology – Shaping the World Atom by Atom* (US National Science and Technology Council 1999). For colour reference turn to page 560.

Furthermore, this shows that nanotechnology is something natural, that it 'is used by' nature, or at least that nanotechnology may copy natural phenomena. This is a powerful message in debates on health, safety and the environment: 'It is already there in nature, and we have a molecular understanding of it: in other words, it is harmless and under control.'

The third type, the combination of nano images with astronomic objects, may appear surprising. Sometimes, the juxtaposition is literal, such as in the front cover of the EC brochure, and the front cover of an older brochure *Nanotechnology – Shaping the World Atom by Atom*, published by the US National Science and Technology Council (1999). Here a nanoscale 'landscape' of atoms appears in front of a dark blue or black sky with a multitude of stars. Sometimes, the sky has some clouds rather than stars (EC 2004, p. 6), or the image of the Earth is just introduced without any obvious reason (EC 2004, p. 7). In other examples, the nanoscale 'landscape pictures' (or photos of laboratories) lack the sky with stars, but have a dark blue sky-like colour tone. There are different ways to interpret the message this conveys. Blue colour may symbolise clean and cool order and control. But why a sky with stars? Philosopher Alfred Nordmann (2004) argued that the implicit message is that nanotechnology should be seen as the sequel to space technology as a "breakthrough technology", defined as "one that breaks through the dam of conventional wisdom and slow progress, opening up prospects for transformative change" (Nordmann 2004, p. 48). Nanotechnology should be compared to space travel: mind-blowing, opening up new worlds for human knowledge, but also for human enterprise. Hence, there is a message for the general public, who should be supportive of nano research out of intellectual curiosity, as well as for investors, who should lift the gaze and recognise the infinity of new natural phenomena and accordingly an equal infinity of opportunities for innovation and business. The historical line underlying such arguments may be drawn from Vannevar Bush (1945) who explained to the US president how science is the 'endless frontier', via Nobel laureate Richard Feynman's famous speech "There is plenty of room at the bottom" (Feynman 1960), which is often held to be an inspiration for nanoscience and nanotechnology. Indeed, the argument should be understood in a broader geopolitical context of expansion and annexation. In the downfall of the colonial empires and with no more blank spots in the world (Earth) map for the pioneers to conquer, economic growth requires new spaces to annex and capitalise upon (Karlsen & Strand 2009). But where to look? The answer is to look up into the telescope and look down into the microscope.

4. Denotation, connotation and construction

So far, we have not really analysed nano images. Rather, we have analysed how they are combined with other types of pictures in non-scientific literature. In the remainder of the chapter we will mainly discuss nano images proper. We will take the perspective of the practising natural scientist and discuss how it relates to the kind of cultural and political reading that we have illustrated above. First, we need to introduce a bit of conceptual apparatus.

In the introduction, we claimed that the scientist is primarily interested in questions such as: 'What nanoscale structures do I see evidence of in this image? Is it a valid and reliable representation of that structure?' Such questions are hardly relevant in the analysis of mixtures of nano images and images of stars, planets and clouds. These pictures are creative inventions and not intended to be representations of reality. Nano images proper, however, are typically intended to represent – *show* – real objects or states of affair, and are interpreted in such terms within the scientific context.

At least since Gottlob Frege, philosophers and logicians have disagreed on the exact nature of reference and representation. The difference between text and image is actually not so important in this question. Philosophy students will be able to confirm that it is surprisingly difficult to state what exactly it means that a term, a theory or an image 'represents' or 'refers to' something in the real world. From a pragmatic point of view, however, a definition is not necessarily needed. It may still be possible to talk meaningfully about reference without solving the philosophical problems of defining it. Such a pragmatic approach is what we propose. If an image A is intended to represent an object B, we may distinguish between two aspects of the representation:

1. The aspect of *denotation* (or reference): What is it an image of? What object in reality is the image made to show? Is it a representation of that object?
2. The aspect of *connotation*: What claims are made in the image about the qualities and properties of the represented object?

We may distinguish these two aspects of representation with the quite different perspective taken above in the chapter:

3. The aspect of *construction*: What information can be found in the image about its makers, in terms of "evidence of how [its] maker or makers have (re-)constructed reality, as evidence of bias, ideologically coloured interpretation, and so on"? (van Leeuwen & Jewitt 2001, p. 6)

For instance, treating Leonardo da Vinci's painting 'Mona Lisa' in representational terms, the aspect of denotation is simply the question of whether the painting depicts Lisa del Giocondo. The aspect of connotation is discussed when we ask what Lisa del Giacondo actually looked like. Did she have long, dark hair, plucked eyebrows and a slight smile? Did the proportions of her skull resemble that of the painting? Art scholars and art lovers might not be so interested in these aspects. Rather, they might approach the picture in terms of the meaning of Leonardo's artistic choices. The aspect of construction appears in claims such as the androgynous features of Mona Lisa's face shed light upon Leonardo's sexuality. Denotation, connotation and construction are not quite independent, though. For instance, the claim that the androgynous features of Mona Lisa's face may shed light upon Leonardo's sexuality, could appear less likely if it were known that Lisa del Giacondo's face actually had such proportions. We shall see a similar kind of "interdisciplinary interference" between these perspectives when turning to nano images.

Let us now consider the famous 'IBM image' made by Donald Eigler working for IBM.[3] What do we see? We are told (in text) that we see 35 xenon atoms that Eigler positioned individually with a scanning tunnelling microscope (STM). Here, many would hold the aspect of denotation is crucially important: Do we see the atoms or not?

Furthermore, what does the image connote? That xenon atoms have a light blue colour and the shape of a cone? That they are darker on their

[3] Figure 3 is credited Donald Eigler and reproduced from http://www.almaden.ibm.com/vis/stm/atomo.html. For the original publication of the underlying scientific work, see Eigler and Schweizer (1990).

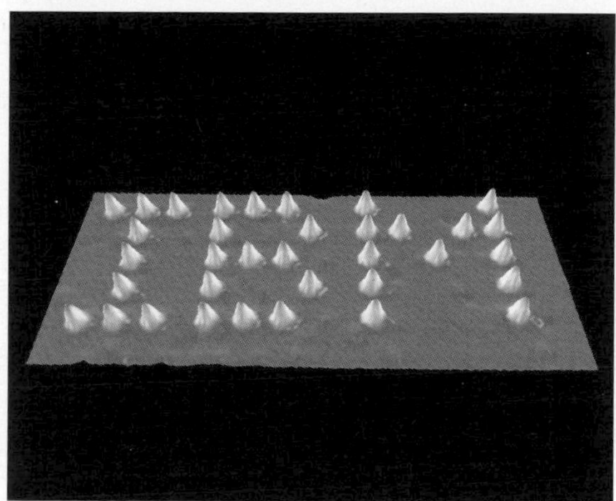

Figure 3. "The Beginning". Image originally created by IBM Corporation. Artwork: Donald Eigler. Reproduction by courtesy of the IBM STM Copyright Policy. For colour reference turn to page 560.

front side and lighter on their back side? Indeed, earlier in the chapter we claimed that the pictures of the EC brochure conveyed messages such as:

- The world of nanotechnology is colourful and beautiful.
- Nanotechnology enables us to see on the nanoscale. The pictures of nanoparticles and atoms are images showing that the particles exist and what they really look like.

At this point, however, knowledge of the physics is necessary to understand the image well. No physicist would see this image as a claim that xenon atoms are blue. Indeed, 'colour' is something that is defined by vision, and there is no vision without electromagnetic radiation with wavelengths within the visual range. The same goes for the colour shades —they are just added to give a sense of three-dimensional depth into the picture. Individual atoms are smaller than the wavelength of visible light, and it is rather an averaged effect from millions of atoms that define properties such as colour and shading. The physicist might summarise his interpretation as follows:

1. The image *denotes*—shows—35 xenon atoms that were positioned with precision in array.
2. The image *connotes* that the density of the signal detected by the STM tip from each atom has the shape of a cone.

The physicist might stop there, disregarding the colour choices as uninteresting and irrelevant, or he may proceed with an analysis of the aspect of construction, asking: Why blue? Why black background? IBM has a blue coloured logo, and is often referred to as 'Big Blue'. Do we perhaps see the resemblance with outer space, nanotechnology being the new breakthrough technology—with IBM as the literal icon of the future? The physicist might think that this is speculation suitable for party small talk but not for work. In that case, we disagree. The important point, in particular for research students in nanotechnology, is that the aspect of construction is there, and may play a political role, one that is either taken seriously or not. This is even more so because people interpret images differently. From a scientific perspective, the blue relief foreground on black background does not connote. No claim is made about blue objects. Within the scientific context, the aspect of construction is simply that a black background gives high contrast to the image, and the human visual system is much better at observing slight curves in a 3D drawing than subtle colour differences, explaining why the image is produced as a 3D model rather than a colour coded 2D image. However, presented out of context to the lay public, there is greater room for confusion and misinterpretations. It is not unlikely that some people will think that xenon atoms are blue cones, implying that atoms are things that we can know the colour of. The makers of brochures and information material accordingly experience (or ought to experience) a moral dilemma (Pitt 2005). Fancy colouring, 3D-effects and other aesthetic effects may help attract the attention of non-scientific readers. However, non-scientific readers may also be more prone to misinterpret such effects in terms of connotation. Accordingly, the more one uses such effects with the well-intended purpose of attracting the intention of citizens to increase knowledge about nanotechnology, the more one may end up misinforming the same citizens. We may take this argument one step further. The makers of brochures often will have to use ready-made nano images produced by scientists, so the moral dilemma should also apply to the scientists themselves.

5. Scientific Interpretation of Nano Images

To focus on two central methods of nanoscience, the basic principle of scanning tunnelling microscopy (STM) and atomic force microscopy (AFM) is that the properties of a specimen surface are studied by moving a probe (tip) across it in an orderly fashion to record measurements of some kind of interaction between the tip and the surface. Keeping track of (x,y)-position of the tip (that is, its spatial position in the plane of the surface), the signals may be displayed in a two-dimensional array that is intended to correspond to the geometry of the specimen surface. It is in this sense that nano images 'show' surfaces. The kind of interaction between the tip and the surface depends upon the choice of technique. In STM, a voltage is set up between the tip and the specimen, and the tunnelling current is measured. In a sense, one measures the conductivity of vacuum near the surface, or how easily electrons are ejected from the specimen to the tip. From quantum mechanics, it is predicted that this current will be extremely sensitive to the distance between the tip and the surface, which makes it possible to (indirectly) detect small variations in distance.

One way of performing STM (called constant current mode) is to vary the height of the tip to keep the measured current between tip and surface constant at every measurement point. This will produce measurements of the height of the tip across the surface, which can be displayed as a 3D 'landscape' picture intended to represent the real 'topography' of the sample (assuming that the conductivity does not vary across the surface). AFM is similar in principle to STM, but instead of using a tunnelling current to measure distance, one relies on the intra atomic forces of repulsion or attraction (such as van der Waals forces) between the tip and the specimen. A laser is then used to measure the deflection of the tip as it is scanned over the surface of the specimen.

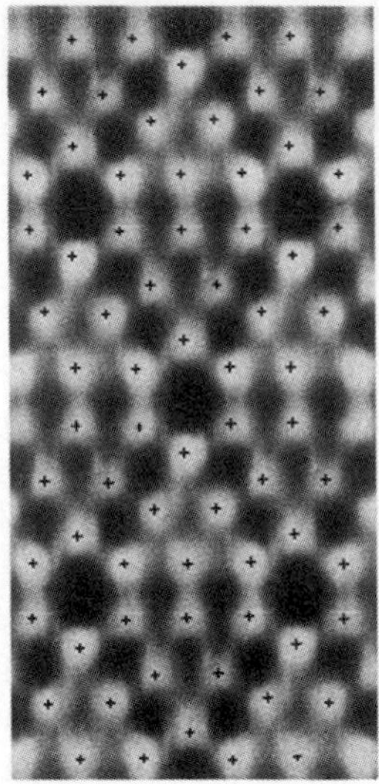

Figure 4. Top view of STM image of Silicon(111) surface. Reproduced from Binnig *et al.* (1983) with permission from authors. Copyright (1983) by The American Physical Society.

It is beyond the scope of this chapter to enter into the scientific details of these techniques. The main point we wish to convey is that the competent scientific attitude towards nano images (such as produced by STMs, AFMs or other scanning microscopes), is that they give valuable information about nanoscale phenomena. In basic science this can be used to test predictions of theories of nano structures. For example, in one of the original STM studies of crystalline silicon, positions of individual atoms are observed indirectly (or more cautiously, inferred) by recognising peaks in the STM signal as "dangling bond positions of the topmost atoms" (Binnig *et al.* 1983; for a profound discussion of these images, see Hennig (2005)). By performing this mapping, a structure is observed that corresponds beautifully to crystalline theory. However, as always when gathering evidence for scientific theories, there is a danger of creating a self-consistent and self-contained ring of theories and experiments that can be completely decoupled from the remainder of scientific knowledge, and thus be relatively worthless in other fields. In the example above, one could argue that using the STM assumes the fact that the tunnelling rate is strongly dependent on distance between the tip and the sample, which is also a prediction of the same atomic theory that was supposedly experimentally verified by the STM image. This is certainly a contrived example, as there is a vast body of theory and other experiments that make atomic theory well justified, which in turn gives credit to STM images as accurately depicting structures on the nanoscale.

Having established the accuracy of nano images, their use for scientific purposes is enabled without regard to whether they philosophically constitute images or not. But one must always carefully take limitations of the imaging process into account when analysing the images. For example, for STM images, the actual measurement is the tunnelling current, and not the distance from the tip to the sample directly. Thus one must take sample specific properties into account in the image analysis. Furthermore, in order to study a sample with an STM, the sample must be conductive and often exposed to vacuum. In order to adhere to these restrictions, samples are often dried and coated in gold or copper, which can potentially alter the sample significantly before starting the imaging procedure. Use of an AFM instead of an STM removes some of these restrictions, as the AFM relies on the atomic forces exerted by the sample on a probe, and requires neither conductivity nor vacuum. On the other hand, the AFM is quite sensitive to the quality of the probe. Also, the AFM can be operated in a number of different modes, measuring atomic forces differently, and thus portraying the sample differently. One can therefore argue that there is not one correct nano image of a given sample, but rather that different images can show different aspects of the sample. This is not a problem for using AFM images in a scientific context, as the scientist will have the necessary background knowledge to understand the images correctly in light of the method used to construct the image. In fact, it can often be crucial to observe different aspects of the sample, in order to get a more complete understanding of its features.

6. Are Nano Images Really Images?

Our hypothetical physicist above summarised his interpretation of the IBM image as follows:

1. The image *denotes*—shows—35 xenon atoms that were positioned with precision in array.
2. The image *connotes* that the density of the signal detected by the STM tip from each atom has the shape of a cone.

Note, however, that there is nothing in the image itself that actually tells us that the cones represent xenon atoms. An alternative interpretation would be:

1. The image *denotes* a record of 35 peaks in the signal recorded with an STM tip moved over a surface.
2. The image *connotes* that the intensity of the signal detected by the STM around each peak has the shape of a cone.
3. In light of the description of the experiment, and in lack of a better explanation, we may postulate that each of the 35 peaks in the conductivity signal are caused by individual xenon atoms, and perhaps even that these atoms were located in a similar or even identical geometrical array as the recorded signals.

Of course, it is also possible to doubt the correctness of the experiment and the results. Some element of faith in the honesty and competence of the scientists who produce such data (or indeed, any data) is inevitable in science.

The first interpretation is a typical realist interpretation. It claims (or takes for granted) that xenon atoms really exist, that an STM allows us to see the atoms, and that we see 35 xenon atoms in this image. The second interpretation is more cautious and insists that what we (hopefully) see, are STM signals and that the best available explanation of the data is that there were 35 atoms there.

Obviously, whatever experimental method one uses at the nanoscale, images are always an assembly of signals recorded with an instrument that records some kind of interaction with the sample other than electromagnetic radiation. For some philosophers, this means that nano images are not images in the ordinary sense, and should not be called as such because they "do not allow us to see atoms in the same way that we see trees" (Pitt 2004, p. 157). Trees we can see for ourselves, we can walk around them and touch them, and we can check whether a particular image is a good representation of the tree is supposed to represent. We do not have to rely on technology to see the tree and check the image.

We have argued above and elsewhere (Birkeland & Strand 2009) that this distinction between seeing directly and seeing through technology is perhaps not so important. Direct vision is also only a partial and fallible source of information about an object, and interpretation is always needed to know what one saw. It is just that one normally sees objects that are so familiar that one does not reflect upon the need for interpretation. For philosophical anti-realists, who think that theoretical concepts and knowledge anyway should not be thought of as expressing the 'Truth' about the universe, but as adequate, useful and reasonable ways to summarise evidence from experiments and experience, the whole distinction between 'real' images and 'not really images' is meaningless. For them, 'atoms' are little more than very useful names anyway.

In this chapter we will not go further into the quarrel about realism, but rather remain agnostic about the real existence of atoms, photons, quarks, electromagnetic fields, etc. Instead, we have indicated how nano images are 'seen' (that is, interpreted) by practising scientists. Our claim is that a scientific interpretation of a nano image (as well as any other image) always assumes that the image only connotes a limited number of important features of what it denotes. In that sense, images are always a kind of shadows of the 'real' objects, at best. Accordingly, to make a correct interpretation, one needs to know the underlying assumptions about the object and the experiment made by the maker of the image, so that one extracts the correct type of information, for instance that the STM image shows conductivity of vacuum close to the conductive surface, and nothing else. This point is relevant also for ELSA research. Let us recall how the examples of the 'Mona Lisa' and the IBM logo showed how there may be interference between the aspects of denotation, connotation and construction. When discussing how "the maker or makers [of the image] have (re-)constructed reality, as evidence of bias, ideologically coloured interpretation, and so on" (van Leeuwen & Jewitt 2001, p. 6), neither the physiognomy of Lisa del Giacondo nor the intended connotations of nano images in terms of properties at the nanoscale are irrelevant.

7. Conclusion

Early in this chapter, we argued how a study of the aspect of construction of images may reveal their political nature. Next, we explained how critical skills and knowledge are needed to properly interpret the denoting and connoting aspects of nano images. Furthermore, we argued that philosophical debates, illustrated by the disagreement between scientific realism and anti-realism, actually may make a difference for the interpretation of nano images. Such philosophical debates may have political implications as well. As seen in chapters 9 and 18, there are heated and difficult discussions about the risks and hazards of nanotechnologies. At the heart of these discussions are questions of knowledge, uncertainty and control. It therefore makes a difference if one believes that it is possible to actually 'see what is going on' at the nanoscale. Images, and interpretations of images, conveying the view that one sees the nanoscale and get a kind of overview of what it consists of, will inevitably supply arguments to those less sceptical of nanotechnology. Can pictures lie?

Pictures may not lie, but a careful reading of them may be necessary. In this chapter, we have presented a very simple procedure when one encounters an image or a picture. We have asked:

1. What does it denote? What is it an image of? Note that there may be different interpretations. What has to be assumed in order to make the claim that the image is an image of real objects?

2. What are the connotations, in terms of qualities and properties of what is denoted? What connotations are "seen"/recognised by competent readers of the image? What assumptions do they make? Consult the accompanying text and context of the image.

3. What other connotations may be made by non-scientific readers? Are they clearly wrong, or is there legitimate doubt?

4. What stories are/can be constructed about the makers of the image, and of the world, based on the use of the image? What are the possible politics of the image? What decisions or arguments may find support in the various interpretations of the image?

Interpretation is no exact science. One's own prejudices, not in the sense of dogmatism, but in the sense of preconceptions that help frame what one sees and understands etc, is unavoidable, necessary and, if dealt with in a systematic, explicit and rigorous way, a resource. Just like in natural science, however, a study of meaning in text and image has to be careful to look for counterexamples so that one does not merely confirm one's own preconceptions. This is also an argument for studying and discussing images together with other people, in particular those coming from other disciplines. Let us add a final argument: It is also more fun!

Questions for Reflection:

1. How important do you think images from scanning probe microscopes (such as the IBM image shown in figure 3) have been for the successful development of the field of 'nanotechnology'? To what extent is this dependent on the image design and style of presentation?
2. What implicit messages, if any, do you see in the EU brochure *Nanotechnology – Innovation for tomorrow's world* (EC 2004)?
3. What is the value of having scientists and ELSA researchers analyse nano-images together? Is the 'method' presented in this chapter a useful way to do this?

Bibliography

Binnig, G., Rohrer, H., Gerber, C. H. and Weibel, E. (1983). 7 x 7 Reconstruction on Si (111) resolved in real space, *Physical Review Letters*, 50, pp. 120-123.

Birkeland, T. and Strand, R. (2009). How to understand nano images, *Techné: Research in Philosophy and Technology*, in press.

Bush, V. (1945). *Science - The Endless Frontier.* A report to the President by Vannevar Bush, Director of the Office of Scientific Research and Development, July 1945, United States Government Printing Office, Washington. http://www.nsf.gov/od/lpa/nsf50/vbush1945.htm (last accessed, April 30 2009).

EC (European Commission) (2004). *Nanotechnology – Innovation for Tomorrow's World,* Brussels: European Commission, Directorate-General for Research. Publication EUR 21151 EN.

Eigler, D. M. and Schweizer, E. K. (1990). Positioning single atoms with a scanning tunneling microscope, *Nature,* 344, pp.524-526.

Feynman, R. P. (1960). There's Plenty of Room at the Bottom, *Engineering and Science* 23, pp.22-36.

Hennig, J. (2005). Changes in the Design of Scanning Tunneling Microscopic Images from 1980 to 1990, *Techné: Research in Philosophy and Technology*, 8(2), pp. 36-55.

Karlsen, J. R. and Strand, R. (in press). Annexation of life: The biopolitics of industrial biology. In Solbakk, J. H., Holm, S. and Hofmann, B. (eds) *The Ethics of Research Biobanking,* Berlin, Springer, pp. 315-330.

National Science and Technology Council (1999). *Nanotechnology – Shaping the World Atom by Atom.* US Government, Washington D.C.

Nordmann, A. (2004). Nanotechnology's worldview: New space for old cosmologies, *Technology and Society Magazine*, IEEE, 23, pp. 48-54.

Pitt, J. C. (2004). The epistemology of the very small. In Baird D., Nordmann, A. and Schummer, J., (eds) *Discovering the Nanoscale,* Amsterdam: IOS Press, pp. 157-163.

Pitt, J. C. (2005). When is an image not an image? *Techné: Research in Philosophy and Technology*, 8(3), pp. 23-33.

Snow, C. P. (1961). *The Two Cultures and the Scientific Revolution.* New York: Cambridge University Press.

Van Leeuven, T. and Jewitt, C. (2001). *A Handbook of Visual Analysis*, London: Sage.

Wickson, F. (2008). Narratives of nature and nanotechnology, *Nature Nanotechnology,* 3, pp. 313-315.

Chapter 4

Poetry from the Laboratory

Hildegard Lee

Cold dark matter tamed
Infinite glow of promise
Two years of funding

*****

There once was a man in a lab
Who did not want to do something bad
He just played with an atom
Unable to fathom
The effect that his actions would have

Nano Meets Macro - Social Perspectives on Nanoscale Sciences and Technologies
by K L Kjølberg & F Wickson

www.panstanford.com
978-981-4267-05-2

Eigler's Eyes 2

Chris Robinson

This is the second of a series of digital drawings parodying an iconic image in nanotechnology: 'Quantum Corral'. The original image shows IBM scientist Don Eigler's use of Scanning Tunneling Microscopy (STM) to arrange 48 iron atoms into a ring in order to 'corral' some surface state electrons, force them into quantum states, and document the resulting electron wave. The image provides access to selected scientific information but it also deceives, in terms of color, shape, scale, orientation etc. While it is interesting to explore ways to generate scientific visualizations that are more accurate, the Visual Arts also enjoy investigating the inaccurate or absurd. Eigler's Eyes 2 makes no pretensions about being true to the nanoscale - popping out self-assembling humans and using a wide array of inaccurate surface imagery – instead it asks how we note and perceive truth.

Chris Robinson is a visual artist interested in the meaning of science and technology and how it assists and influences decision-making. He is currently co-principal investigator on a US National Science Foundation funded project investigating the societal implications and role of images in nanoscience/technology. Robinson is also a member of the nanoCenter at the University of South Carolina and teaches 3D and digital imaging in the Department of Art. Robinson first became interested in nanotechnology through reading Bill Joy's (2001) Wired Magazine article, "Why the Future Doesn't Need Us". He crosses the 'two cultures' of arts and sciences and exhibits, writes, and presents at national and international venues and conferences in both fields.

Question for Reflection:

Since it is not possible to 'see' the nanoscale, how do viewers relate to representations of it? How should they?

For colour reference turn to page 547.

Triangular Masterpiece No. 5

Kristoffer B. Kjølberg

In the recent years I have been working a lot with patterns and shapes used in abstract imagery. One of the fascinating aspects of these pictures is that it's the single bits and pieces they are built from, the 'particles' if you will, that are interesting, more than the design and execution of the picture as a whole. They interact with each other in different ways, and by changing even the smallest variable, the whole perception of the image can be altered in a most dramatic way. This is especially true for repeating patterns, and even though the image I have submitted for this book is not repeating, I think the parable is still valid. I could have changed the way these triangles are juxtaposed in an infinite number of ways, and still the picture would not have been 'finished'. There is no right or wrong way to combine them, but as the artist I get the last say. In this way, I feel like my work sometimes parallels that being done on the nanoscale.

Kristoffer Kjølberg was born in 1980 in Oslo, Norway. He is a graphic designer, cartoonist, illustrator and artist working in a wide range of media. When trying to read up on the theme of nanotechnology for this contribution, by the end of the usually very long articles he often found himself on the verge of a coma.

Question for Reflection:

To what extent can you generate and inspire valuable commentary about nanotechnologies without a detailed understanding of nanoscience?

NanoFireball

Cris Orfescu

'NanoFireball' is a nanosculpture created by the hydrolyzation process of a tiny drop of a titanium metallorganic compound. It was coated with gold to be better visualized with a Scanning Electron Microscope (SEM). The monochromatic electron scan was then digitally painted and manipulated using a technique I call 'digital faux'. Faux painting is an old technique used by decorative painters to recreate the look and feel of natural materials. Like traditional faux, digital faux is done by overlaying translucent layers of color ('digital glazes') to create the perception of depth, volume, and form. The final image was printed on canvas with archival inks.

Nanomaterials have been used for more than 2000 years to create art. One example is the famous Lycurgus cup. It contains gold and silver nanoparticles that make it change color from green to red when light is shone through it. Stained glass (in the 10th-11th centuries) and ceramic glazing (in the 15th-16th centuries) were also based on using metal and metal oxide nanoparticles. 'NanoArt' as a specialised field is, however rather recent. It features both 'nanolandscapes' (natural structures of matter at molecular and atomic scales) and 'nanosculptures' (created by scientists and artists manipulating matter at molecular and atomic scales). These structures are visualized with advanced microscopes, with the resulting images further processed using different artistic techniques.

Cris Orfescu considers NanoArt to be an appealing and effective way to communicate with the general public and raise awareness about nanotechnology and its impact on our lives.

Question for Reflection:

What is your physical and/or emotional response to this image? How might this relate to your attitude towards nanotechnology development?

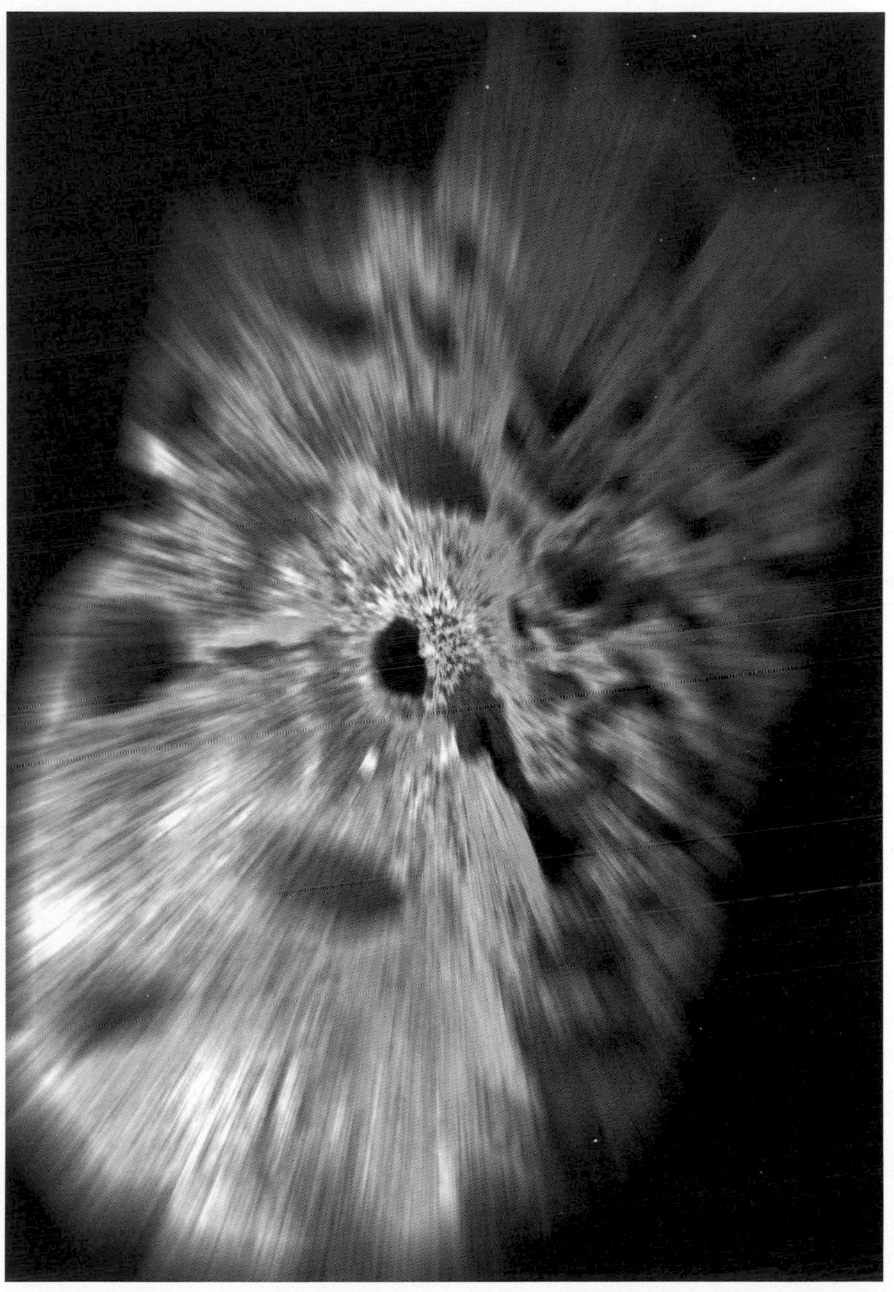

For colour reference turn to page 549.

Nano Meets Macro:

In the Public Eye

This section presents perspectives on how nanoscale sciences and technologies are represented in different social contexts by a diverse set of actors. The field of concern forming this node is how nanoST are understood, presented, talked about and represented in the public sphere. Its central concepts include: perceptions, opinions, rhetoric, metaphors, hype, fear, visions, science fiction, media and representations. There are a number of reasons to be interested in how nanoST appear in the public eye. Most people meet nano for the first time through the arenas discussed in the chapters of this section: through science fiction literature or video games, through the media, and/or through consumer products. They may be introduced to them as a green technology promising global sustainability, or as the newest addition to a long line of commercially driven agriculture applications threatening small scale farming all over the world. They may meet them as an emerging technological revolution, or as an unregulated environmental concern. In the chapters of this section, you will find the different faces of nanoST in the public sphere discussed from the point of view of some of the eyes observing it.

More concretely, this section consists of five academic chapters, one short fiction story and two artworks that all in different ways relate to how nano appears in the public eye. In chapter 5, called 'The Slippery Nature of Nano-Enthusiasm', Rob Sparrow argues that enthusiasts of nanoST use different stories when they wish to promote the technology, and when they are faced with criticism and concern. He points to five areas where he sees this happening and concludes that this inability to stay with the same story makes real public and political debate about consequences impossible. Sparrow wonders: To what extent should scientists be held accountable for their public representations of nanoST? In doing so, the author concludes in a way that also crosses over to the node of 'In the Tough Decisions'.

In chapter 6, Kamilla Lein Kjølberg is interested in how nanoST appear in the media. She has studied newspaper texts over a period of eight years, and found that nanoST are represented mainly as promising and controlled research with great potential for the future. The lack of critical angles and voices, she argues, is a hindrance to the achievement of public engagement in nanoST. In this sense, this chapter can also be seen to relate to the node of tough decision.

Colin Milburn in chapter 7, 'Everyday Nanowars', writes about the ways in which representations of military nanotechnology in computer war games and army promotions inspire each other, both visually and in terms of technological ideas. He makes the argument that representations of military nanotechnology in computer war games help 'normalise' such technologies before they become realities. In addition, he points to a number of ambiguities in the way military nanotechnology is represented and claims that this creates what he sees as a queering. This leads him to conclude that new nanotechnologies – even before they become real – shape how we see and think about ourselves as individuals and as a society. Milburn's reflections in this respect also see his work connect well with the node concerned with the way nano meets macro 'in the big questions'.

Mercy Kamera, in chapter 8, uses the case of nanoST applications in water management to draw attention to the role of technology in the achievement of global sustainability. She asks: Is such a thing as green nanotechnology possible, or is it rather a contradiction in terms? In Kamera's argument, the answer to this depends upon the fundamental frame or the worldview one argues from.

In chapter 9, Lynn Frewer and Arnout Fischer are interested in public perception of risks and benefits related to nanoST in food and agriculture. In the absence of strong public reactions to nanofood so far, they draw on experiences from reactions to previous technological introductions in food and agriculture, and especially to genetically modified food. They argue that communication of risk needs to take into account the ways in which people come to terms with a complex issue like food, which is so intimately related to culture and tradition as well as to physical welfare and health.

Lesley Smith in the science fiction short story 'My Room', included as chapter 10, explores how the public eye itself may become something different with nanoST, and specifically how advances in surveillance technologies may change societies and individuals. What happens if nanotechnologies enable everything to enter into the public eye? The story also explores how people create themselves as virtual individuals, and play with the contrast between a reality where there is no way to hide and a virtual reality of absolute freedom.

In the artwork 'Periodic Table', Teresa Majerus' plays with the way in which images of the nanoscale can tell us stories about what she calls an "imaginary real world". In her work she wants to suggest that as the nanoscale is beyond the level of visual sensory perception, it will always only be visible to us in the form of translations. Through these we not only learn about the nanoscale, but also about life as it appears to us at our own scale.

Tim Fonseca's artwork 'Brainbots', provides a visualization for the often claimed future potential for nanoST to enable integration between the human brain and computational technology. This common representation in the public sphere, whether presented visually or textually, may provoke different reactions and emotions in different people. Sometimes such representations can be manipulative and steer the receiver towards an anticipated sentiment, other times responses may be quite contrary to what was intended.

This section deals with nanoST in the public eye in a number of very different ways, and together the chapters elicit a range of important aspects that we see in the literature relating to this node. There are however a few central aspects of nanoST in the public eye that we were unable to include in this book. One of these areas is the important perspective from consumer research on how nano-products appear in the market. This perspective is concerned with questions such as: Should nanoST consumer goods be labelled, and if so, according to what criteria? How do producers and manufacturers today, in the lack of labelling requirements, announce to consumers that the product stems from a process using nanoST or contains fabricated nano materials? Is the use of nanoST seen as something that promotes sales or something that the producers and marketers would rather the consumers are not aware of, in the sense of the perception that it may scare them away? Another perspective that is absent in this volume is that from literary studies, for instance looking at the rhetoric in use in different types of texts about nanoST. When and how are emotions, like fear and expectation evoked? How do metaphors travel between texts? A third example of a perspective relevant to this node but not included in the book is the existence and use of visions and scenarios of the future in relation to nanoST. The future is an essential part of how nanoST appears in the public eye. How do people relate to the future through these kinds of visions and scenarios? Do people, for instance, differ in their responses and reactions between different stages of the future, from near to far future for example?

NanoST are represented in a number of different ways, and appear in the public eye through a wide range of channels and media. Still, at the time of completing this book most people know little or nothing about nanoST. An underlying theme throughout this section is the tension between the way that nanoST are persistently presented, and yet at the same time, remain remarkably obscure and unintelligible for the vast majority of people. This tension manifests in the public eye and is discussed in this section as an important theme in itself, but it also has significant ramifications for the themes discussed in the three other sections in this book. Does the simultaneous presence and invisibility of nanoST in the public eye influence the development of nanoST and the making of a field of nanoST research? How does it affect the ability and incentive to make tough decisions about, for instance, labelling? And finally, how does it relate to big questions like how nanoST affect the way we understand ourselves and our role in the world?

Chapter 5

The Slippery Nature of Nano-Enthusiasm

Robert Sparrow

This chapter identifies and analyses competing and contradictory claims made by supporters of nanotechnology. Enthusiasts for nanotechnology make one set of claims when they want to advertise and promote this technology and another, often directly opposed, set of claims when sceptics about the technology question their enthusiasm. Working out which of the very different claims made about nanotechnology are true is essential if democratic societies are to be able to make informed decisions about it.[1]

---

[1] An earlier version of this paper appeared on the Friends of the Earth Nanotechnology Project website. I would like to thank Georgia Miller, Peter Binks, and Debra Dudek for helpful comments and discussion over the course of drafting this paper. However, I am, of course, solely responsible for any errors or mistaken judgments which appear here.

Nano Meets Macro - Social Perspectives on Nanoscale Sciences and Technologies
by K L Kjølberg & F Wickson

www.panstanford.com
978-981-4267-05-2

1. Introduction

As a philosopher who writes about the ethics of new technologies, I am often invited to conferences on nanotechnology. At these events, I have usually found myself placed in the position of a critic of nanotechnology, despite the fact that I have no particular concerns about nanotechnology that are not reflections of more general reservations about the pace and direction of technological "progress" (Sparrow 2009). However, what I *have* become concerned about is the way in which public discussion of nanotechnology is being framed. Enthusiasts for nanotechnology make one set of claims when they want to advertise and promote this technology and another, often directly opposed, set of claims when sceptics about the technology question their enthusiasm. As a consequence, the terms of the debate about nanotechnology shift so as to hamper substantial critical engagement about the future of this technology.

It may be that nanotechnology raises no distinctive, ethical, political, or environmental issues. However, it is hard to tell when the debate is plagued by constant equivocation between very different claims by those in favour of nanotechnology. What follows, then, is an examination of the slippery nature of nano-enthusiasm. My hope is that drawing attention to the different and diametrically opposed claims often made about nanotechnology may help those who want to think critically about nanotechnology—and the social decisions we face about it—do so in a clear and sober fashion.

2. Nanotechnology or No Technology?

When researchers and industry spokespeople wish to advertise their products and/or lobby the government for funding, they wax lyrical about the wonders of nanotechnology. Scarcely a week goes by that we do not hear something reported about the promise of nanotechnology. Nanotechnology will make possible marvelous new consumer goods that will improve our lives (Drexler *et al.* 1993; Roco & Bainbridge 2001). Nanotechnology will heal the sick and allow us all to live longer (Roco & Bainbridge 2002). Nanotechnology will be worth so many billions of dollars over the next decade (De Franscesco 2003; Roco 2005; Roco *et al.* 1999).

However, the moment criticism of nanotechnology develops researchers and industry spokespeople often beat a hasty retreat to the position that there is no such thing as nanotechnology: there are only nanotechnolog*ies*—diverse technologies and techniques for manipulating matter at the nanoscale, which have been developed in the fields of chemistry, physics, engineering, and materials science. Any concerns about "nanotechnology" are therefore misplaced.

I would have more sympathy with this demand for terminological precision if it weren't usually made in conference streams on the ethical, legal, and social impacts of nanotechnology, at conferences with nanotechnology in the title, and/or by people who have "nanotechnology" on their business cards. It wasn't, after all, critics of nanotechnology who invented the term, which was coined by a scientist and taken up by other scientists in order to attract attention to their work (Interagency Working Group on Nanoscience, Engineering and Technology 2000; Stix 2001; Taniguchi 1974). Moreover, the effect of shifting the topic of discussion to the question of terminology is usually to divert attention from the original criticism into a tedious and remarkably fruitless debate about the appropriate way of referring to nanotechnology (or nanotechnologies!), which is then repeated at the next forum on nanotechnology.

Regardless of how we choose to refer to them, those technologies that engineer matter at the nanoscale do potentially raise new hazards as well as hold out new promise. The role played by surface chemistry and quantum effects at the nanoscale means that the products of these technologies can have properties that are not possessed by the same materials manufactured at larger scales. The importance of size at the nanoscale arguably justifies grouping these technologies together for the purpose of further developing and investigating the technologies necessary to manipulate matter at this scale. However, it also justifies grouping them together for more critical purposes.

3. Revolutionary or Familiar?

Perhaps the most common claim made in public discussion of nanotechnology is that it represents a "technological revolution". Nanotechnology will, we are told, change the world (Atkinson 2003; Berne 2004; Berube 2006; Interagency Working Group on Nanoscience, Engineering and Technology 2000; Smalley 1999; Roco & Bainbridge 2001). It will make consumer electronics cheap and widely available. It will eliminate pollution and repair damage to the environment. It will even allow "the blind to see and the deaf to hear" (Bond 2007). Indeed, if the writings of some nano-enthusiasts are to be believed, there is little that nanotechnology will not do (Crandall 1997; Drexler *et al.* 1993; McCarthy 2003). Describing nanotechnology as revolutionary draws attention to the novelty and/or power of the technology and consequently assists in attracting funding.

The problem with this rhetoric of revolution is that it also draws attention to the magnitude of the changes nanotechnology promises and to the fact that there are likely to be winners and losers from any such revolution. The rhetoric of revolution also draws attention to the questions of power and democracy involved in technology policy. If there is to be a revolution, it should be a democratic one. Indeed, given that revolutions are dangerous and unsettling, perhaps we don't want a revolution at all (Sparrow 2008)!

When questions about the distribution of benefits of this purported revolution or about who will control it arise, then, enthusiasts for nanotechnology retreat to the contrary claim that nanotechnology is nothing new, that it is in fact entirely familiar (Atkinson 2003; Mody

2006; Royal Society and Royal Academy of Engineering 2004; Schummer 2006). Nanotechnology is merely the latest stage of a continuing process of miniaturisation of technology. Indeed, it is already present in various consumer goods such as paints, sunscreens, and some consumer electronics. Because the technology is familiar, we have nothing to fear from it. We might equally wonder what all the fuss is about and whether the promise of "more of the same" justifies the enormous amount of public money currently being spent on nanotechnology research.

Cynics might note at this point that it is when researchers and corporations want to patent their products that they argue that their products are new and unique. However, when it comes to discussing their possible effects on human health and environment all of a sudden they are moved to argue that these nano-products are nothing new.

The question of whether nanotechnology is revolutionary or familiar is perhaps most important when it comes to evaluating the possible environmental and health impacts of nanoparticles, as the use of engineered nanoparticles for their catalytic or other properties in manufacturing is the nanotechnology which is closest to fruition. In discussions about this topic, it is often pointed out that we are all already regularly exposed to nanoparticles in the form of the exhaust products from diesel combustion engines, soot from forest fires, and salt in sea air (Albrecht *et al.* 2006; Roco 2003; Royal Society and Royal Academy of Engineering 2004; Swiss Re 2004). Exposure to nanoparticles is nothing new and—by implication—nothing to fear. What this observation neglects (besides, bizarrely, the fact that some of these particles are known to be responsible for thousands of deaths each year in modern cities) is that the nanoparticles that have been produced by human activity to this point have been accidental products with large distributions of particle size and shape. Engineered nanoparticles will have been designed to have specific properties by virtue of having particular structures and size distributions. They are therefore likely to behave very differently. It would be extremely foolish indeed to base our assessment of the risks involved in exposure to engineered nanoparticles on our experience with naturally occurring and/or existing anthropogenic nanoparticles (Balbus *et al.* 2005).

4. Inevitable or Precarious?

According to many pundits, the nanotechnological revolution is not only going to change the world, it is going to do so regardless of what you or I think about it. Many writers on nanotechnology seem to feel that technological development has its own dynamic which is effectively beyond human control (Mody 2006). As a result, the development of nanotechnology is, we are told, inevitable. The future is coming and we had better get ready for it (Atkinson 2003; Bond 2007; Drexler 1986; Mulhall 2002).

Yet this certainty that the development of nanotechnology is inevitable seems to be matched by hysteria at the possibility that public hostility to this technology, lack of investment, or a hostile regulatory environment, might prevent it (Bond 2007). Indeed, it sometimes seems that the main function of the claim that the development of nanotechnology is inevitable is to support the argument that we must get ready for it. Unless we direct more money into funding this technology, change our intellectual property law, and educate the public about the benefits of nanotechnology, the nanotechnology revolution will not arrive. Of particular note in this context is the frequency with which consumer hostility to Genetically Modified Organisms (GMOs) is mentioned in discussions of nanotechnology as an example of the way in which public concerns about safety and benefit can remove the incentive to develop certain types of product and thus effectively halt the development of a technology (Balbus *et al.* 2005; Court *et al.* 2004; David & Thompson 2008; Hood 2004; Moore 2002; Nature Biotechnology 2003; Royal Society and Royal Academy of Engineering 2004). Those involved with developing and promoting nanotechnology are terribly concerned to avoid any similar public backlash against nanotechnology. Of course, the possibility that the public might reject nanotechnology suggests that the nanotechnological revolution is not inevitable after all.

5. "Nothing to be Afraid of" or "Cause for Alarm"?

The next contradiction I wish to draw attention to does not appear in public discussions of nanotechnology so much as between the rhetoric and the reality of the regulatory authorities that are likely to be responsible for protecting consumers and the environment from any hazards associated with nanotechnology.

One of the 'big questions' in current discussions of nanotechnology is whether the public is adequately protected from possible hazards associated with exposure to nanotechnology (Albrecht *et al.* 2006; Balbus *et al.* 2005; Friends of the Earth 2006; Hood 2004; Nordman & Holman 2005; Sellers *et al.* 2009; Swiss Re 2004). In order to establish the need for extensions to existing regulations, critics and concerned regulators emphasise our current lack of knowledge about the toxicity or safety of matter engineered at the nanoscale and the gaps in our existing regulatory schemas which mean that materials that are "new" only in relation to their size may not be subject to any special scrutiny (Balbus *et al.* 2005; Royal Society and Royal Academy of Engineering 2004). These studies make a convincing case that the public is currently not adequately protected from possible hazards due to nanoparticles. At the very least, our existing regulatory systems need to be strengthened and modified to ensure that nanoscale particles and materials are evaluated for possible health risks and environmental impacts before being released into the environment (Balbus *et al.* 2005; Royal Society and Royal Academy of Engineering 2004). The fact that the properties of nanomaterials—and of engineered nanoparticles in particular—are so sensitive to their size, shape, and molecular structure suggests that this may be more difficult than first appears (Donaldson *et al.* 2004).

However, these discussions of the need for the extension of existing regulatory systems are striking not just because of what they conclude about the unknown risks involved in nanotechnology but because of what they reveal about our regulators' attitudes towards environmental risks more generally. Having learned that the toxicology of nanoparticles is largely unknown (Albrecht *et al.* 2006; Colvin 2003; Hood 2004), that their movement through environmental systems is poorly understood and extremely difficult to model (Mackay & Henry 2009), that some nanoparticles seem to be able to pass through the skin while others seem to move directly to the brain (Balbus *et al.* 2005), and finally, that cosmetics containing engineered nanoparticles are already on the market (Hood 2004), I naively expect to read expressions of outrage that the public is being exposed to these potentially toxic materials. It seems simple common sense to me that until it can be established that these materials pose no significant threat to human health or to the environment, products containing engineered nanoparticles should not be released on to the market. Instead, it is abundantly clear that many toxicologists, industry figures, and regulators feel that there is nothing untoward in the public being exposed to such risks (Brumfiel 2003; ETC Group 2003). Those involved in regulating chemicals and other possible hazards are well aware that we are all regularly exposed to a myriad of chemicals for which the level of associated risk has never been established; as a result the situation concerning nanotechnology does not —in their minds, at least—cry out for attention.

Existing regulatory systems for environmental protection and health and safety regulation of chemicals turn out to be remarkably unimpressive from the perspective of a concerned citizen. To a large extent, these regulatory systems rely on manufacturers self-regulating and providing data on the chemicals they manufacture and/or import to the relevant regulator. Regulatory agencies are often over-worked, under-staffed, under-funded, and have limited power to investigate and/or punish breaches of the law. It is striking how conservative governments who describe themselves as 'tough on crime' leave corporate individuals to regulate themselves! Hearing that nanotechnology will be regulated in line with existing frameworks therefore does not fill me with confidence.

6. Ethical Issues…What Ethical Issues?

Studies of the ethical, legal, and social issues raised by nanotechnology often conclude by suggesting that nanotechnology "raises many ethical issues" (Roco 2003; Lewenstein 2006). This follows naturally from treating nanotechnology as a revolutionary new technology. It also reflects the tendency in the literature to discuss nanotechnology as though it were a successor to the biotechnology and information revolutions, which clearly have raised many new ethical issues.

My own assessment is that it is in fact difficult to identify any genuinely *new* ethical issues raised by those nanotechnologies that are likely to be developed in the short-to-medium-term future. The most urgent ethical issues associated with nanotechnology concern the relationship between democracy and technology, respect for the environment, risk, privacy, social justice, and the possibility of arms races (Sparrow 2009). All of these issues are already familiar to us as a consequence of existing technologies.

Indeed, the shifting nature of the claims made in the debate about nanotechnology is itself strongly reminiscent of similar phenomena in the debates about the biotechnology and information 'revolutions' (David & Thompson 2008). The history of discussion of these 'revolutions' is also characterised by equivocation about the extent to which they were coherent phenomena, were genuinely new or extensions of what had gone before, and/or were a matter of social choice or historical inevitability. Even the claims made about the implications of this new technology turn out to be old (Edgerton 2006).

However, my concern here is not with the accuracy of the claim that nanotechnology raises new ethical issues but with the apparent ease with which it sits beside the assumption that we should embrace nanotechnology (Whitman 2007). If those developing nanotechnology really believe that it raises so many ethical issues, one would think that this would at least lead them to adopt attitudes of humility and caution regarding this technology. Yet these attitudes are noticeably absent from most discussions of nanotechnology.

Moreover, the idea that the development of nanotechnology is inevitable sometimes produces a shocking and flagrant disregard for the possibility that certain applications of nanotechnology might be unethical. If the development of the technology is inevitable, any negative impacts are equally unavoidable. The only question left is who will profit from this state of affairs; the clear implication is that every nation should work to ensure that it gets its share of the spoils. This argument, of course, also works for the production of opium, selling arms to terrorists, and building weapons of mass destruction. The fact that others are doing, or are likely to do, something wrong, is not itself a compelling reason for us to join them.

In the absence of an acknowledgement of a real possibility that we might choose not to develop nanotechnologies, it is easy to suspect that these gestures towards 'ethical issues' are intended mainly as an advertisement that industry and government are appropriately concerned. A genuine concern for ethical issues would, though, result in less haste in pursuing the profits associated with developing a nanotechnology industry and more reflection and debate on how (and whether!) to proceed.

7. Conclusion

A proper critical assessment of the impacts, costs, and benefits of the adoption of nanotechnology will not be possible until we can clear away some of the hype around it and adjudicate between the competing claims made on its behalf. If there are only different nanotechnologies, if they are already familiar to us, if we have a choice as to whether to develop them, and if they are adequately regulated by existing institutions or something like them, then there may well be nothing to be afraid of and no significant ethical issues that we need to resolve. If, alternatively, nanotechnology is a revolutionary new technology, the development of which appears to be inevitable, and which raises profound challenges to our regulatory systems as well as new ethical issues, then we would do well to proceed cautiously, if at all. Working out which of the very different claims made about nanotechnology are true is therefore essential if we're to be able to make informed decisions about it. It may turn out that each and every claim made about nanotechnology is true of some particular nanotechnology in some particular context—although even this observation presumes the validity of an argument about "nanotechnology vs nanotechnologies"!

However, the real problem arising from the existence of the contradictory claims I have highlighted is not so much that it is hard to work out which of them is true but that the combination of them functions to close down the space in which critical engagement with them might take place. Changing stories allows nano-enthusiasts to avoid having to discuss the full implications of their original claims. When advocates for nanotechnology want to drum up interest in it, or funding for it, they talk about nanotechnology and argue that it is revolutionary; when they want to defuse fears, they insist there are only nanotechnologies which are already familiar. When they want the public to accept nanotechnology, they argue it is inevitable; when they want the government to provide more funding, change the laws, or educate the public to be more enthusiastic about it, then they argue it is precarious (Sparrow 2007). They allow that nanotechnology requires regulation but ignore the problems with the institutions that will be doing the regulating. While they routinely acknowledge the importance of ethical issues, they seldom acknowledge the possibility that these might constitute a reason to turn away from developing nanotechnology. This pattern of claims reflects an attempt by advocates for nanotechnology to have the best of both worlds across these areas. It also functions to continually defer sustained ethical discussion of any of them.

As billions of dollars of public money are poured into nanotechnology research and as the products of nanotechnologies start to be introduced to unwitting consumers and to the environment, we can ill afford to defer discussion of the issues raised by nanotechnology any longer. It is time to hold all those involved in debates about nanotechnology to the claims they make and to highlight and challenge equivocations of the sorts I have identified here. If enthusiasts for nanotechnology try to change their stories when critics respond to their original claims, we should recognize this as a sign that they are more concerned about getting the public to embrace nanotechnology than they are about participating in a genuine debate about it. Yet a genuine, open and vigorous debate is precisely what is required if we want to continue to claim to be a democratic society while pursuing a technology with potentially widespread and profound social and environmental consequences.

Questions for Reflection:

1. Does nanotechnology create any genuinely ‘new’ ethical issues?
2. What is actually required for something to be ‘revolutionary’? Do you think nanotechnology has what it takes?
3. Is it problematic that nano-enthusiasts change their stories in all the ways outlined above? To what extent do you think this phenomenon is restricted to enthusiasts?

Bibliography

Albrecht, M. A., Evans, C. W. and Raston, C. L. (2006). Green chemistry and the health implications of nanoparticles, *Green Chem.*, 8, pp. 417-432.

Atkinson, W. I. (2003). *Nanocosm: Nanotechnology and the Big Changes Coming from the Inconceivably Small*, New York: Amacon.

Balbus, J. M., Denison, R., Florini, K. and Walsh, S. (2005). Getting nanotechnology right the first time, *Issues Sci. and Technol.*, Summer, pp. 65-71.

Berne, R. W. (2004). Towards the conscientious development of ethical nanotechnology, *Sci. Eng. Ethics*, 10, pp. 627-638.

Berube, D. M. (2006). *Nano-Hype: The Truth Behind the Nanotechnology Buzz*, Amherst, New York: Promethus Books.

Bond, P. J. (2007). Preparing the path for nanotechnology. In Roco, M. C. and Bainbridge, W. S. (eds) *Nanotechnology: Societal Implications I*, Dordrecht, The Netherlands: Springer, pp. 21-28.

Brumfiel, G. (2003). A little knowledge... , *Nature*, 424, pp. 247-248.

Colvin, V. L. (2003). The potential environmental impact of engineered nanoparticles, *Nat. Biotechnol.*, pp. 1166-1170.

Court, E., Daar, A. S., Martin, E., Acharya, T. and Singer, P. A. (2004). Will Prince Charles *et al* diminish the opportunities of developing countries in nanotechnology? http://nanotechweb.org/cws/article/indepth/18909 (last accessed Jan 30, 2008).

Crandall, B. C. (ed) (1997). *Nanotechnology: Speculations on Global Abundance*, Cambridge, Massachusetts and London, England: The MIT Press.

David, K. and Thompson, P. B. (eds) (2008). *What can Nanotechnology Learn from Biotechnology?* Burlington, MA: Elsevier.

DeFrancesco, L. (2003). Little science, big bucks, *Nat. Biotechnol.*, 21, pp. 1127-1129.

Donaldson, K., Stone, V., Tran, C. L., Kreyling, W. and Borm, P. J. A. (2004). Nanotoxicology, *Occup. and Environ. Med.*, 61, pp. 727-728.

Drexler, K. (1986). *Engines of Creation*, New York: Anchor Press.

Drexler, K., Peterson, C. and Pergamit, G. (1993). *Unbounding the Future: The Nanotechnology Revolution*, New York: Quill Books.

Edgerton, D. (2006). *The Shock of the Old*, London: Profile Books.

ETC Group. (2003). Size Matters, *ETC Group Occasional Papers Series*, 7, pp. 1-14.

Friends of the Earth (2006). Size does matter. Nanotechnology: Small science - big questions. Special issue (Issue 97) of *Chain Reaction*.

Hood, E. (2004). Nanotechnology: Looking as we leap, *Environ. Health Pers.*, 112, pp. A741-A749.

Interagency Working Group on Nanoscience, Engineering and Technology (2000). *National Nanotechnology Initiative: Leading to the Next Industrial Revolution* , Washington, D.C.: Committee on Technology, National Science and Technology Council.

Lewenstein, B. V. (2006). What counts as a 'social and ethical issue' in nanotechnology? In Schummer, J. and Baird, D. (eds) *Nanotechnology Challenges: Implications for Philosophy, Ethics and Society*, Singapore: World Scientific Publishing, pp. 201-216.

Mackay, C. E. and Hamblen, J. (2009). Toxicology and risk assessment. In Sellers, K., Mackay, C., Bergeson, L. L., Clough, S. R., Hoyt, M., Chen, J., Henry, K. and Hamblen, J. (eds) *Nanotechnology and the Environment*, Boca Raton, FL: CRC Press, pp. 193-224.

Mackay, C. E. and Henry, K. (2009). Environmental fate and transport. In Sellers, K., Mackay, C., Bergeson, L. L., Clough, S. R., Hoyt, M., Chen, J., Henry, K. and Hamblen, J. (eds) *Nanotechnology and the environment*, Boca Raton, FL: CRC Press, pp. 123-144.

McCarthy, W. (2003). *Hacking Matter: Levitating Chairs, Quantum Mirages, and the Infinite Weirdness of Programmable Atoms*, New York: Basic Books.

Mody, C. C. M. (2006). Small, but determined: Technological determinism in nanoscience. In Schummer, J. and Baird, D. (eds) *Nanotechnology Challenges: Implications For Philosophy, Ethics, And Society*, River Edge, NJ: World Scientific Publishing., pp. 95-130.

Moore, F. N. (2002). Implications of nanotechnology applications: Using genetics as a lesson, *Health Law Rev.*, 10, pp. 9-15.

Mulhall, D. (2002). *Our Molecular Future: How Nanotechnology, Robotics, Genetics, and Artificial Intelligence will Transform our World*, Amherst, N.Y.: Prometheus Books.

Nature Biotechnology (2003). Why small matters, *Nat. Biotechnol.*, 21, p. 1113.

Nordman, M. M., and Holman, M. W. (2005). A prudent approach to nanotechnology environmental, health, and safety risks. *Industrial Biotechnol.*, 1, pp. 146-149.

Roco, M. C. (2003). Broader societal issues of nanotechnology, *J. Nanoparticle Res.*, 5, pp. 181-189.

Roco, M. C. (2005). The vision and strategy of the US National Nanotechnology Initiativel. In Schulte, J. (ed) *Nanotechnology: Global Strategies, Industry Trends & Applications*, Hoboken, NJ: Wiley, pp. 79-94.

Roco, M. C. and Bainbridge, W. S. (eds) (2001). *Societal Implications of Nanoscience and Nanotechnology*, New York: Springer.

Roco, M. C. and Bainbridge, W. S. (eds) (2002). *Converging Technologies for Improving Human Performance: Nanotechnology, Biotechnology, Information Technology and Cognitive Science*, Arlington, Virginia: National Science Foundation.

Roco, M. C., Williams, S. and Alivisatos, P. (1999). *Nanotechnology Research Directions: IWGN Workshop Report. Vision for Nanotechnology R&D in the Next Decade*, Maryland: WTEC.

Royal Society and Royal Academy of Engineering (2004). *Nanoscience and Nanotechnologies: Opportunities and Uncertainties*, London: Royal Society & Royal Academy of Engineering.

Schummer, J. (2006). Cultural Diversity in Nanotechnology Ethics, *Interdiscipl. Sci. Rev.*, 31, pp. 217-230.

Sellers, K., Mackay, C., Bergeson, L. L., Clough, S. R., Hoyt, M., Chen, J., Henry, K. and Hamblen, J. (eds) (2009). *Nanotechnology and the Environment*, Boca Raton, FL: CRC Press.

Smalley, R. E. (1999). Testimony to US Congress, p. 1-2. Available via http://www.sc.doe.gov//bes/Senate/smalley.pdf (last accessed Apr 14, 2009).

Sparrow, R. (2007). Revolutionary and familiar, inevitable and precarious: Rhetorical contradictions in enthusiasm for nanotechnology, *NanoEthics* 1, pp. 57-68.

Sparrow, R. (2008). Talkin' 'bout a (nanotechnological) revolution, *IEEE Technol. Soc. Mag.,* 27, pp. 37-43.

Sparrow, R. (2009). The social impacts of nanotechnology: An ethical and political analysis, *J. Bioeth. Inq.*, 6, pp. 13-23.

Stix, G. (2001). Little big science, *Sci. Am.*, 285, pp. 26-31.

Swiss Re (2004). *Nanotechnology: Small Matters, Many Unknowns*. Available via http://www.swissre.com/pws/research%20publications/risk%20and%20expertise/risk%20perception/nanotechnology_small_matter_many_unknowns_pdf_page.html.

Taniguchi, N. (1974). On the basic concept of "Nano-technology". In *Proceedings of the International Conference on Production Engineering, Tokyo, Part II.* Japan Society of Precision Engineering, pp. 18-23

Whitman, J. (2007). The governance of nanotechnology, *Sci. Public Policy*, 34, pp. 273-283.

Chapter 6

Representations and Public Engagement: Nano in Norwegian Newspapers

Kamilla Lein Kjølberg

This chapter deals with the way that nanosciences and nanotechnologies (or just nanotechnology for simplicity) appear in the public sphere. As illustration, a study of representations of nanotechnology in Norwegian newspapers from 2000 to 2007 will be presented. In this study three dominant representations were found: nanotechnology is positive, nanotechnology is under control and nanotechnology is important for the future. The fact that an important part of the public sphere is dominated by these three representations is discussed in relation to public engagement in the development and governance of nanotechnology. The chapter will be concluded by arguing that there seems to be a lack of suitable resources available in the public sphere for meaningful public engagement in nanotechnology, and that the responsibility to provide these resources is shared among everyone that is already engaged in nanotechnology.[1]

[1] A more extensive version of this chapter was published as: Kjølberg, K.L. (2009) Representations of Nanotechnology in Norwegian Newspapers – Implications for Public Participation. *NanoEthics* 3, pp. 61-72.

Nano Meets Macro - Social Perspectives on Nanoscale Sciences and Technologies
by K L Kjølberg & F Wickson

www.panstanford.com
978-981-4267-05-2

1. Introduction

1.1 Public engagement and the public sphere

In this chapter the two terms representation and public sphere will be used repeatedly. Both these terms come with different connotations and associations to people with different backgrounds. Here they will be used quite broadly. The term representation will be used to describe the way that nanotechnology is unavoidably framed, intentionally and unintentionally, whenever it appears for instance in newspaper texts. Public sphere will be used to describe areas of social life and human interaction where issues and problems are discussed and opinions formed. Media is an important part of the public sphere, both in terms of being an arena in which these interactions take place, and by providing information about and attention to certain issues and topics. The way nanotechnology appears in the public sphere is interesting for at least two reasons. In the first instance it tells us something about the debate that has actually taken place, for example in the form of reports from meetings and other public events, as well as letters and opinions in the debate pages. To study newspaper coverage is however equally interesting for another reason: it shapes the present and future public understanding and awareness of what nanotechnology is and what it can do. It is a way to learn about in what ways and how the public has been exposed to information about nanotechnology so far. This again affects the possibilities, hopes, prospects, problems and solutions one sees and takes into account in political, as well as day-to-day, decision-making. In performing and analysing this study, I have tried to keep these two ways to read the newspaper texts apart.

The study that this chapter is built around was motivated by a wish to learn more about to what extent representations of nanotechnology in the media facilitate public engagement, in the sense that the media contributes information and perspectives in the public sphere that can empower critical thinking and reflection. The approach has been to search for intellectual resources, narratives, metaphors and arguments present in the newspapers suitable to facilitate or forestall a lively and meaningful public debate. Two assumptions have been particularly important in shaping this study. The first is that public engagement (in an uninvited form as well as the willingness to arrange or take part in organised activities) stems from a concern or a sense that something is at stake. The second is that certain representations, especially if persistent or reoccurring, are more suited to generate concern and therefore to stimulate engagement. Since the primary aim was to understand the awareness and knowledge of nanotechnology among *the public*, the study was narrowed in its attention to aspects potentially relevant for daily life, work life and education, personal health, environment, military and other prominent societal issues. The issues that were excluded in this way mainly had to do with financial reporting on business and investment.

The outline of this chapter will be as follows: the next section gives a brief background about Norway and its status in relation to nanotechnology and public engagement. Section 2 presents a short introduction to the interest in public engagement for nanotechnology. In section 3 the case study is presented, and section 4 continues to outline its findings and sketch their potential relevance for public engagement. The chapter argues, in section 5, that the dominant representations found in Norwegian newspapers are not particularly suitable to facilitate and inspire engagement.

1.2 Nanotechnology in Norway

Norway, a country with a population of 4.7 million inhabitants, is one of the few European countries outside the European Union (EU). Still, as a member of the European Economic Area (EEA), much European legislation also applies in Norway. Research in nanoscience and nanotechnology has been modest (NFR 2006), although in 2002, the Research Council of Norway established a dedicated funding program for nanotechnology, "NANOMAT", with particular focus on new and functional materials. The three largest universities (those in Oslo, Bergen and Trondheim) all now have nanotechnology research groups and provide education in nanoscience with dedicated bachelor and/or master programmes.

There has not been any public participation exercise arranged in Norway for nanotechnology and the impression of this author, as someone who lives in Norway and works with ethical and social aspects of nanotechnology, is that in general both public awareness and knowledge of nanotechnology is low. A Eurobarometer survey, conducted in 2005, asked inhabitants of European countries about their interest in technologies. Those who answered that they were interested in technologies were asked to state their interest in a number of specific technologies. The survey found nanotechnology to be among the technologies that Norwegians (and most Europeans) were the least interested in, with 8.5 % (of those who had already declared an interest in technologies) answering that they were interested in this technology (EC 2005). Sjøberg and Schreiner (2006) have analysed the results from this survey in a Norwegian context. They propose that the low interest, both in Norway and in Europe, has to do with a lack of knowledge about nanotechnology (Sjøberg & Schreiner 2006).

2. Public Engagement in Nanotechnology

Attention and awareness of the advance of nanotechnology in various strands of social studies of science has overlapped in time with a matured discourse of extended participation in science and technology policy and decision-making (Toumey 2006; Kearnes & Wynne 2007; Rogers-Hayden & Pidgeon 2007). It has been claimed that because of this early awareness, there is a unique opportunity for broader involvement in policy decisions and scientific priorities for nanotechnology, in turn leading to a more democratic and socially robust development (Roco & Bainbridge 2001; Royal Society and Royal Academy of Engineering 2004; Macnaghton *et al.* 2005).

The desire for public involvement to take place early to prevent ethical, social and environmental problems later on, gained increasing ground with controversies around biotechnology in many regions throughout the nineties (Wilsdon and Willis 2004). This move to *upstream public engagement* is commonly described as a way to actively involve groups of the public in democratic processes for science and technology development before both research development and public discourse has become set and entrenched. The call for increased public participation in science and technology decision-making is often grounded in the recognition that the instruments in place for the introduction of new technologies do not adequately address problems of disputed values and unforeseen consequences of ever more powerful technological applications (Wynne 1992; Funtowicz & Ravetz 1993; Gavelin *et al.* 2007). It is the possibility to expose these sorts of issues to a broader range of people at an early stage of development that has caused widespread optimism about the potential to guide nanotechnology development in positive directions. Furthermore, social scientists are not alone anymore in stressing the value of democratisation of science and technology. There is also considerable political will, as evidenced by the two latest research funding frameworks of the European Union: the FP6 (Science and Society) and FP7 (Science-in-Society) (European Commission 2008), as well as interest within the natural sciences (see for example chapter 2).

The problem is just that we see very little of this public participation, engagement and involvement in practice in many regions, among them Norway. This fact has been a subject of research and debate for some time, and there is already an extensive literature evaluating the role, scope and success of public engagement in relation to nanotechnology (Toumey 2006; Kearnes & Wynne 2007; Rogers-Hayden *et al.* 2007; Rogers-Hayden & Pidgeon 2007; Kitcher 2008; Bowman 2007; Gavelin *et al.* 2007; Stilgoe 2007 etc). One of the questions that has been raised in this literature is whether it is too early for members of the public to engage in debates about nanotechnology. To find the right time between *too late* (when technological pathways are set) and *too early* (before necessary knowledge and awareness exist) seems to be something of a paralysing paradox for many proponents of nanotechnology engagement.

The argument that it is *too early* makes sense in many ways. Nanotechnology remains unconsolidated, both scientifically and semantically. The lack of agreement on definitions, scope and visions makes it challenging to discuss its development and governance, in public as in all other settings. The fact that definitions shift and visions vary can also be a factor that may create hesitation and delay in media attention.

3. Nanotechnology in Norwegian Newspapers

In order to see how nanotechnology was represented in the public sphere, a structured reading of all newspaper texts in Norwegian national and local newspapers referring to nanoscience and nanotechnology published between 1st of January 2000 and 31st of December 2007, was conducted. This long timeframe allowed for a study of the development of the newspaper coverage from very early on. The material included all types of newspaper texts, mainly journalistic articles, but also short notices from national and international news agencies, as well as correspondence and debate letters. In this way the newspaper texts give voice to a number of different actors in the public sphere.

A search in the database A-tekst (a comprehensive media archive for newspapers in Norway) in daily newspapers for the Norwegian words for nano-science, nano-technology, nano-ethics, nano-politics and nano-laboratory gave 574 matches. By using as criteria for inclusion in the final material that the text should make claims or give factual information about nanotechnology, the material was reduced to 225 texts. A total of 9 regional/ local newspapers and 7 national newspapers were represented in the material. In a Norwegian context, this includes all main daily newspapers.[2] Once this set of data was defined, the qualitative analysis software Atlas.ti was used to organise the material and to code the information found in the 225 texts. Quotes that contained relevant or interesting information were coded, and one quote could be assigned more than one code. After the coding, the texts were subjected to a systematic textual analysis. The texts were divided into quartiles

---

[2] The newspapers represented in the study are (with number of articles in brackets): Adresseavisen (53), Agderposten (2), Aftenposten morgen (47), Verdens Gang (20), Dagens Næringsliv (19), Dagbladet (15), Stavanger Aftenblad (16), Bergens Tidende (14), Nationen (9), Klassekampen (8), Dagsavisen (8), Nordlys (4), Fædrelandsvennen (4), Harstad Tidende (3), Tromsø (2), Troms Folkeblad (1).

(one quarter of a year), in order to analyze the material for trends and differences over time. All the texts were also registered for their overall tone of general representation of nanotechnology: positive, positive/negative, neutral, negative/positive or negative. The texts were sorted chronologically and read three times. All parts of the texts that presented factual information or made substantive descriptive or normative claims about nanotechnology, science and technology in general or about scientists, were coded in the first two readings. The third reading served to verify the consistency of the use of the codes throughout the 225 texts. In this way, the types of issues present in the quotes were systematized and organized into a set of 84 codes. Examples of such codes are "education", "products and uses", "social aspects", "precaution", "citizens", "convergence" and "institutions". This open and systematic coding allowed for an all encompassing and chronological reading of Norwegian newspaper coverage of individual topics and issues concerning nanotechnology.

4. Three Dominant Representations

4.1 Some general observations

The study showed the emergence of a small, but increasing interest in nanotechnology in Norwegian newspapers. The coverage remained, however, scattered and unfocussed throughout the whole period. There seemed to be a lack of persons or incidents that the interest and coverage could centre around—a function filled for instance by Prince Charles in the UK, after his statement in 2003 about the environmental risk of nanotechnology (Anderson *et al.* 2005). The statement from the prince also gave the British media a natural way to introduce and include aspects of environmental (and social) risks in their coverage about nanotechnology. In Norway there has been no similar anchoring point for relating nanotechnology with broader social concerns in the media.

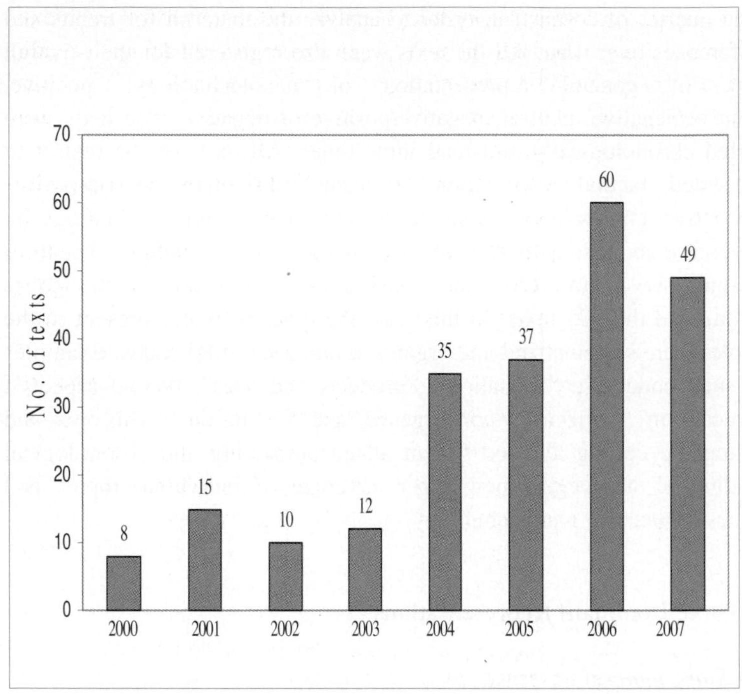

Figure 1: The number of newspaper texts representing nanoscience and nanotechnology per year in Norwegian daily newspapers for the years 2000-2007.

As can be seen from Figure 1, the last four years have had significantly more coverage than the first four. In contrast to many other countries, Norwegian newspapers can not meaningfully be grouped according to the level of education or social status of their readers. Instead, regional differences are far more important, as many Norwegians read their local paper as one of their main newspapers. The 225 texts included 101 texts from local and regional newspapers and 124 texts from national newspapers. Interestingly, one of the regional newspapers (*Adresseavisen*) had 23,6 % of the coverage (53 texts). *Adresseavisen* is the largest regional newspaper of Trøndelag, (including the town of Trondheim where The Norwegian University of Science and Technology, NTNU, is situated). Most of the texts (33 of 53) from *Adresseavisen* were stories with a local angle, entirely or partly focussing on research and teaching activities at the NTNU or the local branch of the research institute SINTEF.[3] In comparison, none of the other local/regional newspapers had more than 14 texts on nanotechnology in this period. This indicates that there may be some regional differences in knowledge and awareness of nanotechnology among the Norwegian public.

Despite the relatively low number of texts, it is clear that nanotechnology is becoming a reality in Norway in this period. The newspapers report from a time when nano-laboratories are being opened; nano-research groups are being formed; university nano BSc- and MSc-study programmes are being started; the Research Council of Norway is opening a funding scheme especially for nanosciences; businesses founded on nanotechnology are put in place; money and prestige are being invested; and products with nanomaterials are entering the market.

In the next sections the three dominant representations of nanotechnology in Norwegian newspapers will be presented. The representations can roughly be classified under the labels: nanotechnology is positive, nanotechnology is important for the future and nanotechnology is under control.

[3] SINTEF (the name is now the organizations full name, although it used to be an acronym for "Stiftelsen for industriell og teknisk forskning ved Norges tekniske høgskole (NTH)") is a private research organisation and the largest independent research institute in Scandinavia.

4.2 Representation one: nanotechnology is positive

As mentioned in the method section, all the texts were registered for their overall tone of general representation of nanotechnology, on a scale from only negative to only positive. Irrespective of topic, more than 60% (143 texts) gave an unequivocally positive representation of nanotechnology. These were for instance stories of useful consumer products on the market, new research initiatives claiming the prospects of new and fantastic applications for medicine as well as simplifying everyday chores, or articles about how our lives in general will change for the better with new nanotechnologies. Two thirds (170) of the texts gave a positive (143) or positive/negative (27) representation. The positive/negative representations were by far mostly positive, but countered with some negative aspects, typically at the end of the text. Asking for instance an interviewed researcher a last question of whether the technology presented in the article could have any ethical concerns, or concluding the text with the statement along the lines of: "Of course one needs to make sure the technology is assessed for adverse effects on health and the environment". The following quote represents an example of the latter (being the very last sentence in a text about the possibilities of nanotechnology):

> *And just like with genetically modified food, another thing that we cannot see, nanotechnology is able to create both justifiable and unjustifiable fear. That is why we need to have a high ethical standard within science and industry. Transparent, publicly funded research has to have high priority in order to recruit the best scientists* (Adresseavisen, 13th October 2004).[4]

In these sorts of statements, the reader is led to think about negative aspects, but at the same time assured that the people working with nanoscience are taking care of this in a satisfactory way.

---

[4] All quotations in this chapter are translated from their original Norwegian by me, with Fern Wickson as a native English speaking consultant.

Considering the body of texts as a whole, the range of questions, concerns and societal issues that were raised was actually quite broad. Most of the concerns were however mentioned only once, and the vast majority of the 82 articles that did include negative aspects, did this in one of the two following ways: 1) the possibility of side-effects on human health and the environment, or 2) technologies out of control threatening human existence as we know it in the far future. The following quote is an example of the latter:

> *Self assembly of course opens for the most frightening scenarios: Billions of nanobots that cannot be stopped. The nice robots that were going to assemble "gammalost"*[5] *continue to work and fill the whole atmosphere with the Norwegian speciality. What possibilities the nanoists in the military industry envision is not healthy to think about.* (Adresseavisen, 8th September 2000).

This quote also shows how apocalyptic visions were sometimes kept satirically at a distance by humorous remarks like this of the nanobots assembling smelly cheese. Broader and possibly less spectacular societal changes and issues in the present or near future were rarely mentioned in the texts.

Stephens (2005, p.196), in a study of the newspaper coverage of nanotechnology in the US, also finds that "[m]ost of the articles, leaning towards a sentiment regarding the benefits or risks of nanotechnology, overwhelmingly lean towards the positive". Anderson and colleagues (2005) have studied the newspaper coverage of nanotechnology in the UK. They find that "[l]ooking across all the coverage in the sample period, the possible benefits to be derived from nanotechnology receive more extensive coverage than do possible risks" (Anderson *et al.* 2005, p. 216).

For the purpose of our interest in this chapter this leads to the question: If the impression most people have of nanotechnology is that it

---

[5] 'Gammalost' is a type of strong Norwegian cheese.

is a positive new technology with few concerns, why would they decide to engage in its development? This first dominant representation is on its own a reason to suspect that the public sphere lacks recourses for raising concern around nanotechnology development. As we shall see, the two other dominant representations pull in the same direction.

4.3 Representation two: nanotechnology is important for the future

The texts included numerous references to the future. First of all, the word *future* itself was used 75 times, and additionally, visions of nano-products that will be realised and scenarios of futures with nanotechnology were common. Words and expressions that drew attention forward in time, rather than to the present, such as robots, self-assembly and science-fiction were used more than 200 times. In this manner, the focus was often quickly shifted towards the future, also in texts about present applications: "Already advanced nanotechnology is being used in diagnostics. But it will be several decades before nano-robots swimming in the veins possibly become a reality" (Aftenposten, 21st September 2001). Even in stories covering something like the launch of a research project with relatively sober ambitions, creative suggestions of how this technology may change our lives in the future were provided. References to science fiction were also often used to point to the potential of today's more or less basic research. "Even if we have a long way to go, this is not science-fiction" (Aftenposten, 15th March 2002).
There were also examples of tendencies towards technological determinism, for instance in the form of headlines such as "Nano is the future" (Dagbladet, 31st December 2006). These sorts of statements were common, and often left unchallenged: "No one has control anymore over the explosive development. People do not have a choice. To choose is an illusion" (Adresseavisen, 7th November 2001). There were however 11 texts which took the position that nano may be nothing but hype and a fashionable term:

> *Nanotechnology is only one of the fashionable words in this context, and one often hears visionary researchers or technocrats seduce the public with the most incredible*

> *scenario of the future where nanobots run around in our bodies and tidy up illness and filth (*Bergens Tidende 7th June 2004).

As many as 40 texts referred to the *revolutionary potential* of nanotechnology and linked this to a description of the future which may be both exciting and scary. The revolutionary potential was exemplified for instance by self-assembly and the ability to build products from the bottom up by rearranging individual atoms. The word *revolution* was also used in quite a few headlines of the type: "Nano-medical revolution" (Aftenposten, 21st September 2001), "The Techno-revolution that will change your everyday life" (Verdens Gang, 21st October 2004), or simply "Nano-revolution" (Klassekampen, 8th April 2005). The use of the term revolution was clearly linked to the future in these texts. These texts typically stated that nanotechnology has given us useful consumer products today, and has potential for fantastic applications in the future. The many consumer products that were described in the texts were presented as useful and as making boring everyday chores, such as washing, much easier, if not superfluous.

Represented as something basically making the far future brighter, it is hard to see how publics would see any reason to engage. In competition with issues on the current agenda that are presented as dramatically urgent, such as global climate change and loss of biodiversity, why would anyone prioritise to use their limited time and energy on discussing the development of nanotechnology?

4.4 Representation three: nanotechnology is under control

The third dominant representation found in the emerging discourse of nanotechnology in Norwegian newspapers was that of control. The notion of control was used in two different ways that are important for our context. The first relates to the researcher's knowledge and control over the *scientific* processes, and the second to control over how technology will affect society. Both representations are interesting for the later discussion of public engagement.

The representation of technological control presents the researcher as having the ability to completely master the physical properties of materials or to fully predict the output of a manipulation or a process at the nanoscale. As many as 46 quotes dealt with this, and the following is an example: "But it is only recently that scientists have been able to control materials all the way down to the level of the single atom." (Aftenposten, 20th September 2001). With this language, the reader is given the impression that the researchers working at the nanoscale can move the atoms around with perfect precision. In addition to the explicit use of the word control, the expression *building blocks* was used in 33 of the texts to explain how nanoscience works by moving the atoms directly. "Scientists believe that by using atoms as building blocks, we can create technical solutions beyond most people's imagination" (Aftenposten, 18th September 2001). This language gives associations to children's toys like LEGO®, and may create an impression that atoms can easily be put together and taken apart and moved around with perfect precision and control. The following two quotes are examples of this way of describing nanotechnology: "The foundation for nanotechnology is to build with natures own building blocks, the atoms." (Adresseavisen, 8th September 2000), "Nanotechnology is about developing techniques to direct the smallest building blocks of nature, which is atoms and molecules"(Dagsavisen, 8th March 2004).

To indicate control over the technological *development* was also common in the texts. This refers to quotes portraying the ability of society to direct the development of science and how technology and society will interact. This was found in as many as 76 quotes, and this is one example: "Whether you like it or not, nanotechnology is here to stay. It is up to us and future generations to decide how we want to use it"(Aftenposten, 7th February 2007). The lack of acknowledgement of past unexpected uses of technology and surprises in the interactions between new technologies and society is of particular interest for the matter of concern in this chapter.

Interestingly, a small number of texts contrasted the dominant representation of control by presenting the possibility that society might completely lose control over new and powerful technologies. Two individuals were reported to raise this concern. The only international story that got coverage more than once was Bill Joy's article "Why the future doesn't need us" (Joy 2000). Joy's article was mentioned in 5 texts. His concern is that new technologies, and he mentions nanotechnology as one of three examples, may start irreversible processes too complex to understand or predict. His article was mainly read as arguing that science should not necessarily do everything that it can do, and that there may exist scientific paths that responsible researchers should not follow. The other person to warn about a lack of control was Terje Berg. He is a Norwegian who was referred to in a text from 2001 as having a background in philosophy and astronomy and as a Senior Associate Fellow at the Foresight Institute in California. Berg pointed to the low awareness of nanotechnology among politicians and decision-makers in Norway, and stressed that it was a democratic problem that nanotechnology was being developed without any form of political control. Researchers and corporate boards made important decisions, without the politicians even knowing or caring, he claimed. This was in 2001, but this remarkably strong claim was never followed up in the newspapers.

With the important exception of the above mentioned six stories referring to Joy and Berg, the dominant representation was that society has control, both in the technological sense, and in the sense of actively directing the course of the development. To the extent that risks of undesirable effects and consequences were mentioned, these were often represented as under control because funding is now also allocated to research into ethical and social aspects. A representative from the Research Council of Norway is quoted in this way: "This is why we also study the effects on ethics, environment and society, so that the substances that are developed won't be dangerous" (Verdens Gang, 21st October 2004). Most researchers with funding to study 'the effects on ethics, environment and society' probably do not see their role as someone who shall guarantee that nothing will be dangerous. On the contrary, quite a few see it as part of their job to point out exactly that such guarantees are impossible due to the scientific uncertainty associated with novel technologies.

That nanotechnology is represented in the public sphere as something that is under control is not at all surprising. It is nevertheless not particularly suitable to triggering public engagement. When what the public know about nanotechnology is that it means that scientists move individual atoms with perfect precision and regulators and ethicists actively direct the scientific development to steer it away from potential risks and concern, there appears no reason for them to be concerned, far less to take a more active role. In this representation all aspects of nanotechnology appear as clearly under the responsibility of relevant and competent bodies and individuals.

5. Implications for Public Engagement

The years between 2000 and 2007 are a period when nanotechnology becomes a reality in Norway in terms of political and scientific priorities, as well as in consumer products on the market and investment of capital. This study has shown that there is a certain awareness of these processes in the written media, even though in terms of number of texts the coverage is rather low. The question that initiated this study was: why has this period of increased activity and awareness in Norway not seen more public engagement? Uninvited public engagement, as well as participation in arranged dialogue and consultation, stem from a concern or a sense that something is at stake. One of the most important places where this concern may be triggered is through the media. In the newspaper texts it may take the form of either journalistic articles that ask critical questions, or letters of correspondence from organizations or individuals raising concerns.

The approach in this chapter has been to search for intellectual resources, narratives, representations and arguments present in the newspapers (as part of the public sphere) suitable to facilitate or forestall a lively and meaningful public debate. In this study, however, the written press was found to be dominated by three representations; nanotechnology as positive, nanotechnology as under control and nanotechnology as important for the future. Based on this finding it is not surprising that the public is unconcerned about nanotechnology development. There is reason to believe that those that have a general idea about the emerging existence of nanotechnology in Norway will have an impression of it as something that allows scientists to work with more precision and control to develop spectacular and attractive applications that will benefit us all (or our descendents) in the far future. There are specialised social researchers paid to worry about ethical and social concerns, and the present applications are useful and harmless.

By studying the texts, information is obtained about what the newspapers actually write about, and how they write about it. In addition to describing the representations of nanotechnology, however, the structured reading also offeres an opportunity to reflect upon the issues not included or covered in the newspaper texts. It is striking, for instance, that actual and potential conflicts, as well as scientific uncertainties were hardly explored. Hornmoen (2006), who has studied science journalism in Norway, sees the lack of newspaper articles exposing the limitations of science as a general problem for the public sphere in Norway He describes what he misses in this way: "Journalism that succeeds in showing how different interpretations, sets of minds, values and interests are associated with new fields of research invites the reader to critically reflect and debate science and its role in society" (Hornmoen *et al.* 2006, p. 65).

In the first section of this chapter the argument that it is too early for public engagement was introduced. The study showed no notable change in the dominant representations of nanotechnology in the public sphere across the eight year period. At the same time, this period has brought scientific and political awareness. Even though nanotechnology is still to a large degree scientifically, legally and semantically unconsolidated, it is a reality in the sense of being able to influence scientific and political priorities and attract investment. Nanotechnologies also exist in the sense of products in the market, even though the fine lines of what exactly falls inside and outside the definition are still up for discussion. In this sense it is getting increasingly difficult to claim that it is too early for public debate. How do we move from a situation of *too early* to one that is ready to engage in the issue of nanotechnology while it is still upstream in its development?

The main line of argument in this chapter has been to analyse the representations and arguments present in the public sphere in relation to their ability to facilitate debate. As mentioned in section two, the role, scope and success of public engagement in relation to nanotechnology is currently the subject of extensive debate in academic circles. In these debates engagement is normally taken as a suitable object for social scientists and governmental institutions to plan, initiate and organize. Still, the problem of framing creates hesitation for many institutions when wanting to approach the public. When the knowledge and awareness is so low, and the resources and time available for consultation limited, one of the main challenges is to empower the public efficiently without framing the issue too much. One way for those worried about the lack of public engagement in nanotechnology to approach this challenge is to provide the media, and other parts of the public sphere, with a broader range of perspectives. This is important in addition to for instance designated webpages with information about nanotechnology, because those that are not already engaged are unlikely to come across such pages. To get access to the media, or to gain attention in other parts of the public sphere, with words of sober caution or slow unspectacular social change is not at all an easy task. To get these sorts of perspectives into the public sphere may however be necessary for public engagement in nanotechnology to take place upstream of potential harm. If harm or controversy takes place, it will of course easily reach newspaper headlines. The availability of a plurality of perspectives and resources in the public sphere suitable for generating upstream public engagement in nanotechnology is the shared responsibility of everyone already engaged in nanotechnology development.

6. Conclusion

This chapter has shown that for the last eight years, there has been an unfortunately one sided coverage of nanotechnology in Norwegian newspapers, which does little to stimulate public engagement. For public participation to take place for nanotechnology, be it spontaneous or arranged, up-, mid- or downstream, it is a good start to call for a joint effort between responsible natural scientists, nano-ethicists and aware journalists, as well as other voices, to present a broader range of stories and representations in the public sphere. Particularly it is crucial that science is presented in such a way that it becomes visible how different interpretations, values and interests shape a new field of research.

Questions for Reflection:

1. Is it, in your opinion, too early, too late or just the right time for public engagement in the development of nanotechnology?
2. Apart from the media, what other actors are important for creating attitudes and awareness about nanotechnology?
3. How is nanotechnology represented in the media in your country?

Bibliography

Anderson, A., Allan, S. Petersen, A. and Wilkinson, C. (2005). The framing of Nanotechnologies in the British newspaper press, *Science Communication*, 27, pp. 200-220.

Bowman, D. M. (2007). Nanotechnology and public interest dialogue: Some international observations, *Bulletin of Science, Technology & Society*, 27, pp. 18-132.

European Commission (2005). *Social Values, Science and Technology Special Eurobarometer 225*, Brussels: EC. Available from http://europa.eu.int/comm/public_opinion/index_en.htm.

European Commission (2008). Socio-economic sciences & humanities and science in society in 2007, *Highlights of the Year*, EUR 23172.

Escobar, A. and Alvarez, S. E. (eds) (1992). *The Making of Social Movements in Latin America: Identity, Strategy, and Democracy,* Boulder: Westview Press.

Funtowicz, S. O. and Ravetz, J. (1993). Science for the post-normal age, *Futures*, 25, pp. 739-755.

Gavelin, K., Wilson, R. and Doubleday, R. (2007). *Democratic Technologies? The final report of the Nanotechnology Engagement Group (*NEG), London: Involve.

Hornmoen, H., Meyer, G. and Sylwan, P. (2006). *Fornuften har flere stemmer*, Oslo: Cappelen Akademisk.

Joy, B. (2000). Why the future doesn't need us, *Wired*, 8 (4).

Kearnes, M. and Wynne, B. (2007). On nanotechnology and ambivalence: The politics of enthusiasm, *NanoEthics*, 1, pp. 131-142.

Kitcher, P. (2007). Scientific research–Who should govern? *NanoEtchics*, 1, pp. 177–184.

Macnaghten, P., Kearnes, M. and Wynne, B. (2005). Nanotechnology, governance, and public deliberation: What role for the social sciences? *Science Communication*, 27, pp. 268-291.

Norges forskningsråd (2006). *Nasjonal strategi for nanovitenskap og nanoteknologi.* Report from the Research Council of Norway, Oslo.

Roco, M. C. and Bainbridge, W. S. (2001*). Societal Implications of Nanoscience and Nanotechnology*, Boston: Kluwer.

Royal Society and Royal Academy of Engineering, RE/RAE (2004). *Nanoscience and Nanotechnologies: Opportunities and Uncertainties*, London: Royal Society.

Rogers-Hayden, T., Mohr, A. and Pidgeon, N. (2007). Introduction: Engaging with nanotechnologies – Engaging differently? *NanoEthics*, 1, pp.123-130.

Rogers-Hayden, T and Pidgeon, N. (2007). Moving engagement "upstream"? Nanotechnologies and the Royal Society and Royal Academy of Engineering's inquiry, *Public Understand. Science*, 16, pp. 345–364.

Sjøberg, S. and Schreiner, C. (2006). *Holdninger til og forestillinger om vitenskap og teknologi i Norge - En framstilling basert på data fra Eurobarometer og ROSE*, Oslo: University of Oslo.

Stephens, L. F. (2005). News narratives about nanotechnology in major U.S. and non-U.S. newspapers, *Science Communication*, 27, pp.175-199.

Stilgoe, J. (2007). *Nanodialogues – Experiments in Public Engagement with Science*, London: Demos.

Toumey, C. (2006). National discourses in democratizing nanotechnology, *Quaderni*, 61, pp. 81-101.

Willis, R. and Wilsdon, J. (2004). *See-through Science*, London: Demos.

Wynne, B. (1992). Uncertainty and environmental learning: Reconceiving science and policy in the preventive paradigm, *Global Environmental Change*, 2, pp. 111-127.

List of Norwegian newspaper articles referred to in the text:

Adresseavisen, 8. September 2000: "Nano for liten, nano for stor"
Adresseavisen, 7. November 2001: "Noen grunner til nøkternhet"
Adresseavisen, 13. October 2004: "Det vi ikke kan se"
Aftenposten, 18. September 2001: "Førstesiden"
Aftenposten, 20. September 2001: "Vår nye ferd innover Rør til 5000 kr. Grammet"
Aftenposten, 21. September2001: "Nano-medisinsk revolusjon"
Aftenposten, 15. March 2002: "Dette er USAs fremtidssoldat"
Aftenposten, 19. September 2004: "Forskningsdager for alle"
Aftenposten, 15. June 2005: "REPLIKK Nanomaterialer - et fremtidig miljøproblem?"
Aftenposten, 7. February 2007: "Fremtidens forskning"
Bergens Tidende, 7. June 2004: "Fysikk og teknologi i Afrika og på Mars"
Bergens Tidende, 22. June 2006: "Fred Kavli kommandør"
Klassekampen, 8. April 2005: "Nano-revolusjon"
Dagbladet 31. December 2006. "Nano er Fremtida"
Dagsavisen, 8. March.2004: "Selvvaskende vinduer, Klær som ikke blir skitne, Billakk som ikke får riper, Nye behandlingsformer - Din nye mirakelhverdag"
Verdens Gang, 21.October 2004: "TEKNO-REVOLUSJONEN som vil forandre din hverdag"
Verdens Gang, 21.October 2004: "Utreder etikk"

Chapter 7

Everyday Nanowars: Video Games and the Crisis of the Digital Battlefield

Colin Milburn

Military scientists frequently claim that nanotechnology will transform the future of warfare, integrating soldiers and machines into the 'digital battlefield' at a molecular level and making combat programmable at every scale: war becomes a video game. At the same time, consumer video games increasingly feature simulations of 'nanowar' based on the research agendas of real scientific institutions. This chapter examines the convergence of military nanotechnology with video game culture. Focusing on the recent game *Crysis*, which follows the adventures of a soldier equipped with a nanotechnology battlesuit (based loosely on US army prototypes), this chapter investigates how players navigate the condition of nanowar. *Crysis* players typically adapt to the official rhetoric of military nanotechnology, which presents nanotechnology as a form of masculine empowerment. At the same time, these players often comprehend the functioning of military nanotechnology as occasioning a state of crisis: a crisis of gender and human embodiment. Drawing upon queer theory (the field of critical analysis that studies the construction and deconstruction of normative assumptions about gender and sexuality) this chapter shows that, even as video game players adjust to the concepts of nanowar, their engagement with the digital battlefield as an everyday playspace simultaneously opens the discourse of military nanotechnology up to other politics, other genders, and other futures.

Nano Meets Macro - Social Perspectives on Nanoscale Sciences and Technologies
by K L Kjølberg & F Wickson

www.panstanford.com
978-981-4267-05-2

1. The Digital Battlefield

The digital battlefield: an immense network of computers, sensors, and communications systems linking soldiers and machines into common channels of data, where every vehicle, weapon, and combat trooper is rendered a component in the fully integrated circuits of command and control. Under the logic of 'network centric warfare'—a defining feature of militarism in the wake of postmodernity (Gray 1997; De Landa 1991; Singer 2009)—the digital battlefield emerges as a shared awareness of combat space, a hypermedia environment generated by the flows of information streaming from various mobile units, intelligence sources, and global positioning satellites, analyzed and computed in real time through massively-parallel processes, generating the zone of warfare as a virtual reality to be navigated and coordinated on a computer monitor or a vehicular heads-up display.

The completely digital theatre of war appears just off screen but closer than ever, an emergent development of the military-entertainment complex (Lenoir 2003; Stockwell & Muir 2003; Terry 2007; Der Derian 2009). The promise of programmable war (where everything from the operation of unmanned ground and aerial vehicles to the bio-monitoring of soldier physiologies is algorithmically synchronized) exemplifies the condition of virtuality we see operating everywhere today, where material bodies become discretely atomized nodes in computational phase-space. For instance, Jean-Louis 'Dutch' DeGay, an equipment specialist at the US army's Natick Soldier Center, describes the digital battlefield as depending on two specific engineering objectives:

> *One is a new suite of vehicles and the network that those vehicles will operate in, and the other is the next-generation soldier who will be a node, if you will, to plug into that network and interact with those vehicles...We look at the soldier at the next-generation platform. The Star Trek analogy is the Borg, a group of people who are plugged into a supercomputer and part of the collective, so they can share information and push data back and forth; what one person knows, everybody knows* (DeGay quoted in Howard 2006).

This modular soldier of the future, plugged into virtual combat zones born from the dreams of science fiction, currently evolves in military programs all over the world, including Germany's IdZ (Infanterist der Zukunft, or 'Infantryman of the Future'), India's F-INSAS (Futuristic Infantry Soldier as a System), the United Kingdom's FIST (Future Integrated Soldier Technology), France's FELIN (Fantassin à Equipement et Liaisons Integrées, or 'Foot Soldier with Integrated Equipment and Links') and BOA (Bulle Operationnelle Aéroterrestre, the network-enabled 'Aeroterrestrial Operational Bubble' combat system), Singapore's ACMS (Advanced Combat Man System), Norway's NORMANS (Norwegian Modular Network Soldier), Australia's Land 125, the United States' Future Combat Systems and Future Warrior, and many others. These programs instantiate the protocological and topological production of decentralized control networks wherein, as the critical theorists Alexander Galloway and Eugene Thacker have written, "the scale is fractal in nature, meaning that it is locally similar at all resolutions, both macroscopic and microscopic. ...[Such] networks are elemental, in the sense that their dynamics operate at levels 'above' and 'below' that of the human subject." (Galloway & Thacker 2007, p.155-157). They operate fundamentally, in other words, as elemental interactions of bits and atoms.

The digital battlefield, involving the total integration of bits of matter with bits of code, necessarily extends from the macroscale down to the nanoscale. According to the US Army Research Office, the final realization of this vision for future warfare will depend upon innovative military science that looks into "creation and utilization of materials, devices, and systems through the control of matter on the nanometer-length scale and into the ability to engineer matter at the level of atoms, molecules, and supramolecular structures." This research will contribute to "increasing command and control, lethality, mobility, survivability, and sustainability of systems in the field...[It will ultimately enable] a strategically mobile force capable of handling the full spectrum of future operations from stability and support operations through major theater war" (US Army Research Office 2001, p. 5).

Nanotechnology would seem to provide revolutionary solutions for integrating soldiers and information systems in the battlespaces of the future (Altmann 2006). As US Navy Lieutenant Shannon L. Callahan explains, if the ultimate goal of this projected revolution in military affairs involves "integrating the infantryman's capabilities into the digitized battlefield without adversely affecting his performance, thereby multiplying his lethality through an ability to communicate what he sees and knows up to higher headquarters," then nanotechnology and its characteristic "tiny devices could be the revolution's enabling technology" (Callahan 2000, p.25, 20, referencing Gourley 1998). To be sure, this revolutionary notion informed the US army's decision in 2001 to create the Institute for Soldier Nanotechnologies at MIT:

> *The individual soldier [of the future] ...will require systems revolutionary in their capabilities. Recent advances in the field of nanoscience suggest that it may be possible to provide the soldier with radically new capabilities in full-spectrum threat protection without incurring significant weight or volume penalties. Such soldier systems will only be realized by directing additional resources to the Army's Science and Technology Program in the emerging field of nanoscience. For that reason, the Army's Science and Technology Program in the emerging and assigns arena is being extended ...to create a University Affiliated Research Center (UARC) entitled the 'Institute for Soldier Nanotechnologies'.* (US Army Research Office 2001, p. 3).

When the Army Research Office (ARO) first disseminated its Broad Agency Announcement about the Institute for Soldier Nanotechnologies, the solicitation for research proposals appeared on the ARO webpage attached to a cartoon of the digital battlefield (Figure 1). Futuristic soldiers—what the army has dubbed 'Future Force Warriors'—charge across a hyperbolically digitized combat zone. The computational grid or matrix on which the Future Force Warriors wage battle materializes directly from a field of green-tinted binary code, 1s and 0s from the antique past of monochrome monitors now morphing seamlessly into the scene of high-tech warfare: a nod to *The Matrix* films, perhaps, where a virtual world indistinguishable from reality emerges from the electronic translations of computer language. This digital cartoon advertising the US army's vision for soldier nanotechnologies makes one thing perfectly clear: the anticipated digital battlefield of the future would seem to be at last realized in the collision of cyberspace with nanospace.

Figure 1. The Digital Battlefield: US ARO Final Solicitation, Institute for Solder Nanotechnologies (2001). For colour reference turn to page 561.

This image, visually enacting the military investment in nanotechnology as enabling the digital battlefield, redeploys and vivifies the widespread discourse of 'digital matter' circulating in the fields of nanoscience. 'Digital matter' is the prevalent notion among advocates of radical nanotechnology that, as the nanoscientist J. Storrs Hall has put it, "Nanotechnology will make matter into software" (Hall 2005, p.271). For military science, the implication becomes: 'Nanotechnology will make war into a video game.'

The notion that nanotechnology will empower military science to transform the field of combat, the constitution of weaponry, and even soldiers' bodies with the ease of toying with pixels on a screen, appears frequently in the research agendas and media artifacts of nanoculture. Shortly after establishing its operations at MIT, the Institute for Soldier Nanotechnologies (ISN) produced a series of publicity videos entitled 'Soldier of the Future' that presented the potential of nano to reprogram the shape of warfare. The videos depict this potential by remediating the visual conventions of action-adventure video games. For example, the first ISN video, created in 2004 by North Bridge Productions (a division DigiNovations) in collaboration with the video game company Boston Animation, splices interviews of real ISN scientists together with fictive animations of military nanotechnology in action. The animated sequences draw upon the same 3D graphics architecture used in other Boston Animation video games like *Darkened Skye* (2003). These animated vignettes focus on the lightweight, skintight exoskeleton that the ISN aims to engineer: a nanotechnology battlesuit, or nanosuit.

According to the vignettes, the nanosuit will harden instantaneously to stop sniper bullets, rapidly synthesize antitoxins in response to chemical weapon explosions, and administer first aid to wounded soldiers by "applying little electric currents to systems [of 'exomuscle'] that are unimaginably small and light" (Soldier of the Future 2004). Rendering biometric data visible on the 'battlefield network', the nanosuit will enable soldiers and commanders alike to observe its molecular operations on the body. In one of the vignettes, the nanosuit worn by a stricken soldier injects nanoparticles into his bloodstream to combat deadly enemy toxins. The soldier's companion, watching this microscopic process take place via the monitor in his own nanosuit, says:

"The drama unfolding in my viewscreen was riveting" (Soldier of the Future 2004). Such histrionics of military nanotechnology, turning the chemical interior of the soldier body into yet another setting for the theatre of war, would appear no mere fantasy. For the Soldier of the Future video announces that the nano-enabled digital battlefield is almost here: "Nanotechnology research is taking this out of the realm of dreams and, within in a couple of decades, into the field" (Soldier of the Future 2004).

A later ISN video depicts the future of soldier nanotechnology through an introductory first-person action sequence (Figure 2) whose visual field is isomorphic with the heads-up displays (HUDs) featured in 'first-person shooter' video games like *Doom* (1993). The narrative here concerns a small squad of nanosuited soldiers infiltrating an enemy bioweapons bunker. During this search-and-destroy mission, an airborne bioagent (dubbed "bad stuff" by the squad leader) infects Jones, one of the soldiers. But before Jones is even aware of it, the nanosystems that constantly monitor his blood chemistry immediately go into action to contain the contamination.

In this video, the digital battlefield—the networked informatic space produced by the various recording devices and nanosensors in each of the soldiers' uniforms—is available to the troops as well as to the military scientists coordinating the operation through their command terminals: the first-person shooter perspective is identical for the soldiers and the military scientists. When Jones' bio-contamination occurs, the distributed HUD indicates that a man has been exposed and therefore must be sent out, dropped from the mission—identical to the convention in video games where the HUD inventories 'remaining lives' or 'extra men'. Subsequently, the video offers us a molecular view of "what happens inside the suit," showing the subdermal functions of the onboard "med-surveillance systems": nano-syringes constantly puncture the soldier's skin, sucking samples of his blood through "micro-blenders", filtering his pureed RNA molecules through lab-on-a-chip biometric monitors, which then beam the resulting data back into the digital network. This vignette closes by showing that the nanosuit's embedded systems have intervened in time to save the soldier's life, so that he can go on fighting on the digital battlefield.

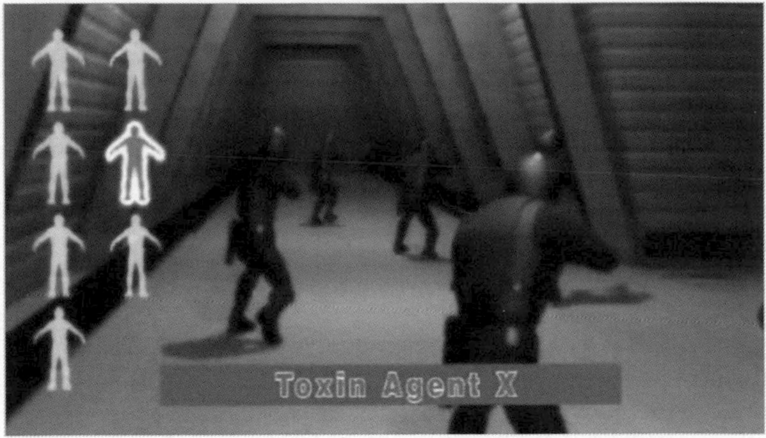

Figure 2. First-Person Shooter: Soldier of the Future video, Institute for Soldier Nanotechnologies (2005). For colour reference turn to page 561.

Video game traditions here shape the way that military nanoscience presents itself to the public, as striving for a digital future where wars are rebootable and soldiers' lives are replayable, thanks to the struggles of intrepid researchers. As the first Soldier of the Future video tells us, ISN scientists are currently "mounting an assault on that challenge [of improving soldier survivability] with tools that were unimaginable just a few years ago" (Soldier of the Future 2004). We are presented with the vision of a very near future where very real wars will be indistinguishable from video games precisely because nanotechnology gives us digital command and control even at the atomic scale, the ability to program warfare right down to the molecular level. Given that we can now rigorously imagine military nanotechnologies of even the most radical and far-out varieties, given that we can make digital animations right now of what, according to the ISN, was literally "unimaginable just a few years ago," that very possibility for the future already virtualizes the present. We engage with this virtuality and navigate its potentials through interactions with militarized media that do not simply represent, or speculate, but rather simulate the actualization of digital matter. In other words, video game animations animate the technology, literally, and in every way.

Even before it gets here, radical nanotechnology is already being shaped at a social level through video games of military science. For at the same time as various military institutions project the nanotechnology battlefield in the form of a video game, popular culture now consumes certain commercial video games as projections of the nanotechnology battlefield. By virtue of this constitutive transcoding or feedback between popular games and military science, the discourse of military nanotechnology emerges through a collective process that Henry Jenkins (2006) has called 'convergence culture': a mode of cultural production "where old and new media collide, where grassroots and corporate media intersect, where the power of the media producer and the power of the media consumer intersect in unpredictable ways" (Jenkins 2006, p.2). In converging at the site of the video game, military scientists and video game players together engage in unpredictable contests over the shape of the things to come in the nanotechnology era. For even as institutionalized Soldier of the Future programs advance particular technocratic visions of totalizing, programmable warfare, popular engagements with military nanotechnology frequently open onto virtual futures of an altogether different order.

2. Crysis Mode

A growing number of consumer video games today animate the technical concepts and political discourses of military nanotechnology, making the digital battlefield less a dreamscape of future wars than an everyday playspace, easily accessible and endlessly reloadable. Games such as *Deus Ex* (2000) and *Deus Ex: Invisible War* (2003), *PlanetSide* (2003), the *Red Faction* games (2001, 2002), *Nano Breaker* (2005), *Project: Snowblind* (2005), the *Metal Gear Solid* series (1998—2008), *Heroes of War: Nanowarrior* (2009), and several dozen others turn speculative nanoscience and military engineering diagrams into *recreational* experiences. Like the ISN publicity videos, these games contribute to the incursion of futuristic military technologies into everyday life by simulating the conditions for advanced nanowarfare as both present and playable. In doing so, such games variously participate in the militarization of popular culture, making the state of perpetual armed conflict into a form of consumable pleasure and often naturalizing militaristic values of imperialism, xenophobia, misogyny, and aggressive masculinity in the process (Power 2007; Nieborg 2006; Ottosen 2009). But this is only part of the story. Indeed, for many players of these games, such quotidian engagements with the digital battlefield may simultaneously produce perceptions and sensations of nanowarfare radically divergent from those of the official military programs on which these games are based. After all, as Michel de Certeau (1984, p. xii) has famously argued, "Everyday life invents itself by *poaching* in countless ways on the property of others," tactically appropriating, transforming, or *recreating* cultural materials in ways that might profoundly subvert their 'proper' meanings, affects, and visions of the future. So even as nanowar video games normalize the digital battlefield as clear, present, and inevitable, they also enact its *recreational potential.*

Consider the blockbuster game *Crysis*, released in 2007 for the Windows operating system by the German game company Crytek. *Crysis* enables the player to operate as an elite US Special Forces soldier, equipped with a nanosuit that mimics prototypes from the Institute for Soldier Nanotechnologies and the US Future Warrior program. According to Bernd Diemer, the senior designer on the *Crysis* project:

> *Taking inspiration from the Future Warrior 2020 program, we developed the Nano Fibre Suit [a.k.a. Nanosuit] that can enhance strength, speed and armour levels. The player can max the speed to dash across an open field, change to the strength setting and silently punch out a sentry* (Diemer quoted in Booker 2006).

In many ways, *Crysis* is nothing less than a playable version of the scenarios depicted in the ISN Soldier of the Future videos. The game takes place on a tropical island in the South China Sea, where the North

Korean military has commandeered a mysterious alien artifact discovered by US archeologists. Quickly, the island becomes a stage for the performance of globalization as militarization. The plot unfolds through open-ended 'sandbox' gameplay. The player can proceed through the various military objectives with a large degree of freedom, selecting missions according to individual preference rather than any prescribed order, thanks to the logistics of network-centric warfare that transform the island into a fully rendered digital battlefield. But whether battling North Korean soldiers in the opening segments of the game, or surviving the onslaught of alien creatures in the later chapters of the story, the narrative, such as it is, takes a back seat to the real focus of the game: namely, the playability of the nanosuit itself, and its relation to the figure of the male soldier inside (James Dunn, identified in the game under the codename of 'Nomad').

The nanosuit has several functions that the player can activate at will. Whenever we initiate these nano-functions, an ominous male voice booms into our headset, mechanically announcing the outcome of our selection: "MAXIMUM ARMOR," "MAXIMUM STRENGTH," "MAXIMUM SPEED," and so forth. This voice would seem to be the programmed rhetoric of military science as such, built into the operating system of the suit that we inhabit in first-person perspective. The voice, along with the incessant text material that pops onto our HUD, reminds us, comforts us, about our invulnerability—our virtual impenetrability—in the embrace of the nanosuit boasting maximum armor, maximum strength, maximum hardness.

The game's ludic insistence on the hardness of the nanosuited soldier channels the standard masculinist rhetoric now well entrenched in the discourse of military nano (Milburn 2008). This rhetoric of maximum hardness would seem to be some kind of conceptual protection against an otherwise emasculating notion of the soldier body being thoroughly penetrated at all times by invasive nano enhancements. Indeed, various long-term visions for military nanotechnology, involving nanowires in soldier's brains, nanodevices circulating in their bloodstreams, nano-actuators embedded in their muscles, and so forth, are common in DARPA (US Defense Advanced Research Projects Agency) programs and human enhancement discourse—not to mention most military science fiction (Gray 2001; Garreau 2005; Hantke 1998; Roco & Bainbridge 2003). Such visions are so completely predicated on the opening of the body, the rendering of the soldier body as a porous membrane where materials pass back and forth with abandon, a site of endless penetrations, that it is as if the hardness and hardening processes of nano must be emphasized, rhetorically if nothing else, in order to psychologically protect the 'soldier male mentality' (Theweleit 1987) from collapsing under fear of disintegration, the anxiety of being opened up by tiny probing technologies.

For even while insisting on maximum hardness and maximum impenetrability, the *Crysis* multimedia experience offers several third-person perspectives on the operations of the nanosuit: non-playable scenes from the game trailer that give an illicit vantage point on the high-tech war machine Nomad has become, showing us the extent to which nano enhancement, even in rendering the solider body maximally hard, simultaneously renders it maximally fluidic (Figure 3) (Figure 4). Such images of nanoparticles and nanodevices dropping into the bloodstream, embedded into muscle fibres, infiltrating the epidermis (similar to the ISN images of nano-needles puncturing the skin inside the protective suit) expose a conundrum in the imaginary of military nanotechnology. This is a discursive contradiction whereby the soldier is made nanotechnologically hard only by virtue of being nanotechnologically soft. The soldier becomes penetrated, invaded, conscripted—not a closed off, armored body, but an open, wet, and humble body encased in a machinic system whose operations are entirely invisible. Invisible, at least, save for these privileged digital animations that games like *Crysis* and the ISN videos give us, situating us initially in the first-person perspective and then, ecstatically, in the third-person perspective directly at the nanoscale, so that we can watch the body's molecular opening and perceive its nanoscale fluidity, even while being bombarded by the official rhetoric of hardness. By viewing or interacting with such renditions of the battlefield of digital matter, we are suspended between these two perspectives, unable to resolve them because they are both insisted on, graphically and semiotically.

This is what I would call the "crysis mode" of the nanowarrior, strung out between the poles of maximally hard and maximally soft, armored and fluidic, meaty and machinic, self and other. This is a psychological identity crisis made possible, even mandated, by the advent of military nanotechnology already, right now, even before nanotechnology of this nature has become a reality.

Concerns about this crysis mode circulate both overtly and covertly among self-professed 'hardcore' male gamers who actively contribute to discussions of *Crysis* in online forums, blogs, video-posting websites, and gaming magazines. Between 2006 (when pre-release materials for the title first appeared) and 2009, I monitored more than ten thousand of these public discussions. The vast majority of the players involved seemed to support the notion that *Crysis* appeals almost exclusively to "we with the Y-Chromosome" (AkumaX 2008) (and hence, providing another meaning for the 'y' in *Crysis*, beyond the obvious reference to Crytek and its first game, *Far Cry* [2006]). As a prerequisite for playing the game, in addition to owning high-end computer equipment, some players would "even add a new requirement: a penis. *Crysis* is the manliest game on any system" (Roher 2007). Such notions conform to video game industry stereotypes, considering the extent to which certain game genres—especially military first-person shooters like *Crysis*—have traditionally targeted male consumers, discounting female gamers, as well as transgender and intersex gamers, from consideration in game design and marketing strategy (a situation that has only recently started to change) (Cassell & Jenkins 1998; Kafai *et al.* 2008; Deuber-Mankowsky 2005). Certainly, there are many female gamers who do play *Crysis*. But in public discussions of the game, hardcore gamers usually assume by default that all the other discussants are male. This is so much the case that, in one conversation where a player named "Anna" announced her female identity ("I'm a girl gamer, would like to chat, but I'm busy playing *Crysis*"), the immediate response was incredulity, with several respondents insisting that this 'girl gamer' must really be a 'boy gamer' in disguise (Anna 2007). Most players involved in *Crysis* discussions make a point of their maleness, either in specific postings or in their author profiles. Whether or not they are actually male as they profess to be (and it is conceivable that some female gamers might seek to covertly 'pass' as male in these public spaces, for example, to

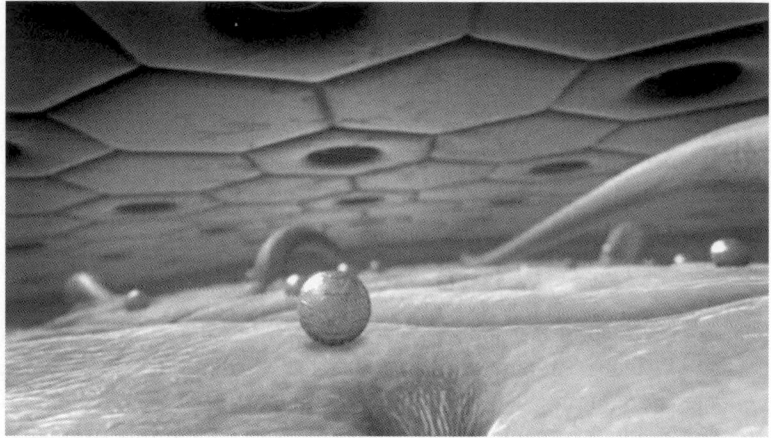

Figure 3. Inside the Nanosuit: Dropping from holes in the exoskeleton, nanospheres infiltrate the pores of Nomad's skin. ©2007 Crytek. For colour reference turn to page 562.

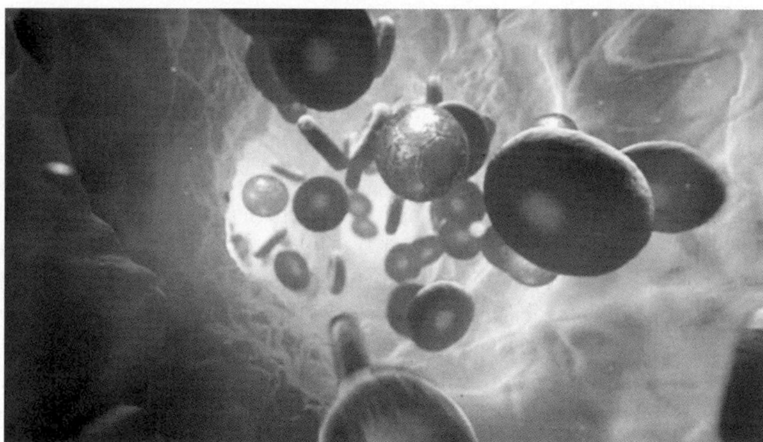

Figure 4. Wet Nanotechnology: Power-boosting nanospheres flow through Nomad's bloodstream. ©2007 Crytek. For colour reference turn to page 562.

participate without harassment), the intensely gendered discourse surrounding *Crysis* nevertheless plays out the stakes and the standards of militarized masculinity in the age of nanotechnology. Moreover, to whatever degree there might be slippage between the *offline* gender identity and the *online* persona of a *Crysis* player—that is, to whatever extent female gamers might be performing as male when playing or discussing *Crysis*—this kind of transgender recreation (or drag) would therefore *already enact* the problematics of the crysis mode (Turkle 1995; Stone 1995; Case 1996; McRae 1997; Danet 1998). For the relationship between the hypermasculine image of military nanotechnology in the game (the nanosuit's baritone assurance of maximum power enveloping the tumescent bodies of Nomad and his fellow soldiers, Psycho, Prophet, Aztec, and Jester) and the gendered body of the solider-player who inhabits this image, is precisely what is rendered precarious by the discourse of *Crysis* itself.

For many players who work through *Crysis*, military nanotechnology emerges simply as a mode of empowerment without imperilment, without risk to masculinity. The game cultivates an affirmative and affective response to the nanosuit and its capabilities, generating fantasies of individual superiority and invulnerability under the regime of nanotechnology. The narrative of the game disappears in relation to the affective force of the playable nanosuit, the imaginary transformation of the player into an indomitable nanowarrior, backed up by the insistence that all of this is scientifically "plausible." As one player writes, the game "manages to make you feel like a badass thanks to the high-tech nano-suit, which has four settings to help with combat situations...By the end of [the] single player [game narrative] it'll be second nature...[T]he nano-suit really helps you feel superior for a plausible reason [namely, the science of nanotechnology]" (Adams 2007).

Yet, the fantasy of absolute power, absolute impenetrability, on stage here stands in paradoxical contrast with the fact that such super powers are only enabled by the molecular opening and invasion of the soldier body. Even as some players identify the nanosuit as fundamentally protective of subjective integrity (in accord with the rhetoric of 'maximum hardness'), other players instead are made aware of the fact that the nanosuit "allows you to use enhanced abilities to supplement your battle prowess...by releasing nanobots into your bloodstream to heighten a particular trait" (Bertz 2008a). Some, finding the disconnect between 'hard' rhetoric and 'wet' imagery confusing, have chosen to interpret the playable nanosystems as actually operating at two distinct levels, dividing the armoring functions from the penetrating functions: "In fact, I believe in the game the 'nano suit' is two part, the part that covers the users body and provides armour / cloaking, and the free flowing 'capsules' within the bloodstream that boost strength and speed upon command" (Shadonic0 2008). Others express a clear preference for the exoskeletal functions that might be made distinct from suit's transvenous nanospheres: "I suppose the balls [passing from the suit into the bloodstream] are things using nano technology to speed up the body. I like the idea of [exomuscle] carbon nanotubes conforming to the muscles to make them stronger better myself" (Broadsword 2007). Yet certain players alternatively go so far as to perceive all functions of the nanosuit as transforming soldiers at the level of biomolecules, as actually getting into the DNA and altering the soldier's genetic code:

> *[T]he coolest thing about this game, the Nano Suit. The Nano Suit is a very high tech piece of military property. It has the power to alter your genetic code and will give you 4 different powers, Defense which will make you invulnerable to bullets and more for short time, Speed which will make you ten times faster for short time, Cloak which makes you invisible for short time, and my favorite Strength which give you super strength to pick enemys [sic] up and throw them and such* (Edmunds 2007).

For those players who recognize some fundamental contradiction in the promotion of maximum hardness through nanotechnologies that function only by making the soldier body into a weeping membrane, or molecular sponge, the condition of crysis is often perceived as a failure of normative gender: "little nano guys go into your bloodstream to make you faster!!...I'm not even kidding man, these little thingies like balls or something went inside the guy's bloodstream and he dyked out. Like, ran fast" (peacemaker898 2007). The penetrability of the male soldier ('the guy'), the porosity of his body relative to 'these little thingies like balls' that get inside him, here provokes a player-response in which the enhanced power of 'maximum speed' is understood as a queering. The onscreen nanowarrior's ability to 'run fast' by virtue of being corporeally invaded by nanospheres is seen as identical to sexual chaos, gender slippage: the guy 'dykes out', the male becomes female, lesbian. Absolute hardness and absolute fluidity, absolute armor and absolute penetrability, absolute masculinity and absolute femininity, give way when confronted with the functioning of military nanosystems. Certainly, many players have commented on a sense of gender trouble (Butler 1999) and queering suggested by the nanosuited soliders in the game: "JAKE DUNN (a.k.a. Nomad) is a raving lesbian trapped in a man's body…and Psycho ha[s] a fetish for guys in nano suits" (Milos 2007).

Crysis players incessantly discuss such ambivalences immanent to the technical condition of the nanosuited soldier, smeared between maximum hardness and maximum fluidity, precisely because the game presents it as going both ways, as a form of queering, channeling the ambiguities inherent in the broader discourse of military nanotechnology and rendering visible their insolubility. Thus the 'crysis mode', the condition of irresolvable tension enabling and mobilizing the nanowarrior, is turned into playable format by the games of military nanotechnology. In working through games like *Crysis*, players become accustomed to the concepts and the operations at stake, and begin to inhabit this peculiarly militarized version of the nanotechnology future.

3. Hardware Fetish

The power fantasies animated by *Crysis* might seem localized to the irreal spaces of science fiction, purely virtual. And yet, inhabitation of the digital battlefield of the gameworld, adaptation to the functioning of the soldier figure in the grip of the nanosuit, quickly translates into a tacit sense of personal investment in the as-yet fictive future, an embodied response to what one player aptly calls the game's "unsatiable hardware fetish" (redwinedrummer 2007).

The *Crysis* program demands so much computational capacity just to run, that only very powerful PCs (equipped with top-of-the-line processors, graphics cards, and sound systems) are capable of serving the game at its optimum settings. With its resource-intensive engine (Crytek's CryEngine 2), which provides robust and vivid 3D graphics, sophisticated AI behaviors, and a responsive open-world environment allowing gamers to explore the island warzone with few limitations, *Crysis* requires advanced computer hardware with maximized tech-specs simply to be played: maximum hardware. The boundary-pushing hardware requirements of *Crysis* have become legendary: "*Crysis* demanded so much graphical horsepower it crippled most PCs" (Bertz 2008b, p.126). These outrageous requirements make the game itself appear as an artifact transported backwards in time from some kind of science fiction future: "It's too demanding today, that's just the simple truth. Nothing runs it" (Emil 2007). Another gamer concurs: "It may be cheaper to travel forward in time to play *Crysis* than it is to build a machine capable of bending the game to its will" (The Top 50 Games of 2007: *Crysis* 2008, p.42). This idea of future-ladenness echoes throughout the gaming community: "*Crysis* is the future. The present isn't ready for it" (Eric 2007).

Regularly invoking the same militarized rhetoric and science fiction idioms that characterize the narrative of *Crysis* itself—the 'crippling' of lesser PCs, the struggle for technoscientific power, and the notion that you might need a time machine to operate the game—players of *Crysis* express a mixture of frustration and swagger in describing their gaming experiences. The frustration appears when they are unable to make the game work: "I get annoying *screen tearing* constantly . . . that gives me a headache . . . this bastaaard [sic] of a game. I hate it, I want Crysis dead" (l88bastard 2008). But many players start to swagger as soon they are able to make the game run at all, even imperfectly: "I did play Crysis on its very high settings and I did lag a bit at times and I couldn't put the anti-aliasing up all the way but no computer can run Crysis on very high ...but all in all my computer is awesome and is the envy of everyone I know" (Pat M. 2009).

Although the game 'cripples most PCs', those players equipped with superior hardware have comparatively little trouble. These better-equipped players therefore identify their relationship to their own awesome computers in terms that evoke the in-game relationship between Nomad and his awesome nanosuit, while warning other players that, without such a computer, the full awesomeness of the nanosuit will remain elusive. For example:

> *Crysis [offers] a high-tech nanosuit that augments natural human qualities like speed, strength, and fortitude. . . . But all of this comes at a price: the cost of a state-of-the-art gaming machine. Play without a technologically advanced rig and you'll have no choice but to run the game at the lowest visual settings* (Sapieha 2008).

It would appear that the logic of the game narrative and the rhetoric of 'maximum power' surrounding the nanosuit maps onto players' own relationships to their gaming systems.

It is as if the military fantasy depicted inside the game of a soldier engaged in international (and interplanetary) warfare, pumped up to the max by virtue of his nanotechnology exoskeleton, enables players to comprehend their technical ability to navigate the game's digital battlefield in equivalent terms: superior computer hardware becomes the player's own high-tech exoskeleton. Like Nomad's nanosuit, the player's 'state-of-the-art gaming machine' becomes a hardware prosthesis, a 'technologically advanced rig' practically imported from the future, transported backwards in time into the player's living room and grafted to the player's body at the moment of picking up the controls. To own such a computer is to be directly in touch with the future depicted in the game. Which is to say, the computer now appears as a fetish, an embodied symbol of the military nanotechnology inside the game, a hardware appendage of the player identified as source of high-tech power, maximum hardness, maximum strength.

For some players, this fetishism has sexual implications, responding to and playing with the overt phallocentrism of the game itself. Indeed, a number of players have noted that the game's incessant incantation of 'maximum strength' and 'maximum armor' seems really to mean, as one player puts it: "MAXIMUM penis length" (dudeworld1 2009). Players of

Crysis therefore often measure and compare their relative computing powers in phallic terms: "Those are some pretty heavy processor requirements [to run the game], glad I have a larger e-penis than the rest of you" (polystethylene 2008). These gamers regularly post digital photos of their computers and lists of their tech specs to online message boards, sometimes even distributing close-up shots of their CPUs or graphics cards in ways that self-consciously mimic the pornographic conventions of 'cruising for sex' websites. Occasionally, fellow players of *Crysis* appear to be quite impressed with such displays of hardware prowess: "[T]hat rig will give you a HUGE e-penis. Congrats on that" (jazzzy 2008). Or: "Damn. That rig is sweet dude. You easily have the best rig here. That pic of your rig you emailed me was sweet. I love your rig" (alexander224 2009). Yet just as frequently, other players respond with distain or hostility towards those players who advertise pornographic details about their *Crysis*-mastering hardware: "Congratulations, you can show off your computer on the internet. Your penis probably got larger because of what a good computer you have" (Billy Bob Joe 2008). But whether admiring or ridiculing such displays of 'e-penis' power, the communal discourse of *Crysis* players routinely affirms a linkage between the symbolism of 'maximum hardness' in the game, the technical capacities required to play the game at maximum settings, and the image of the male body.

We see here a strong instance of what Sherri Turkle has described as the tendency of computer-users to identify the computer as a 'second self', to see the computer as a prosthesis or a mirror: "This kind of identification is a powerful source of computer holding power. People are able to identify physically with what is happening inside the machine. It makes the machine feel like a part of oneself" (Turkle 2005, p.171). This sensation of bodily extension or identification with the computer by no means requires physical connection with the hardware, but instead simply relies on symbolic interaction; as Turkle writes, "The sense of physical relationship depends on symbolic contact" (Turkle 2005, p.176). In the case of military nanotechnology games like *Crysis*, symbolic contact with the digital battlefield produces a feeling of inhabitation, a merger with the hardware as the physical site of everyday nanowars.

Players regularly see the predicament of Nomad in the game as equivalent to the predicament of the computer as it struggles with the demanding *Crysis* software: "Crysis: Not just a game, but also the state your PC is in trying to run this game!" (txtmstrjoe 2008). Frequently, this translation gets discursively rendered as a *corporeal* predicament, a symbolic crysis of the player's own body. For by promoting an anthropomorphism of the gaming system as an extension or reflection of the player (Lupton 2007), the discursive field of *Crysis* effectively 'e-masculates' the player, where 'e-masculate'—like the gamer vocabulary of 'e-penis'—signifies, on the one hand, the *electronic performance of machismo*, the conference of an inflated soldier-male superiority through the virtuosity of gaming hardware, and, on the other hand, the electronic *emasculation* or castration of the player as soldier-male through the very same gaming hardware. For example, as one player puts it: "*Crysis* may very well kick your computer in the balls" (Adams 2007). We see here the multiple and irresolvable tensions inherent to the crysis mode of the nanowarrior, where the ability to swagger about e-penis power is only enabled by a simultaneous risk of e-penis wounding, a deflation caused by the very measure of electronic-masculation: the software of *Crysis* itself. Here, the power enhancement associated with those 'little thingies like balls' that drop into your bloodstream to make you run fast is nearly identical to the experience of getting 'kicked in the balls': nano-empowerment only through nano-imperilment. In other words, full-on crysis mode.

4. Queer Adventures

Players experience the crysis mode in the paradoxical identification of their own gaming systems as the enabling condition to enter the digital battlefield of *Crysis* and, simultaneously, as the very site of limitation or insufficiency relative to the demands of that battlefield. Some players, desperately wanting to join in the fun and live the dream, nevertheless fear for the safety of their computers, and likewise, their own security. One writes that if he dared to install *Crysis*, the game would "rape my computer in its ass aaaaall [sic] night long man. And then the next morning my computer would wake up in a street without its shell on and be completely and utterly violated. RAPED" (afrospezz 2008). For yet another wannabe *Crysis* player, the game's thuggish hardware demands appear as an alluring danger, a fantasy of 'rough trade': "Just thinking about it makes my anus bleed" (Steakslim 2007). The power fantasies of *Crysis* come alongside recognition of the technical limits of average graphics processors to handle the game, ubiquitously expressed as a bodily—and, more specifically, an anal—vulnerability. Such transference of anxiety between hardware integrity and personal security, of course, is at odds with the game's own rhetoric of 'maximum armor'. But as we see, this is the nature of the crysis mode: protection enabled only by total exposure to the risks of technology, hardness enabled only by total molecular penetration. In crysis mode, pleasure in extreme militarism comes simultaneously with being forced to one's knees and made to service the fetish object: "*Crysis* can bring even the manliest of rigs to their knees" (Swertfeger 2008). No doubt about it: "Even top end systems were brought to their knees by this game" (SightseeMC 2007).

Playing within the crysis mode—this volatile condition where even the manliest 'tops' might be made into 'bottoms'—thus appears for some gamers as a queering of normative gender and heterosexual imperatives, or more specifically, as a failure of masculinity in their own embodied relations to the game. Homophobic and misogynistic responses to such experiences are unfortunately frequent within the community of *Crysis* players. A few identify their computational shortcomings as sexual debasement, as an involuntary servitude to the phallic demands of the game: "I wish my pc didn't suck total penis filth" (astromario 2008). Others try to reject the near-universal fetishism of *Crysis* as the measurement of hardware power and the benchmark for assessing 'MAXIMUM penis length', because, according to this logic, everyone inevitably comes up short in comparison to *Crysis* and therefore can only see themselves as the effeminate recipients of its awesome technology: "Come on guys, lets all take our lips off Crysis's dick" (SuperLuigiBros 2006).

In many ways, then, *Crysis* would seem to animate a queerness of the digital battlefield—a latent queerness identified by several players in the 'fetish for guys in nano suits' that exists at the core of today's speculative militarism, both in video games and in 'real life'. For example, one gamer describes Nomad's exoskeleton in *Crysis* (Figure 5) as "some gay nanosuit…Note of the use of the word gay, not as an insult but as a descriptive word for the sexual status of the nanosuit, which resembles a dude's ballet costume" (gatman 2006). Yet another gamer, referring specifically to the 'Soldier of the Future' prototype exoskeleton designed by the US Army's Natick Soldier Center and featured in the ISN 'Soldier of the Future' videos (Figure 6), writes: "Looks like it's for the homosexual type" (Led_poison 2008). Such assessments of futuristic military wear as a form of gay fetishism suggest that eroticized images of 'guys in nanosuits' become signifiers for a future-in-the-present that intersects both military utopias and gay fantasias. *Crysis* exposes the extent to which the fetishistic projection of the future of military technology onto the playable form of the Soldier of the Future—whether a costume mock-up or a video game character—is a fundamentally queer practice. For on the digital battlefield, there is no natural or normative future, only that future in crisis.

Crysis therefore vivifies a future in the present that is entirely, and in every sense, queer. Some players simply cannot handle this realization: "Yea im done with *Crysis*, too many gay things happened to me in the game…nano suit my ass" (xXxKEVxXx 2008). The failure of the nanosuit to guarantee heterosexual male identity in the future is charged with 'gayness'. Many gamers see the nanosuit itself in this way and seem to understand that not only success but also failure to meet the game's standards for virtuosity means adaptation to the crysis mode: "I'm not qualified to review this game. I couldn't play through this whole mess. But I'm an arrogant bitch who has to give his two cents to everything, so here they are: The whole beginning looks like a gay latex fetish porn" (CoinMatze 2008). In agreeing to the terms of *Crysis*, that is, in attempting to play the game and even temporarily inhabiting its worldview, the player lives the future of everyday nanowars that seem indistinguishable from 'gay latex fetish porn'—where the fetishistic qualities conveyed upon the nanosuit are indistinguishable from the fetishistic qualities conveyed upon the player's hardware attributes and the technical skills needed to complete the game, and where failure to do

Figure 5. Cruising in his nanosuit, Nomad throws us a smoldering, over-the-shoulder glance. ©2007 Crytek. For colour reference turn to page 563.

Figure 6. Future Warrior, US Army Natick Soldier Center. 2003. A member of Natick's Operational Forces Interface Group poses in a mock nanosuit costume. For colour reference turn to page 563.

so, as much as any success in doing so, destabilizes conventionally gendered notions of subjectivity. Indeed, in playing nanowar porn, even if in the end to turn away from it in frustrated expression of insufficiency ('I'm not qualified'), the male player recognizes himself as becoming-queen, an 'arrogant bitch'.

Others instead find nothing but pleasure in inhabiting this fetishistic symbolic field where, under the regime of military nanotechnology, gender becomes fluid and sexuality opens to play. As one player attests: "I've take a huge slap…Crysis killed me once again. He even rape me so hard, I want a 9800 GTX [Nvidia GeForce graphics card] now! I wasn't expect the game to be SO good. I really had some much fun, some much gorgeous moments [sic throughout]" (svenminoda 2007). The imagined phallic demands of the *Crysis* software would seem similar—yet all in good fun?—to 'slapping' and 'raping' the player himself. But such "violation" is surprisingly expressed as affirmative pleasure ('SO good'). Rather than protection against further penetration, purchasing a more powerful Nvidia 9800 GTX graphics card would instead seem to mean more thoroughly opening oneself to repeat performances, to meet the game's own appetites and desires on more equal terms, and thereby to receive yet more pleasure from the intensity of the experience (albeit, perhaps, without suffering so much 'screen tearing'). To be sure, several male players describe their pleasure in the game with similar masochistic joy, with similar fluidity of gender categories, and with similar desire to repeat again and again: "I'm the biggest Crysis whore in the world" (rott112 2008). With little discomfort or anxiety, these particular players find that the sexual ambiguities animated by *Crysis* as completely convergent with the playability of the nanosuit that makes the game so much fun in the first place. One player (self-identified as a straight married man and a military engineer in the US Air Force), in confessing his fanboy desire to "get on this game's dick now," imagines an alternative sexual response, a male femininity: "And the Nano-suit makes me wet . . . sooo much fun" (easy_bake 2007).

Whether trying to distance themselves from the queer future of everyday nanowars through a vocabulary of homophobic panic, or instead joyously enjoying the ride and looking forward to further adventures, players of *Crysis* appear to accommodate themselves to the logics of the game, inhabiting the nanosuit as an incarnation of their own vortical relationship to their computers, and navigating the concomitant crysis mode without ultimate resolution. If there is resistance here to the future of nanowarfare, it is a resistance produced from within, an internal tension between the force objectives of soldier-male militarism and the nanotechnological queering that destabilizes those force objectives from the inside. For recognition that the dominant rhetoric of the game is actually at odds with its own representations of the technical details of nanowarfare—that is, recognition of the crysis mode—opens players' eyes to the other self-critical and self-queering dimensions of the game. In their communal discussions of the internal resistances or tensions experienced in playing with Nomad, many players become evermore attentive to the failures of the in-game military rhetoric to live up to its own hype, thus opening a space for critical analysis of the overt military objectives of *Crysis*, providing a subversive perspective on the political ideologies at stake. For example, one player writes:

> *I also liked that the game tried to be topical. You played as a technically superior enemy to the Koreans, hunting them, which obviously had a relation to ideas about the major Western powers and their foreign policies in mind. This was a battle to win a major resource, possibly a power source. (Much like oil or nuclear power).*
>
> *The arrival of the aliens can be read in several ways too. The destructive power unleashed by war and violence perhaps? Or more likely given the nature of the aliens, it was representative of the earth rebelling against mankind's destruction, or even more simply, destruction from ecological change. The earth essentially erupts like a volcano of ice, smothering the tropical jungles. (Referring to current fears about global warming and the melting of the ice-caps.) So it was nice to see some thought went into it.* (Matt 2007).

So amidst all the enthusiastic chatter about maximum power and e-penises, *Crysis* players become capable of intuiting their own experience of the crysis mode as opportunity to question, to evaluate, and to self-reflect on their own recreational participation in everyday nanowars.

5. Conclusion

The video games of military nanotechnology (animating real-life research programs like the ISN, as well as popular fictions like *Crysis*) make the vision offered by Dutch DeGay for the digital battlefield of the future an everyday reality. Channeling the Borg from Star Trek, and implicitly, their notorious motto—"Resistance is futile"—DeGay's vision reveals the conscriptive logic of the digital battlefield. This is the logic by which soldiers are totally 'plugged into' the data network of C^4ISR (Command, Control, Communications, Computers, Intelligence, Surveillance, and Reconnaissance), thanks to their hyperreal merger with nanotechnology and other emergent sciences that would seem to make war into a video game. Considering that video game technologies are already intricately woven into today's systems of militarism (Lenoir 2000), games like *Crysis* that make the digital battlefield into an intensely playable experience themselves become part of the apparatus of training and induction.

As DeGay has said, "it's no shock to see that technology being used for training [soldiers]...It has always been a challenge to get the time, space and resources to train soldiers, so video games give leaders and commanders an easier portable training option." And likewise, a recruitment option: "It's not a stretch of the imagination that people who have the chance to see the capabilities the next generation of soldier [in a video game]...might find the idea of being a soldier more appealing. Or at the very least give them a better idea of what it might be like to be a soldier." Accordingly, DeGay has served as technical advisor for two recent 'soldier of the future' games: *Ghost Recon: Advanced Warfighter* (Ubisoft, 2006) and *EndWar* (Ubisoft, 2008). 'Authenticity' is key in such games, according to DeGay, in order to give players "as real a military experience possible in a video game" (DeGay quoted in Gaudiosi 2006). As we see, the video games that stage everyday nanowars in our own living rooms—*Crysis*, certainly, but also dozens of others—attend directly to the visions of research programs in military nanoscience that seek to transform the soldier of the future from science fiction into reality, making these visions as immediate and immersive as possible, as something that we can all enjoy. Perhaps resistance is futile. Indeed, as one player has said of *Crysis*: "Now that I've tried it I can't resist it" (Nocutius 2007).

And yet, in making nanowars playable, these video games stage the soldier's conscription into the networks of the digital battlefield as a situation of interminable, internal crysis, where resistance and accommodation to the logics of the digital battlefield are experienced simultaneously: an effect of the dynamic tension between the molar and the molecular, the digital and the material, the straight and the queer, the male and the female, the core and the network, the solid and the fluid, the 'e-masculation' function and the 'e-penis' function, presented as immanent to the high-tech systems of nanowar. By performing the play of sexual difference, morphing sexualities, gender trouble, and molecular fluidity all at the site of the nanowarrior body, mapping directly onto the player's own 'wet' relationship with high-tech hardware, games like *Crysis* spectacularly self-deconstruct, liquidating the foundations of the androcentrism upon which their own militaristic power fantasies are erected. Indeed, the power struggles of military games that depend on enforcing constructions of normative masculinity seem to regularly undermine their own metaphysics of gender at every level, thereby rendering the serious notion of 'everyday nanowars' entirely non-serious, laughable, something of a pleasurable joke, providing players with a critical form of resistance *from the inside* against the otherwise overwhelming futility of resistance: in other words, a pleasurable queering of the digital battlefield itself.

Upon inserting *Crysis* into any computer, the pre-credits launch sequence to the game—featuring the familiar male voice of the nanosuit —announces that players of this game will experience the ultimate in high-tech digital warfare: "MAXIMUM GAME". But as many players have commented upon viewing this opening clip, what the voice of the nanosuit actually says is rather ambiguous. One player sums up the inherent indistinction, the irresolvable double-speak and double-think that is nothing but the crysis mode in all its playfulness: "I swear to god that I thought he said 'MAXIMUM GAY'" (OldParr 2007).

Let's all play.

Questions for Reflection:

1. In what ways do popular culture and military science influence one another? What are the social/political consequences of this?
2. How do our concepts of gender and sexuality shape visions of future nanotechnologies, and vice versa?
3. What does engaging in war as everyday entertainment tell us about ourselves and society?

Bibliography

Anna (2007). Response to Polkastripe, 'Girl Gamers', *GameSpot Forums*, 17 November 2007. http://www.gamespot.com/pages/forums/show_msgs.php?topic_id=26040943 (last accessed Nov 11, 2008).

Adams, D. (2007). Crysis review: Crytek blasts back onto the PC with the most beautiful kind of violence, *IGN PC*, 12 November 2007. http://pc.ign.com/articles/834/834614p1.html (last accessed Feb 11, 2009).

Afrospezz (2008). Response to Koroushghazi, 'Crysis Demo - Destructible Vegetation DX9', *YouTube*, October 2008. http://www.youtube.com/watch?v=8SkXuoElvXY (last accessed Feb 3, 2009).

AkumaX (2008). Why do we play videogames? *Giant Bomb*, 17 September 2008. http://www.giantbomb.com/profile/akumax/why-do-we-play-videogames/30-11549/ (last accessed Nov 11, 2008).

Alexander224 (2009). Response to Hektikninja, 'What Are the Specs of Your Rig?' (Post #60), *Zoklet.net*, 5 February 2009. http://bbs.zoklet.net/showthread.php?t=4253 (last accessed Mar 15, 2009).

Altmann, J. (2006). *Military Nanotechnology: Potential Applications and Preventative Arms Control*, London: Routledge.

Astromario (2008). Response to '*Crysis*: Oceanic Onslaught Gameplay', *GameTrailers.com*, 6 July 2008. http://www.gametrailers.com/player/27855.html (last accessed Jan 18, 2009).

Bertz, M. (2008a). Farcrysis, Review of *Crysis*. *Game Informer,* 18(1): 96.

Bertz, M. (2008b). Meanwhile, across the Island . . . , Review of *Crysis Warhead*. *Game Informer,* 18(12): 126.

Billy Bob Joe (2008). Response to Spiffymarlin, 'How Well Would This Run?', *GameSpot GameFAQs—Crysis Message Board for PC*, 13 July 2008. http://www.gamefaqs.com/boards/genmessage.php?board=931665&topic=44239785 (last accessed Feb 2, 2009).

Booker, L. (2006). Inside Crysis, *Atomic: Maximum Power Computing*, 21 September 2006. http://www.atomicmpc.com.au/Feature/60160,inside-crysis.aspx (last accessed Mar 6, 2009).

Broadsword (2007). Response to Fanboi69, 'Nano Suit: What Does The "Nano" Suit Do That We Haven't Already Seen?', *inCrysis—Crysis Forums*, 22 August 2007. http://incrysis.com/forums/viewtopic.php?pid=147758 (last accessed Feb 18, 2009).

Butler, J. (1999 [1990]). *Gender Trouble: Feminism and the Subversion of Identity*, New York: Routledge.

Callahan, S. L. (2000). Nanotechnology in a new era of strategic competition, *Joint Force Quarterly*, 26, pp. 20-26.

Case, S. E. (1996). *The Domain-Matrix: Performing Lesbian at the End of Print Culture*, Bloomington: Indiana University Press.

Cassell, J. and H. Jenkins. (1998). *From Barbie to Mortal Kombat: Gender and Computer Games*, Cambridge, MA.: MIT Press.

Certeau, M. D. (1984). *The Practice of Everyday Life*, Translated by S. Rendall, Berkeley: University of California Press.

CoinMatze (2008). Like That Hot, Dumb Girl in School, *Giant Bomb—Crysis Reviews*, 21 July 2008. http://www.giantbomb.com/crysis/61-11757/user_reviews/?review=779 (last accessed Feb 1, 2009).

Crysis (2007). Video game, Crytek (developer), Redwood City: Electronic Arts.

Danet, B. (1998). Text as mask: Gender, play, and performance on the intemet. In Jones, S. G. (ed) *Cybersociety 2.0: Revisiting Computer-Mediated Communication and Community*, Thousand Oaks: Sage, pp. 129-158.

De Landa, M. (1991). *War in the Age of Intelligent Machines*, New York: Zone Books.

Der Derian, J. (2009). *Virtuous War: Mapping the Military-Industrial-Media-Entertainment-Network*, 2nd ed, New York: Routledge.

Deuber-Mankowsky, A. (2005). *Lara Croft: Cyber Heroine*, Translated by D. J. Bonfiglio, Minneapolis: University of Minnesota Press.

Dudeworld1 (2009). Comment on '*Crysis* Cinematic Intro', *You Tube*, 13 January 2009. http://www.youtube.com/watch?v=5QzKIeLRVtM (last accessed Feb 23, 2009).

Easy_bake (2007). Response to Tangojorg, 'Btw (Crysis: Demo and Beta)', *The Box Heads Forums*, 29 October 2007. http://www.theboxheads.net/modules.php?name=Forums&file=viewtopic&t=4410 (last accessed Apr 3, 2009).

Edmunds, D. (2007). Crysis, *I'm Having a Crysis*, 1 November 2007. http://dustball21d.blogspot.com/2007/11/crysis.html (last accessed Dec 13 2008).

Emil (2007). Response to Alec Meer, 'Crysis in Crisis, Unreal in Real Trouble', *Rock, Paper, Shotgun*, 15 December 2007. http://www.rockpapershotgun.com/2007/12/15/crysis-in-crisis-unreal-really-suffering/ (last accessed Dec 17, 2007).

Eric (2007). Response to Alec Meer, 'Crysis in Crisis, Unreal in Real Trouble', *Rock, Paper, Shotgun*, 17 December 2007. http://www.rockpapershotgun.com/2007/12/15/crysis-in-crisis-unreal-really-suffering/ (last accessed Dec 17, 2007).

Galloway, A. R. and Thacker, E. (2007). *The Exploit: A Theory of Networks*, Minneapolis: University of Minnesota Press.

Garreau, J. (2005). *Radical Evolution: The Promise and Peril of Enhancing Our Minds, Our Bodies — and What It Means to Be Human*, New York: Doubleday.

Gatman. (2006). Response to Pe2, 'Things You'd Like to See Changed/Added/Removed in Crysis So Far' (Reply #161), *In Crysis—Crysis Forums*, 14 September 2006. http://incrysis.com/forums/viewtopic.php?id=292 (last accessed Jan 20, 2009).

Gaudiosi, J. (2006). Red Storm Entertainment Enlists Army Specialist for GRAW, *Xbox.com | Tom Clancy's Ghost Recon Advanced Warfighter*™, 16 May 2006. http://www.xbox.com/en-US/games/t/tomclancysghostreconadvancedwarfighter xbox360/20060516-soldierinterview2.htm (last accessed Jan 7, 2009).

Gourley, S. (1998). Lethal Combination, *Jane's Defence Weekly*, 30(14): pp. 39-42.

Gray, C. H. (1997). *Postmodern War: The New Politics of Conflict*, New York: Guilford Press.

Gray, C. H. (2001). *Cyborg Citizen: Politics in the Posthuman Age*, New York: Routledge.

Hall, J. S. (2005). *Nanofuture: What's Next for Nanotechnology*, Amherst, New York: Prometheus Books.

Hantke, S. (1998). Surgical strikes and prosthetic warriors: the Soldier's body in contemporary science fiction, *Science-Fiction Studies*, 76, pp. 495-509.

Howard, C. E. (2006). Fueling the Future, *Military & Aerospace Electronics*, 7 (7): 26-31.

Jazzzy (2008). Response to Poobah, 'Buying a Rig to Run Crysis' (Reply #19 & #22), *Crysis Online—Forum*, 17-18 August 2008. http://www.crysis-online.com/forum/index.php/topic,25098.0.html (last accessed Jan 12, 2009).

Jenkins, H. (2006). *Convergence Culture: Where Old and New Media Collide*, New York: New York University Press.

Kafai, Y. B., Heeter, C., Denner, J. and Sun, J. Y. (eds) (2008). *Beyond Barbie and Mortal Kombat: New Perspectives on Gender and Gaming*, Cambridge, MA: MIT Press.

L88bastard (2008). I Hate Crysis, *Nvidia SLI Zone*, 3 December 2008. http://forums.slizone.com/index.php?showtopic=30865 (last accessed Dec 15, 2008).

Led_poison (2008). Response to Knowledge-Funk, 'New US Military Technology - The Exoskeleton', *GameSpot Forums*, 9 March 2008. http://www.gamespot.com/pages/forums/show_msgs.php?topic_id=26280598 (last accessed Apr 16, 2008).

Lenoir, T. (2000). All but War Is Simulation: The Military-Entertainment Complex, *Configurations*, 8, pp. 289-335.

Lenoir, T. (2003). Programming theaters of war: Gamemakers as soldiers. In Latham, R. (ed) *Bombs and Bandwidth: The Emerging Relationship between Information Technology and Security*, New York: New Press, pp. 175-198.

Lupton, D. (2007). The embodied computer/user. In Bell, D., and Kennedy, B. M. (eds) *The Cybercultures Reader*, London and New York: Routledge, pp. 423-432.

Matt (2007). Response to Alec Meer, 'Crysis in Crisis, Unreal in Real Trouble', *Rock, Paper, Shotgun*, 15 December 2007. http://www.rockpapershotgun.com/2007/12/15/crysis-in-crisis-unreal-really-suffering/ (last accessed Dec 17, 2007).

McRae, S. (1997). Flesh made word: Sex, text and the virtual body. In Porter, D. (ed) *Internet Culture*, New York: Routledge, pp. 73-86.

Milburn, C. (2008). *Nanovision: Engineering the Future*, Durham: Duke University Press.

Milos (2007). Response to Imacker, 'Is It Just Me or Is Crysis Boring?', *Crysis Online—Forum*, 19 November 2007. http://www.crysis-online.com/forum/index.php?topic=14615 (last accessed Feb 2, 2008).

Nieborg, D. B. (2006). Mods, Nay! Tournaments, Yay!—The Appropriation of Contemporary Game Culture by the US Military, *Fibreculture,* 8. http://journal.fibreculture.org/issue8/issue8_nieborg.html.

Nocutius (2007). Reply to Werty316, 'Crysis Demo - The Countdown Is Over!' (Reply #21), *Bjorn3d.com—Forums*, 27 October 2007. http://www.bjorn3d.com/forum/showthread.php?t=18951 (last accessed Mar 13, 2008).

OldParr (2007). Response to M337ing, '*Crysis* Multiplayer Walkthrough', *GameTrailers.com*, 18 September 2007. http://www.gametrailers.com/player/usermovies/107487.html (last accessed Mar 3, 2009).

Ottosen, R. (2009). The military-industrial complex revisited: Computer games as war propaganda, *Television & New Media,* 10, pp. 122-125.

Pat M. (2009). Customer Testimonial, *BuyDirectPC*, 8 January 2009. http://www.buydirectpc.com/reviews.html (last accessed Feb 2, 2009).

Peacemaker898 (2007). Response to Fanboi69, 'Nano Suit: What Does The "Nano" Suit Do That We Haven't Already Seen?', *in Crysis—Crysis Forums*, 22 August 2007. http://incrysis.com/forums/viewtopic.php?pid=147758 (last accessed Feb 18, 2009).

Polystethylene (2008). Response to Bot-IGN, 'Crysis Warhead System Requirements Revealed', *IGN.com—Message Boards*, 14 August 2008. http://boards.ign.com/Boards/Message.aspx?brd=9108&topic=169131636&page=1 (last accessed Feb 7, 2009).

Power, M. (2007). Digitized virtuosity: Video war games and post-9/11 cyber-deterrence, *Security Dialogue,* 38(2), pp. 271-288.

Redwinedrummer (2007). Crysis Alert. Crysis Alert., *Reverse-Engineering the Cross-Com: The Technological Endeavors of an Advanced Warfighter,* 29 November 2007. http://redwinedrummer.wordpress.com/2007/11/29/crysis-alert-crysis-alert/ (last accessed Jan 15, 2008).

Roco, M. C., and Bainbridge, W. S. (eds) (2003). *Converging Technologies for Improving Human Performance: Nanotechnology, Biotechnology, Information Technology and Cognitive Science*, Dordrecht: Kluwer.

Roher, J. (2007). Fantesticle Penisula Adventure, *Not Clickable*, 28 October 2007. http://www.notclickable.com/blog/fantesticle-penisula-adventure/ (last accessed Jan 8, 2009).

Rott112 (2008). Response to Rot112, 'Dox 180.70...Thank You Dox' (Reply #3), *NotebookReview.com—Notebook Forums and Laptop Discussion*, 28 November 2008. http://forum.notebookreview.com/showthread.php?t=325733 (last accessed Feb 23, 2009).

Sapieha, C. (2008). *Crysis* (M): Super-Realistic Violent Sci-Fi Shooter, *Common Sense Media.* http://www.commonsensemedia.org/game-reviews/Crysis.html (last accessed Sep 13, 2008).

Shadonic0 (2008). Response to Jpear5000, 'So We Have All Learned Something from Crysis', *GameSpot GameFAQs—Crysis Message Board for PC*, 4 July 2008. http://www.gamefaqs.com/boards/detail.php?board=931665&topic=44051199 (last accessed Mar 1, 2009).

SightseeMC (2007). Response to Alec Meer, 'Crysis in Crisis, Unreal in Real Trouble', *Rock, Paper, Shotgun*, 15 December 2007. http://www.rockpapershotgun.com/2007/12/15/crysis-in-crisis-unreal-really-suffering/ (last accessed Dec 17, 2007).

Singer, P. W. (2009). *Wired for War: The Robotics Revolution and Conflict in the Twenty-First Century*, New York: Penguin Press.

Soldier of the Future (2004). Video, MIT Institute for Soldier Nanotechnologies, Boston: North Bridge Productions and Boston Animation.

Soldier of the Future (2005). Video, MIT Institute for Soldier Nanotechnologies, Boston: North Bridge Productions and Boston Animation.

Steakslim (2007). Response to Curium244, '780i Asus P5n-T Deluxe' (Reply #11), *Crysis Online—Forum*, 12 December 2007. http://www.crysis-online.com/forum/index.php?topic=16845 (last accessed May 5, 2008).

Stockwell, S., and Muir, A. (2003). The Military-Entertainment Complex: A New Facet of Information Warfare, *Fibreculture,* 1: http://journal.fibreculture.org/issue1/issue1_stockwellmuir.html.

Stone, A. R. (1995). *The War of Desire and Technology at the Close of the Mechanical Age*, Cambridge, MA.: MIT Press.

SuperLuigiBros (2006). Response to Cornholio12, '*Crysis* Graphics Overrated!', *E-mpire Forums*, 30 May 2006. http://forums.e-mpire.com/archive/index.php/t-57239.html (last accessed Mar 13, 2009).

Svenminoda (2007). Response to Washd123, 'My Final Crysis Review+ Lvl16', *GameTrailers.com—washd123's Game Pad*, 1 December 2007. http://washd123.gametrailers.com/gamepad/index.php?action=viewblog&id=206855 (last accessed Nov 11, 2008).

Swertfeger, C. (2008). Intel E7200 2.53ghz Dual Core Processor, *Techware Labs*, 20 November 2008. http://www.techwarelabs.com/reviews/processors/intel_e7200_2_53ghz_dual_core_processor/index.shtml (last accessed Apr 17, 2009).

Terry, J. with Kelly, R. (2007). Killer Entertainments, *Vectors* (1). http://www.vectorsjournal.org/index.php?page=7&projectId=86.

The Top 50 Games of 2007: *Crysis*. (2008). *Game Informer,* 18(1), pp. 42.

Theweleit, K. (1987). *Male Fantasies: Volume 2: Male Bodies: Psychoanalyzing the White Terror*, Translated by S. Conway in collaboration with E. Carter and C. Turner, Minneapolis: University of Minnesota Press.

Turkle, S. (1995). *Life on the Screen: Identity in the Age of the Internet*, New York: Simon and Schuster.

Turkle, S. (2005 [1984]). *The Second Self: Computers and the Human Spirit*, 20th anniversary ed., Cambridge, MA.: MIT Press.

Txtmstrjoe (2008). The Crysis Controversy, *Overclock.net*, 18 July 2008. http://www.overclock.net/blogs/txtmstrjoe/591-crysis-controversy.html (last accessed Mar 1, 2009).

U.S. Army Research Office (2001). *Institute for Soldier Nanotechnologies [Broad Agency Announcement DAAD19-02-R-0001], October 2001*. Formerly available at: http://www.aro.army.mil/soldiernano/finalsolicit.pdf (now available through the Internet Archive's Way Back Machine) (last accessed Apr 1, 2009).

XXxKEVxXx. (2008). Response to Aua_Violator, 'Patch 1.1 Has Finally Killed Crysis for Me' (Reply #20), *Crysis Online—Forum*, 15 January 2008. http://www.crysis-online.com/forum/index.php?topic=19081 (last accessed Jan 30, 2008).

Chapter 8

Nanotechnology and Global Sustainability: The case of water management

Mercy W. Kamara

Water is life. During the last few years, a freshwater crisis has been predicted and listed among pressing global problems. Nano innovations have been proposed as instruments for curbing the predicted freshwater crisis and for ameliorating water pollution problems. Whereas proponents of nano-based water management argue that nano innovations will further a sustainable development, critics argue that nano innovations will promote unsustainable development. This chapter engages the ensuing conflict concerning the sustainability of nano innovations. It presents this conflict as emanating from worldview differences between nano proponents and critics. It demonstrates how those differences in worldviews create, rework, and sustain competing meanings of novel phenomena and for such information as nano innovations. It further explains how processes of socialization and internalization of a particular worldview and its underlying belief systems influence nano proponents' or critics' thoughts and practices, including what they envision as possible and doable or how they interpret and seek solutions to both new and old problems. Accordingly, the chapter argues that what nano innovations are and what their effects on the environment and society are depend on the worldview, local context, and interests of the affected community.

Nano Meets Macro - Social Perspectives on Nanoscale Sciences and Technologies
by K L Kjølberg & F Wickson

www.panstanford.com
978-981-4267-05-2

1. Introduction

Freshwater crisis, scarcity, or stress (Chenoweth 2008a; Savage *et al.* 2009; Theron *et al.* 2008) has been predicted and listed among such pressing global problems as energy and poverty. Water is life. Freshwater is a basic need that is critical to human health and survival, social welfare, and economic and industrial development. It is little wonder that access to clean and freshwater is a key priority of the sustainable development agenda. It is anchored in chapter 18 of the United Nations' Agenda 21, which is the United Nations (UN) action for alleviating negative human impacts on the environment (UN Department of Economic and Social Affairs 2004); in the Millennium Declaration of the UN General Assembly of 2000; and in the Johannesburg's Declaration for a Sustainable Development of 2002 (Chenoweth 2008b). Thus, the need to manage the risks of the predicted water crisis cannot be overstated (Sustich *et al.* 2009).

A notable dimension of the predicted water crisis is that, while on the one hand the UN predicts a looming water crisis, on the other hand, it observes that "there is enough water for everyone" (Duncan et al. 2009, p. 468). Chenoweth (2008a) observes that "globally, there is adequate freshwater available, and that looks set to continue in the long term" (Chenoweth 2008a, p. 29). According to Chenoweth, figures "indicate that we currently extract less than 10 per cent of the 43,750 cubic kilometres of freshwater returned each year to the Earth's rivers, lakes and aquifers" (Chenoweth 2008a, p. 29). In light of these figures, one could rightly wonder why we have a looming water crisis.

A number of factors have informed the prediction of the water crisis. One of these factors is uneven distribution of rain and precipitation, spatially, temporarily, and geographically (Chenoweth 2008a). A second factor, especially in such water rich developing regions as East Africa or the Amazon, is due to poor public water management and distribution infrastructures, corruption, and political-economic hurdles that impede proper water governance, management, and supply (McDonald & Ruiters 2005). A third factor, especially in developed countries such as Denmark, is due in part to concentrated urbanization and in part to over consumption of freshwater (personal communication with a Danish water engineer). Globally, perhaps the major cause of water problems is increased pollution of groundwater and surface water from industrial, municipal, and agricultural waste disposal, which release harmful chemicals from detergents, fertilizers, and pesticides or harmful heavy metals into available freshwater for human use. This pollution can be caused by inadequate sanitation or poor infrastructures for waste disposal, and sewage treatment. It can also be caused by salinity from widespread and inefficient irrigation or from soil erosion (Theron *et al.* 2008; Strathmann *et al.* 2009).

As increased industrialisation has been accompanied by increased water pollution that poses unwanted risks to human health and environment, there have been calls for developing innovative ways of mitigating this pollution. Nanosciences and nanotechnologies (hereafter nano) have been proposed as instruments for curbing the predicted freshwater crisis and for ameliorating water pollution problems (Theron *et al.* 2008; Sustich *et al.* 2009; Roco 2009; Hillie & Hlophe 2007). As nano application in water purification, treatment, cleaning, or remediation is gaining currency, nano-based water management tools are starting to include the use of nano-structured and nano-reactive membranes or polymers in water filtration, including the use of metallic or magnetic nanoparticles and nanomaterials ranging from nano zeolites, carbon fibres, and single-walled or multi-walled carbon nanotubes to iron, zinc, or titanium oxides' nano metal catalysts. They also include nanomaterial-based sensors, enzyme nano particles, and self-assembled monolayers on mesoporous supports (SAMMs) (Theron *et al.* 2008; Savage *et al.* 2009).

The attractiveness of nano-based water management is unsurprising. During the last three decades, nano developments have been accompanied by visionary speculations, promises, and expectations about their potential benefits and opportunities for society (Roco 2009; Theron *et al.* 2008). Products emanating from nano research have been hailed as exciting emerging and transformative innovations of our time (Amato 1999). In fact, proponents of nano innovations frame real or imagined nano products and deployment in terms of "sustainable development"[1] (Masciangioli 2002; Ministeriet for Videnskab, Teknologi og Udvikling 2004; Juma & Yee Cheong 2005). That is, nano innovations will contribute to the reduction and treatment of pollution or solve environmental and ecological problems. They will promote the development of cheaper, durable, and more efficient products and augment industrial development, national economic growth, and global competitiveness (Jørgensen *et al.* 2006; Chemical Industry Vision2020 Technology Partnership 2007; Malanowski 2001; European Commission 2004a; 2004b; National Science and Technology Council 2003). In this spirit, many nano research projects, policies, programs, or proposals have been presented as contributions to sustainable development endeavors (cf. Majewski & Chan 2008; Nelson & Strano 2006). However, the invitation to see nano innovations as sustainable development has not been accepted by everyone.

Skeptics argue that the deployment of nano-based products pose known environmental, ecological, or human health hazards, including inflammation, oxidative stress, toxicity, immune suppression, carcinogenicity, or cyto-toxicity hazards[2] to human, animal, or aquatic organisms. Critics point to scientific ignorance stemming from the indeterminacy or ambiguity of ubiquitous physical-chemical properties and functions that may accompany nano-particle exposure and their unwanted effects on complex, non-linear, and self-organizing environmental, ecological, biological, or aquatic systems (cf. Oberdorster *et al.* 1992; 2005; Hoet, *et al.* 2004; Royal Commission on Environmental Pollution [RCEP] 2008; Barnard 2006; Maynard *et al.* 2006; Uskokovic 2007). In light of known or unknown uncertainties, some critics have gone as far as questioning the unexpressed underlying assumptions that are driving and shaping nano innovations and trajectories, sometimes claiming that these unspoken assumptions are

---

[1] Technically, sustainability and sustainable development are not necessarily the same, since sustainability refers to any given system whose performance can be sustained, and sustainable development refers to the extent to which development, in particular, can be sustained. In everyday language and in this chapter, however, the two terms are used interchangeably to denote a harmonious environmental, social, and economic system whose performance can be sustained at the present time and in the future.

[2] These include risks from both natural and human-engineered nano-particles.

precluding critical reflection and even benign alternative visions and innovations (Wilsdon *et al.* 2005; Kearnes *et al.* 2006; Felt *et al.* 2007). Thus, it seems that nano proponents and critics are divided over the issue of the meaning of the innovations emanating from nano research and development (R&D). On a closer look, they are also divided over the meaning of *sustainability*, which may explain why proponents and critics of nano innovations do not conceive nano products in the same manner. These meaning differences are provoking tensions and conflicts about nano innovations, in particular concerning the sustainability or unsustainability of nano innovations.

This chapter engages the conflict concerning the sustainability of nano innovations. It presents this conflict as emanating from worldview differences between nano proponents and critics and demonstrates how those differences in worldviews create, rework, and sustain competing meanings of novel phenomena and for such information as nano innovations. A worldview is a term I apply throughout this chapter to signify ideas or beliefs that shape and are closely intertwined with a given community's actions, rendering them intelligible—a social fact or reality (Barnes 1983). I seek to examine the underlying goals, priorities, and values (hereafter referred to broadly as *interests*) that these different worldviews put forth and sustain in scientific or public arenas and how they represent nano innovations as sustainable or unsustainable. Particularly, the chapter seeks to draw attention to how the processes of socialization and internalization of a particular worldview influence nano proponents' or critics' thoughts and practices, including what they envision as possible and doable.

The chapter begins by summarizing the sustainable development idea (hereafter SD)—to show why SD permeates nano debates in scientific and public spheres. Section 3.1 examines the framing of nano innovations in water management as a SD, arguing that this dominating frame is informed by a deep-rooted collectively endorsed belief in science-driven technology, particularly, the power of science to predict and control complex social and natural phenomena. It argues that this belief system—indeed, a 'blind faith' (Anderson et al. 2009; Felt et al. 2007) in science—has, over the last thirty years or so, been complexly synthesized with a 'blind faith' in the productive forces of commercializing, privatizing, and commoditizing public and common resources at a global level. Together, these beliefs combine to create and sustain a powerful and dominating nano sustainability frame. A frame (Gamson & Modigliani 1989) is understood throughout this chapter to be a "central organizing idea or storyline to a controversy that provides meaning to an unfolding of a series of events, suggesting what the controversy is about and the essence of an issue" (Nisbet & Huge 2006, p. 10). Frames are "thought organizers, devices for packing complex issues in persuasive ways by focusing on certain interpretations over others, suggesting what is relevant about an issue and what should be ignored…to structure social meaning" (Nisbet & Huge 2006, p. 10–11). Section 3.2 examines the framing of nano innovations in water management as an unsustainable development. It argues that this framing is grounded in a deep-rooted collectively endorsed belief in precautionary science, which is complexly synthesized with a belief in a democratized science by a particular unit group in contemporary society. Together, these two beliefs combine to create a compelling counter-dominant nano frame that calls for alternative innovations—nano or non-nano. Dominancy here and throughout this chapter is understood in both a political and a cultural sense. Politically, it refers to the supremacy or control of one social group over others due to unbalanced access and possession of economic, social, or symbolic capital. Culturally, it refers to the control and sustenance of power by one social group over others' imaginations, ideas, thoughts, practices, and actions. This control occurs, in part, due to the powerful group's communication prowess or access to leadership positions. In part, it occurs through the ability to persuade others to accept the dominant group's visions and ideas as the only possible or worthwhile ones, indeed, as the way or only way to think or do things. Counter-dominancy is the alternative force that tries to create counter visions and ethical leadership (McLeish 1993). In the final section, the chapter argues that the creation, reworking, and sustaining of nano meanings—including the sustainability or unsustainability of nano innovations—must be understood in light of the worldview and local context in which they occur and the interests that this worldview seeks to further and maintain.

2. Sustainable Development

The 1972 United Nations Stockholm Conference represented the first world appeal to recognize and mitigate harmful effects that accompany science and technological developments. A conference that ushered in the era of environmentalism and public awareness of the need to protect our environment conceived the "sustainable development" idea (Baylis & Smith 2005). However, it was not until 1987 that the SD idea won international recognition, when the UN World Commission of Environment and Development published the Brundtland report, which gave equal emphasis to ecological, social, and economic developments. The basic normative and ethical ideas of SD as defined by the Brundtland report are listed below (Kamara *et al.* 2006; Kamara 2008):

1. **Meeting needs**: Present society should take care of the *needs* of the poor and future generations and respect their rights to a human existence based on reasonable standards of welfare. Security of food, work, energy, water, and health care are major concerns.
2. **Social fairness**: A fair distribution of resources, such as money, information, health, etc., within global populations is important in itself, as well as for the development of environmental sustainability.
3. **Maintenance of natural resources and nature**: Scarcity of natural resources and nature set *limits* on the exploitation of resources and nature. Care should be taken not to disrupt the regenerative capacity of nature. Biodiversity should be preserved and the use of renewable energy enhanced.
4. **Sustainable economy:** Revitalization of economic growth based on new qualities, such as fair distribution and producing more with less use of resources, is paramount to fighting poverty and environmental degradation.

What was revolutionary about the Brundtland report was that it questioned the classical European idea of development that is founded on a belief in limitless economic growth, national well-being, and competitiveness through the powers of scientific and technological progress (Kragh 2005; Kjaergaard 2006). The Brundtland report idea of SD was recognized during the 1992 UN Rio Earth Summit, where the Rio Declaration was adopted. Both the Brundtland report and the Rio Declaration conveyed the clear message that the adoption and implementation of precautionary and public participation principles were crucial tools for achieving SD. Since then, the idea of SD has been welcomed and represented globally. However, that the Brundtland report idea of SD was presented does not mean that its initial intention was accepted indiscriminately; it was reworked and modified by global communities and organizations with reference to the given community's underlying belief system and the interests it serves. Nano innovations were developed in the wake of this SD consciousness. As this consciousness permeated public imaginations and government policies, support for science-driven innovations came to be tied with their assumed effects on SD. Granted, nano innovations in themselves have no inherent meaning; they acquire meaning through use and with reference to the worldview and local context in which they are justified and deployed.

3. Framing Nano Innovations in Water Management

Like earlier science-driven innovations, nano innovations cannot be understood outside the worldview in which they were developed: modern developed society and its deep-rooted collectively endorsed belief in the power of science and technology. This belief can be traced from the history of classical European thought to contemporary visions of a knowledge-based society. This belief—in fact, a 'blind faith' (Anderson *et al.* 2009; Felt *et al.* 2007) in science and technology—is, according to sociologists of knowledge, a social reality in modern developed societies: the way to do things, the ways things are, the way things are done, the way everybody sees and does things, the way it is. According to sociologists of knowledge, a social reality is created and sustained by coordinated collective human thoughts, talk, and practices in a given community or society. It is sustained and reinforced through education, training, socialization, coercion, mutual symbolic sanctioning, or habituation. For our purposes, the social reality concept helps us to capture how a community's worldview comes to be created and sustained and to serve as a powerful background for the following: first, what is imaginable, thinkable, and doable; second, solutions to old or new problems that are sought; third, the way new and novel information or phenomenon are interpreted; fourth, the imaginations, thoughts, or actions that are open or closed to people by virtue of their membership in a given community or society and its consensual reality (Barnes 1983; Mazzotti 2008). When we examine nano deployment in water management through SD, we gain fascinating insight into competing social realities that rework and frame nano innovations accordingly—as outlined in Table 1 below.

3.1 The nano sustainability frame

It is plausible to say that the nano SD frame in water management is created and sustained by mainstream scientific experts, government agencies, corporate organizations, and mainstream natural and social science community groups' talks, thoughts, and practices. It is now accepted that nano proponents subscribe to a blind faith in science and technology, seeing science-driven technological solutions to social, human, and natural problems as the only game in town (cf. Anderson *et al.* 2009, p. 145; Felt *et al.* 2007; Wilsdon & Willis 2004; Wilsdon *et al.* 2005; Kearnes et al. 2006). However, it is now accepted that nano proponents have learned from nuclear power, the mad-cow disease, and genetically modified food controversies that ignoring known or unknown nano concerns can only derail nano support or innovations (Anderson *et al.* 2009; Giles 2003; Editor 2003; Royal Society and Royal Academy of Engineering 2004). Accordingly, in the field of water management, proactive nano risk and ethical governance is of concern to **both** critics and proponents of nano innovations (cf. Savage et al. 2009; Anderson *et al.* 2009; RCEP 2008; Theron *et al.* 2008; Scheufele et al. 2007; Berube 2006; Colvin 2003; Maynard *et al.* 2006; Royal Society 2004).

Table 1 shows that those claiming nano innovations as contributing to sustainable development tend to hold a worldview that is normally associated with unsustainable development.

Table 1: Competing frames for understanding nano and sustainability

Frame	Nano seen as Sustainable Development	Nano seen as Unsustainable Development
Problem	slow, complex, inefficient, or ineffective existing inputs and services	undemocratic institutions; short-term profit-driven innovations; lack of political will
Engines	multinational corporations	a balance between industrial corporations and locally based innovations or micro-industries
Tagged value on natural ecosystems, including human	capital to be exploited; assets providing environmental services	assets have intrinsic value and should be protected; common/public good to be shared, nurtured, and sustained
Dreams	learn and imitate nature's methodology; perfect, predict, and control nature; tame nature's complexity, uncertainty, and ambiguity	learn from nature; accept, live with, and respect nature's complexity and uncertainty
Goal/visions	economic growth, global competitiveness, and dominance	fair and just society, trade, distribution of goods, and wealth
Ethical world	new market niche; mass industrial production of simple, fast, effective, efficient, and targeted short-term techno-fixes; capital increase or payback on investment; sustainable business and capital; exploit anything and everything	sustainable tools and services defined in a time horizon of many human generations; defend public good
Drivers	principle of "sound science"	precautionary science and public participation principles

An examination of nano proponents' practices, talks, and claims about nano innovations indicate that a blind faith in techno-science driven progress permeates nano water management investments, research, and development. Informed by scientific studies, proponents of nano-based water management agree that the predicted water problems are caused by human induced pollution (cf. Theron *et al.* 2008; Savage *et al.* 2009), overconsumption of freshwater, concentrated urbanization growth, or poor urban planning. Interestingly, nano proponents see water overconsumption as an unavoidable side effect of improved lifestyles and economic endeavors and view pollution as emanating from aging water infrastructures (cf. Sustich *et al.* 2009). While the former is supported and encouraged, the latter is seen as a technical problem that can be fixed by advanced technologies—now presented by nano innovations (Sustich *et al.* 2009). In fact, proponents of nano deployment see the predicted water crisis as an opportunity, a business niche for nano products, with all eyes focused on a staggering global market. In this view, nano innovations are seen as drivers for national economic growth and global competitiveness (Sustich *et al.* 2009).

Further, an examination of nano proponents' talks and practices show that nano proponents are driven by the dream to perfect, tame, or alleviate perceived shortcomings of nature (Wickson 2008). Thus, although the UN and others observe that there is enough freshwater in the world albeit unevenly distributed (Chenoweth 2008a; 2008b), proponents of nano-based water management insist that there is a limited global freshwater supply (Loncto *et al.* 2007). They observe that global resources for freshwater represent 3% of the world's freshwater supply, with the rest, 97%, being salt water. Observing that 70% of the available freshwater is used for desirable agricultural or industrial production, interestingly, nano proponents invite us to see the predicted water crisis as "the new oil of the 21st Century" (Choi 2005; Loncto *et al.* 2007). Viewing the predicted water crisis as the new oil has two effects. On the one hand, it triggers public imagination and a genuine fear of global terrorism or war (akin to the wars and terrorism triggered by oil conflicts) and threat to national and international security. Therefore, describing the predicted water crisis as the new oil crisis is meant to persuade the public to see nano as the tool for mitigating terrorism or wars and for promoting national well-being and security. On the other hand, viewing the predicted water crisis as the new oil crisis is meant to trigger risk takers', industrialists', and investors' imaginations concerning a lucrative business and opportunity for global dominance and competitiveness, which the oil business is famous for. Thus, this view persuades risk takers to invest in the new business opportunities provided by nano deployment in water management. Consequently, the application of nano-based catalysts, filters, nano-filtration membranes, carbon nanotube membranes, nano-sensors, and magnetic nanoparticles—among others in water management—come to create, on the one hand, a public good but, on the other hand, a private good.

Furthermore, there already exist conventional, sustainable, and cheap water management technologies (cf. Sobsey *et al.* 2008; Meridian Institute 2006; Hillie *et al.* 2006). These include conventional water filters (based on ceramic, biosand, fabric filters, etc.), conventional heat and UV radiation, conventional chemical treatment, desalination, and arsenic removal technologies. However, proponents of nano-based water management invite us to see these conventional technologies as inefficient or obsolete. They insist that nano-based management is more advanced and high tech (Tahaikt *et al.* 2008; Majewski & Chan 2008; Koratkar *et al.* 2007; 2003; 2005; Xue *et al.* 2008; Hillie & Hlophe 2007; Theron *et al.* 2008). Perhaps the flagship devices for nano-based water management are carbon nanotube membranes and self-assembled monolayers (SAMs) of titanium nanoparticles, silver nanoparticles, silica nanoparticles (Koratkar 2007; 2005; 2003; Brown 2006; Majewski & Chan 2008). While nano proponents argue that carbon nanotubes' membranes provide advanced micro-filtration of unwanted elements from salt or waste water, SAMs are hailed as superior at removing unwanted molecules and pathogens on-site — including proteins, viruses, bacteria, and oocytes — in difficult-to-reach remote areas (Majewski & Chan 2008; Koratkar 2007; Xue *et al.* 2008).

Although much has been made of these flagship innovations,[3] carbon nanotubes and SAMs are still in the early stages of product R & D. It is

---

[3] These innovations have been developed by both private and public university scientists, the latter in collaboration with government and industrial partners.

expected that their success will depend on their performance at the commercial stage of product development, which will involve huge industrial capital investment to transform laboratory prototypes into mass-produced goods. Those who understand the world of corporate business accept that, once on the market, nano products will compete with existing conventional and cheaper means of water management (Berger 2007; Loncto *et al.* 2007). Therefore, the success of nano-based water management products will involve more than their novelty. They will need intensive marketing and advertisement. Successful public relations activities involve huge economic budgets, which are feasible for big corporations (Berger 2007; Loncto *et al.* 2007; Cornelissen, 2004). It is this need for mass-producing laboratory prototypes and marketing and advertisement that ties nano-innovations in water management—both at public universities and start-up research laboratories—to big corporations and their obvious interest and drive to explore and expand their global-market niche (Berube 2006). Corporations' desire for new markets has no bounds. If allowed, this drive could involve commoditizing everything, including human beings and natural or publicly owned resources and infrastructure, such as water or public water infrastructure. The temptation to commoditize everything is well exemplified by the American company CNI, which has more than 100 nano patents and applications with no less than 5,000 claims (Berube 2006) or the recent patenting of the gene of the Maasai people of East Africa by American scientists (Gathura 2009).

However, it is the real need for capital investment that persuades society to allow such ownership that, by default, strips previous public goods (such as sea, ground, or surface water) of their public ownership status, only to tag them as private or capital assets whose exploitation by big corporations is justified or legalized (AzoNanotechnology 2006). In addition, this real need for capital investment justifies big corporations' access and acquisition of exclusive rights to exploit public goods for private gain, in part, due to corporations' monopoly and ownership of nano innovations or processes. This ownership occurs through patents rights, corporations' ability to mass produce lab prototypes, and marketing prowess. By tying water management to nano innovations, investors, by default, transform publicly owned water resources that include rivers, sea, or ground water into highly priced property: "the new oil of the 21st Century" (Berube 2006, p. 191; see also Choi 2005 or Berger 2007).

Nano innovators invite risk takers, business entrepreneurs, and investors to see water as "already the world's third-largest industry; only oil and electricity are larger" (Loncto *et al.* 2007, p. 158). They remind investors that the "current size of the global water market is $287 billion right now, and expected to be $413 billion by 2010" (Choi 2005, p. 5). They remind investors, too, that the demand for freshwater is expected to grow by 70% over the next few years and contend that investments in nano-based water purification devices would reap benefits from the "new oil" market (Choi 2005; Roco 2009, p. xxiii), which is an irresistible opportunity:

> *[E]stimated costs [of improving aging water infrastructure] through 2019 for infrastructure replacement in the United States alone ranges from $485 billion to $896 billion (excluding operations and maintenance using current technologies). As staggering as this sum appears, it is most likely only a fraction of the true cost* (Sustich *et al.* 2009, p. xxxii).

Persuaded that nano innovations will play a key role in water infrastructure replacement improvements, risk takers have invested millions of dollars in no fewer than 44 companies developing nano-based water management products, based in countries ranging from Morocco, to South Africa, Israel, New Zealand, Singapore, Australia, and the USA (Berube 2007; World Business Council for Sustainable Development [WBCSD] 2008). Thus, nano-based water management is transformed into an engine of business expansion and a key for opening up new global business opportunities that might have been otherwise impossible to access or justify.

This development may not be accidental. Business theorists have observed how the developed world markets are flooded (cf. Prahalad & Hammond 2002; London & Hart 2004; Berube, 2007) and have invited big corporations to see "the fortune at the bottom of the pyramid, [as an opportunity for] eradicating poverty through profits" (Prahalad 2006) in developing countries.[4] These theorists invite big corporations to see the basic needs of the world's poor— including freshwater supply—both as an unexploited market niche for big corporations' interests and profits and as a noble cause: "enabling dignity and choice" (Prahalad, 2006). Accepting such claims, nano researchers in water management remind us that, in developing countries, "6,000 [people] are dying every day due to water related diseases" (Majewski & Chan 2008, p. 292). They remind us that "in 2002, 1.1 billion people did not have access to a reliable water supply and 2.6 billion people—the majority of them living in Africa and Asia—lacked access to adequate sanitation" (Hillie & Hlophe 2007, p. 663). They also remind us that "poverty and water are closely linked, and access to water resources has become widely equated with ensuring that basic human needs are met" (Hillie & Hlophe 2007, p. 663). Such claims have become common in high-impact scientific journals.

Optimistically, these claims can be read as scientists' attempts to underscore the social virtues and motivations of their work, to be good global citizens, or to adhere to funding agencies stipulations that scientists demonstrate the social relevance of their research. However, a skeptic may read these claims as examples of how science and scientists are politicizing science, indeed, how science and scientists are used as instruments for authorizing and justifying injustices. The symbolic authority of claims made in high-impact journals becomes a powerful instrument for stripping poor people's vital requirement for freshwater of its status as an intrinsic human need, one that should be met through political will, existing conventional water purification, and management technologies (Sobsey *et al.* 2008), indeed, through public managed-water infrastructures (Chenoweth 2008a; 2008b). Such a pessimistic reading can be extrapolated from the fact that most nano-based water management research projects are undertaken as university-government-industry collaborations (cf., Tahaikt *et al.* 2008; WBCSD 2008). By implicitly coupling freshwater supply with nano deployment, a capital value is placed on poor people's intrinsic need for water, a value that

[4] It should be noted here that commoditization of water is already occurring in water-rich countries. For example, various multinational companies are acquiring water supply dominance in water rich countries such as Germany (cf. Wissen & Naumann 2006).

stems from marrying the provision of freshwater with nano technoscience devices, investments, and management (Juma & Yee-Cheong 2005). However, when seen through new business strategy theories (cf. Prahalad & Hammond 2002; London & Hart 2004; Prahalad 2006), the world's four billion poor, as a whole, become an attractive object of big corporations' desires. Through the authority of natural scientists or social scientists, the poor come to lose their intrinsic human value and are viewed as a capital value that can be exploited in the interest and profit of nano researchers and big corporate industries.

It is little wonder that the coupling of nano innovations with water management has coincided with the competition for global economic dominance and leadership. This coupling has been a strong driving force behind nano investments in Europe and North America. For example, in the US and EU, nano proponents lobby for increased public and corporate funding for nano R & D in water management (among other applications) by invoking metaphoric imageries of a threatening "nano race" and alleged "nano wars" among "nano powers": Russia, China, India, and South America (Hullman 2006, p. 13). Viewing nano innovations and investment as war or race engagement calls upon governments and investors to join that race or war or risk losing the nano race or war in the "third industry" (water). European and North American governments and industries have been called upon to see nano support and investment as being about losing or winning US and EU global economic dominance, leadership, and competitiveness. EU or US governments have been called upon "to harness nanotechnology to drive the new global economy" (Goel 2008, p. 1) in order to maximize EU and US global impact and competitiveness in the "third industry" (Sargent 2008; Hullmann 2006). Key to winning the "nano-race," supporters claim, is the ability to be the first to develop effective high-tech water purification tools and devices. According to nano proponents, however, it is the ability to move from laboratory prototypes to mass industrial products that would enable EU and US industries to scale up or reduce prices, which is critical for breaking markets, acquiring dominance, and maintaining a competitive edge in the global market. More importantly, it is the ability to innovate and develop quick-fix replaceable devices—

well exemplified by SAMs—that will be critical for the sustainability of nano business and dominance (Sustich *et al.* 2009; Roco 2009).

3.2 The nano unsustainability frame

The above described nano sustainability frame has not been without controversy. It has been accompanied by and is in tension with a nano unsustainability frame that is created and sustained by nano critics' talk, thoughts, and practices (Giles 2003; Organic Inform 2008). Nano critics are heterogeneous. They include such national or international environmental organizations as Friends of the Earth, Greenpeace, and ETC group (Canada); general stakeholder groups such as consumer organization; Soil Association (UK); and organic associations. Nano critics include dissident, critical, or activist nano researchers, as well as nano social science researchers. Because of nano critics' diverse institutional, social-cultural, and ideological backgrounds, the nano unsustainability frame is informed by diverse modalities of knowledge and representations of nano innovations. However, it is plausible to say that the core of the unsustainability frame is a belief in precautionary science, which is synthesized with the belief in the need to democratize science, nano innovations policy, and the decision-making process.

Accordingly, most nano critics, like nano proponents, are individuals or groups who believe in science and technology, as evidenced by their selective invocation of scientific studies and authority as tools for defending their nano unsustainability claims. However, although nano critics accept that developments in science and technology have generated products and innovations that they embrace and celebrate, they see these greater 'goods' as accompanied by the production and distribution of unacceptable greater 'bads' (Beck 1992). It is the latter that disconcert nano critics and have come to influence and shape the way they see and think about advances in nano innovations. Perhaps more important, nano critics are influenced by the belief that science is deeply social, political, and value-laden (Wynne 2006; Pielke 2007). Accordingly, nano critics are not anti-science, nor do they want less science; they want more science—diverse and multiple—as resources for furthering heterogeneous social interests (Felt et al. 2007).

Consequently, perhaps one of the overarching concerns of nano critics is a perceived democratic deficit in nano research, development, and deployment. Nano critics indicate that nano industries, including water management industries, have not been open about the presence, functions, and nature of the properties of deployed nano components and that, when industries do provide information, very little of it is trustworthy (cf. Colvin 2003; Mueller & Nowack 2008). For this reason, nano critics argue that it is impossible for governments, scientists, or critics to characterize the types, properties, qualities, and functions of nanoparticles in consumer products or to assess their longer term effects on non-linear ecological, environmental and biological ecosystems. In the main, this is because industries flag such information as confidential and protected from public disclosure under trade secret regulations (Mueller & Nowack 2008). The democratic deficit concern is heightened by the fact that governments, both in developed and developing world, have allowed nano-based products to enter the market place or consumer products with little or no regulatory surveillance and control or robust mechanisms for public participation, consultation, and information (Hansen *et al.* 2008). Consequently, nano critics invite us to see governments as lacking the political will to protect citizens from the predation of monopolistic multinational companies:

> *Three years ago, scientists advised the Government [UK] that the release of nanoparticles should be avoided as far as possible. . . . Though the government acknowledged the risks, no action has been taken to impose controls...the government is ignoring the initial indications of risk and giving the benefit of the doubt to commercial interest rather than the protection of human health and the environment* (Soil Association, 2008).

Perhaps the issue of ownership and control of nano innovations is the real concern for nano critics. As argued above, the deployment of nano-based technologies in water management is intimately paired with commoditization and commercialization of water services and supply worldwide (cf. McDonald & Ruiters 2005; Naumann & Wissen 2006).

Indeed, since the ability to invest and mass-produce cheap commercial nano products from invented prototypes is dependent on huge capital investments, which are feasible for big multinational organizations, the protection of common water resources from big corporations' predation is one of the greatest concerns of nano critics, a concern that is deeply rooted in the democratic questions of justice and fairness. For nano critics, natural water resources are considered a public good that should be accessible to all human beings, irrespective of their economic situation and power. Accordingly, nano critics invite us to see big corporations that have heavily invested in and developed nano-based water management technologies as being driven solely by the business bottom line. Indeed, we are reminded that "not a single product out there advertises to be cheaper because it is nanotechnology enabled" (Berger 2007). Others argue that such products would not be affordable for the world's poor (Invernizzi *et al.* 2008). Accordingly, nano critics invite us to see nano-based water management technologies as mere tools for the intensification and expansion of an already existing trillion dollar water commercialization market (McDonald & Ruiters 2005) or as part of a larger effort to explore new business ventures or consumer markets. It is because of this stance that some nano critics foresee future scenarios in which nano may accelerate corporate takeover, control, and monopoly of formerly public-owned common-good resources in developed and developing countries as big corporations "accelerate existing trends of patent monopolies over life—making new corporations life lords" (Miller 2008).

A related democratic concern, albeit at the global level, is that of human rights and justice. Akin to environmental racism or environmental classism, the fear of nano racism and classism has been raised by some nano critics who invite us not to lose sight of this risk as more and more nano particles enter the marketplace. This reasoning contends that certain nano decisions, actions, and policies would constitute racial discrimination or class discrimination or would give rise to the creation of certain advantaged races or classes of people. While the former becomes a hot spot for nano release and exposure by way of placing nano particles or nano industry production facilities in certain countries, areas, or communities, the latter reaps premium benefits from nano investments and development, coupled with premium protection from nano exposures and hazards (Berube 2007, p. 301; Invernizzi *et al.* 2008; Foladori & Invernizzi 2007).

Accompanying the democratic and public resource ownership questions is an increasing concern over longer term effects of nano particles on the environment, human health, and ecosystems. According to nano critics, there is a disconcerting lack of research funding for ecological and environmental research projects to examine unwanted negative effects that may accompany the use of nano particles in water management (Organic Inform 2008). Critics note that the impacts of nanoparticles on human health, the environment, wastewater disposal channels, incineration plants, landfills, and sewage treatment plants are largely unknown (International Center for Technology Assessment 2008). Pointing at how existing risk-assessment models are limited (see Chapter 18)—in that they are based on unreliable extrapolations and analogies from their bulk versions or natural nano particles—, nano critics press for more research that examines the effects of engineered nano particles on lakes, rivers, groundwater, sediments, surface erosion, and irrigation (Bass 2008; Benn & Westerhoff 2008). Through the invocation of the "new asbestos" and "late lessons from early warning" metaphors (Hansen *et al.* 2008, p. 444), critics draw our attention to existing studies that have raised concerns over the unwanted environmental effects of carbon nanotubes and titanium dioxide, silver, or iron nano particles in various consumer products: "a number of studies have demonstrated the toxicity of nanoparticle silver to bacteria, suggesting that the antimicrobial effects of silver may be detrimental to aquatic ecosystems" (Benn & Westerhoff 2008, p. 4133). Critics call for the need to support diverse, heterogeneous, and alternative science.

Perhaps the ironic twist of nano innovations as a tool for water management, which is a growing concern for nano critics, is the downstream polluting effects of nano particles on ground and surface water. Critics are inviting us to consider how nano particles in cosmetics or clothing are readily washed down the drain—to be released into sewer systems and wastewater treatment plants (Mueller & Nowack 2008; Bass 2008). They point out that nano particles present in sewage may remain in wastewater biomass that is collected at wastewater treatment plants and, thus, end up in the environment or human food because of the dispersion of wastewater biosolids on agricultural land or through waste disposal in surface or sea water. This dispersal may occur directly or through trophic level exposure (through various stages within the food chain, for example, human consumption of fish that have been exposed to nano particles) and transfer of nano particles through complex non-linear eco-system networks or food webs. Pointing at an existing inability to remove some of the nano particles from wastewater treatment systems, these critics point at how remnants of nano particles in treated effluent find their way into streams, rivers, and ground or sea water environments—the very water that nano innovations were meant to purify, treat, and manage—with unknown risks to complex environmental and biological systems (cf. Borm & Berube 2008; Mueller & Nowack 2008; Weisner *et al.* 2007; Holbrook *et al.* 2008; Maynard 2006; Barnard 2006).

4. Conclusion

This chapter has engaged existing conflicts and tensions concerning the sustainability of nano innovations by focusing on nano-based water management. It assumes a bird's eye view—albeit an imperfect one—of nano innovations in scientific and public spheres, mapping competing frames for sustainability and unsustainability of nano innovations. It develops a typology of nano sustainability and unsustainability frames that are in tension in creating and sustaining competing nano meanings. It understands these frames as stemming from different deep-rooted beliefs that proponents and critics have been socialized into and have internalized. These deep-rooted beliefs are understood as repertoires through which nano proponents and critics see and rework new ideas, phenomena, or information such as SD and nano innovations. The resulting frames of reference magnify or diminish certain elements of nano innovations or their effects on the environment and society. Critically, this chapter argues that the meaning of the effects of nano innovation in water management, environment, and society are created, mediated, and sustained by the recipient community's belief systems—

and the interests they protect, further, and sustain. Accordingly, the chapter argues that what nano innovations are and what their effects to the environment and society are depend on the worldview, local context, and interest of the affected community. Nano innovations may mean a sustainable business and economic development. They may mean unsustainable exploitation of the worlds' poor human and natural resources.

Therefore, what nano innovations mean and entail depends on the local context and circumstances of their application and reception. In democratic 21st century societies, all these meanings must be recited, respected, and recognized in order to enrich the decision-making process. In the policy-making and implementation process, these differences in meaning will continue to intensify. A consensus may be impossible and unrealistic to hope for and could possibly be detrimental to society and the human endeavor to deepen our knowledge and to embrace diversity. These differences should continue in a creative and productive tension. In conclusion, this chapter can offer only hope that proponents and critics of nano innovations can seek to see what they cannot see by way of learning to learn: seeking to see things from myriad perspectives, listening to and doubting their own voices and meanings, and listening to and doubting the voices and meanings of others. Thus, they should engage in processes of mutual learning and co-construction of each others' meanings in an open, dynamic, and respectful enquiry.

Questions for Reflection:

1. What underlies different positions on the question of whether nanotechnology contributes to sustainability? How do these manifest in the example of nano-based water management?
2. Provide arguments for and against nano-based water management in: a) developing countries, b) developed countries, and c) a global context.
3. What is your contribution to global sustainability?

Bibliography

Amato, I. (1999). *Nanotechnology—Shaping the World Atom*, Washington, D.C.: National Science and Technology Council.

Anderson, A., Petersen A., Wilkinson, C. and Allan, S. (2009). *Nanotechnology, Risk and Communication*, UK: Palgrave Macmillan.

AZoNanotechnology (2006). *Carbon Nanotubes May Offer a Cheap Technique for Desalination*. http://www.azonano.com/news.asp?newsID=2322 (last accessed Mar 26, 2009).

Barnard, A. (2006). Nanohazards: Knowledge is our first defense, *Nature Materials*, 4, pp. 245–248.

Barnes, B. (1983). Social life as bootstrapped induction, *Sociology*, 17, pp. 524-545.

Bass, C. (2008). The (Nano) Silver Bullet, *The New Republic*. http://www.tnr.com/politics/story.html?id=21ebcaab-34c6-4279-882d-63cff63d964d (last accessed Mar 26, 2009).

Baylis, J. and Smith, S. (2005). *The globalization of World Politics* (3rd ed), Oxford: Oxford University Press, pp. 454–455.

Beck, U. (1992). *Risk Society: Towards a New Modernity*, London: Sage.

Benn, T. M. and Westerhoff, P. (2008). Nanoparticle silver released into water from commercially available sock fabrics, *Environ. Sci. Technol.*, 42, pp. 4133-4139.

Berger, M. (2007). Water, Nanotechnology's Promises, and Economic Reality, *Nanowerk*. http://www.nanowerk.com/spotlight/spotid=2372.php (last accessed Mar 26, 2009).

Berube, D. M. (2006). Nano Hype. The Truth Behind the Nanotechnology Buzz, Amherst: Prometheus Books.

Borm, J. A. and Berube, D. (2008). A tale of opportunities, uncertainties, and risks, *Nanotoday*, 3, pp. 56-59.

Brown, S. (2006). Water, Water Everywhere, *Science NOW Daily News*, May 25. http://sciencenow.sciencemag.org/cgi/content/full/2006/525/1 (last accessed Mar 26, 2009).

Chemical Industry Vision2020 Technology Partnership (2007). *Vision2020 Goals*. http://www.chemicalvision2020.org/ (last accessed Mar 26, 2009).

Chenoweth, J. (2008a). Looming water crisis simply a management problem, *New Science*, 2670, pp. 28-32.

Chenoweth, J. (2008b). Minimum water requirement for social and economic development, *Desalination*, 229, pp. 245-256.

Choi, C. Q. (2005). Water, Water Everywhere Nano, *Space Daily,* March 18. http://www.spacedaily.com/news/nanotech-05zb.html (last accessed Mar 26, 2009).

Colvin, V. (2003). The potential environmental impact of engineered nanomaterials, *Nature Biotechnology*, 21, pp. 1166–70.

Cornelissen, J. (2004). *Corporate Communications: Theory and Practice*, London: Sage.

Editor (2003). Don't believe the hype, *Nature*, 424, p. 237.

Duncan, J. S., Savage, N. and Street, A. (2009). Competition for water. In Savage, N., Diallo, M., Duncan, J., Street, A. and Sustich, R. (eds) *Nanotechnology Applications for Clean Water,* Norwich: Andrew, pp. 463-489.

ETC Group (2004). Down on the farm: The Impact of Nano-Scale Technologies on Food and Agriculture, Ottawa: ETC Group.

European Commission (2004a). *Towards a European Strategy for Nanotechnology*, Luxembourg: Office for Official Publications of the European Communities.

European Commission (2004b). Nanotechnologies: A preliminary risk analysis on the basis of a workshop organized in Brussels on 1-2 March 2004 by the Health and Consumer Protection Directorate General of the European Commission, Luxembourg: Office for Official Publications of the European Communities.

Felt, U. and Wynne, B. (2007). *Taking European Knowledge Society Seriously*, Report of the Expert Group on Science and Governance to the Science, Economy and Society Directorate, Directorate-General for Research, European Commission.

Foladori, G. and Invernizzi, N. (2007). Agriculture and Food Workers Challenge Nanotechnologies, *Rel-UITA*. http://www.rel-uita.org/nanotecnologia/trabajadores_cuestionan_nano-eng.htm (last accessed Mar 26, 2009).

Gamson, W. and Modigliani, A. (1989). Media discourse and public opinion on nuclear power: A constructionist approach, *American Journal of Sociology*, 95, pp. 1-37.

Gathura, G. (2009, March 1). Row as Maasai Gene Patented, *Sunday Nation*. http://www.nation.co.ke/News/-/1056/540420/-/u31ino/-/index.html (last accessed Mar 26, 2009).

Giles, J. (2003). What is there to fear from something small? *Nature*, 426, p. 750.

Goel, A. (2008). Harnessing Nanotechnology to Drive the New Global Economy, *Xconomy*. http://www.xconomy.com/boston/2008/04/28/harnessingnanotechnology-to-drive-the-new-global-economy/ (last accessed Mar 26, 2009).

Hansen, S. F. Maynard, A., Baun, A. and Tickner, J. A. (2008). Late Lessons from early warnings for nanotechnology, *Nature Nanotechnology*, 3, pp. 444–447.

Hillie, T. and Hlophe, M. (2007). Nanotechnology and the challenge of clean water, *Nature Nanotechnology*, 2, pp. 663–664.

Hillie, T. Munasinghe, M., Hlope, M. and Deraniyagala, Y. (2006). *Nanotechnology, Water and Development*, Meridian Institute.

Hoet, P. H. M. Brüske-Hohlfeld, I and Salata, O.V (2004). Review: Nanoparticles—Known and unknown health risks, *Journal of Nanobiotechnology*, 2(12), pp. 1-15

Holbrook, R. D. Murphy K. E., Morrow, J. B and Cole, K. E. (2008). Trophic transfer of nano particles in a simplified invertebrate food web, *Nature Biotechnology*, 3, pp. 352-355.

Hullmann, A. (2006). *The Economic Development of Nanotechnology - An Indicators Based Analysis*, European Commission, DG Research, Nano S&T - Convergent Science and Technologies. ftp://ftp.cordis.europa.eu/pub/nanotechnology/docs/nanoarticle_hullmann_nov2006.pdf (last accessed Mar 26, 2009).

International Center for Technology Assessment (2008). Groups Demand EPA Stop Sale Of 200+ Potentially Dangerous Nano-Silver Products (press release), *Organic Consumers Association*, May 1. http://www.organicconsumers.org/articles/article_11955.cfm(last accessed 26 Mar 2009).

Invernizzi, N. Foladori, G and Maclurcan, D. (2008). Research note. Nanotechnology's controversial role for the south, *Science, Technology and Society*, 13, pp. 123-148.

Jørgensen, M.S, Munch Andersen. M., Hansen, A., Wenzel, H., Pedersen, T.T., Jørgense.n, U., Falch, M., Rasmussen, B., Olsen, S. L, and Willum, O.

(2006). Green Technology Foresight about Environmentally Friendly Products and Materials – Challenges from Nanotechnology, Biotechnology and ICT, Environmental Project No. 30, Miljøstyrelsen, Copenhagen.

Juma, C. and Yee-Cheong, L. (2005). *Innovation: Applying Knowledge in Development, UN Millennium Project,* London: Earthscan. http://www.unmillenniumproject.org/documents/Science-complete.pdf (last accessed Mar 26, 2009).

Kamara, W. M, Coff, C. and Wynne, B. (2006). *GMO and Sustainability: Contested Visions, Routes and Drivers.* Report Prepared for the Danish Council of Ethics.

Kamara, M. W. (2008). *Public Participation in African Biosafety Regulations and Policies*, Issue Paper prepared for the African Union.

Kearnes, M., Macnaghten, P. and Wilsdon, J. (2006). *Governing at the Nanoscale. People, Policies and Emerging Technologies*, London: Demos.

Kjaergaard, P. C. (2006). Kundskab og Magt (Science and Power). In Kjaergaard (ed) *Lys over Landet* (*Light over the Land*). *Dansk Naturvi denskabs Historie*, Bind 3. Aarhus Universitetsforlag, pp. 11-42.

Koratkar, N., Modi, A., Lass, E., Wei, B. and Ajayan, P. M. (2003). Miniaturized gas ionization sensors using carbon nanotubes, *Nature*, 424, pp. 171-174.

Koratkar, N., Suhr, J., Keblinski, P. and Ajayan, P. (2005). Viscoelasticity in carbon nanotubes composites, *Nature Materials*, 4, pp. 134-137.|

Koratkar, N, Wang, Z., Ci, L., Chen, L., Nayak, S., and Ajayan, P. M (2007). Polarity-dependent electrochemically controlled transport of water through carbon nanotube membranes, *Nano Lett*, 7, pp. 697-702.

Kragh, H. (2005). Signalement af Perioden: 1730-1850 (Description of the Period 1730-1850). In *Natur, Nytte og Aand: 1730-1850* (*Nature, Usefulness and the Spirit: 1730-1850*). *Dansk Naturvidenskabs Historie,* Bind 2, Aarhus Universitetsforlag, pp. 9-22.

Loncto, J., Walker, M. and Foster, L. (2007). Nanotechnology in the water industry, *Nanotechnology Law and Business*, June, pp. 157-159.

London, T. and Hart, S. L. (2004). Reinventing strategies for emerging markets: Beyond the transnational model, *Journal of International Business Studies*, 35, pp. 350-370.

Majeswski, P. and Chan, C. P. (2008). Water purification by functionalised self-assembled monolaters on silica particles, *Int. J. Nanotechnol.*, 5, pp. 2–3.

Malanowski, N. (2001). Study for an Innovation—and Technological Analysis (ITA) on Nanotechnology, *VDI Technology Center, Future Technologies Division. Future Technologies No. 35*, Duesseldorf 2001.

Masciangioli, T. (2002). *Nanotechnology for the Environment*, Presentation at National Center for Environmental Research (NCER), US EPA 11 March 2002.

Maynard, A., Aitken, R.J., Butz, T., Colvin, V., Donaldson, K., Oberdorster, G., Philbert, M.A., Ryan, R., Seaton, A., Stone, V., Tinkle, S.S., Tran, L., Walker, N.J. and Warheit, D.B. (2006). Safe handling of nanotechnology, *Nature*, 444(16), pp. 267-269.

Mazzotti, M. (2008). Introduction. In Mazzotti, M. (ed) *Knowledge as Social Order. Rethinking the Sociology of Barry Barnes*, Padstod: Ashgate, pp. 1-13.

McDonald, D. A. and Ruiters, G. (2005). *The Age Commodity. Water Privitization in Southern Africa*, London: Earthscan.

McLeish, K. (1993). Guide to Human Thought - Ideas that shaped the world, Bloomsbury.

Meridian Institute (2006). *Overview and Comparison of Conventional Water Treatment and Water Nanobased Treatment Technologies*, Background paper for the International Workshop on Nanotechnology, Water and Development, Chenai India, 10-12 October.

Miller, G. (2008). *Nanotechnology in Food and Agriculture*, FoE Presentation. Hosted by Reclaim the Food Chain, the sustainable food campaign of Friends of the Earth Adelaide and The Bob Hawke Prime Ministerial Centre, UniSA. http://www.unisa.edu.au/hawkecentre/events/2008events/Nanotech_GeorgiaMiller.pdf. (last accessed 26.03.2009).

Ministeriet for Videnskab, Teknologi og Udvikling (2004). *Teknologisk fremsyn om dansk nanovidenskab og nanoteknologi.* http://www.teknologiskfremsyn.dk/link.php?folder_id=22 (last accessed 26.03.2009).

Mueller, N. C. and Nowack, B. (2008). Exposure modeling of engineered nanoparticles in the environment, *Environ. Sci. Technol.*, 42, pp. 4447-4453.

National Science and Technology Council (2003). *National Nanotechnology Initiative: Research and Development Supporting the Next Industrial Revolution*, Supplement to the President's FY 2004 Budget, Washington, D.C.

Naumann, M. and Wissen, M. (2006). A new logic of infrastructure supply: The commercialization of water and the transformation of urban governance in Germany, *Social Justice*, 33, pp. 20–37.

Nelson, D. J. and Strano, M. (2006). Richard Smalley: Saving the world with nanotechnology, *Nature Nanotechnology*, 1, pp. 96–97.

Nisbet, M. C. and Huge, M. (2006). Attention Cycle and Frames in the Plant Biotechnology Debate: Managing Power of Participation Through the Press/Policy Connection, *The Harvard International Journal of Press/Politics*, 11, pp. 3-40.

Oberdorster, G. Ferin, J., Gelein, R., Soderholm, S. C. and Finkelstein, J. (1992). The role of the alveolar macrophage in lung injury: Studies with ultrafine particles, *Environmental Health Perspective*, 97, pp. 193-199.

Oberdörster, G., Maynard, A., Donaldson. K, Castranova, V., Fitzpatrick, J., Ausman, K., Carter, J, Karn, B., Kreyling, W., Lai, D., Olin, S., Monteiro-Riviere, N., Warheit, D. and Yang, H (2005). Principles for Characterizing the Potential Human Health effects from exposure to nanomaterials: Elements of a Screening Strategy, *Particle and Fibre Toxicology*, 2, p. 8.

Organic Inform @ Elm Farm (2008, February 28). *Nanotechnology in Water Purification.* http://www.organicinform.org/newsitem.aspx?id=425 (last accessed Mar 26, 2009).

Pielke, R. (2007). *The Honest Broker. Making Sense of Science in Politics*, Cambridge University.

Prahalad, C. K. (2006). The Fortune at the Bottom of the Pyramid: Eradicating Poverty Through Profits, Upper Saddle River: Pearson.

Prahalad, C. K. and Hammond, A. (2002). Serving the world's poor profitably, *Harvard Business Review*, 80, pp. 4–11.

Roco, C. R. (2009). Foreword: The potential of nanotechnology for clean water resources. In Savage, N., Diallo, M., Duncan, J., Street, A. and Sustich, R. *Nanotechnology Applications for Clean Water*, Norwich, NY, USA: William Andrew, pp. xxiii-iv.

Royal Commission of Environmental Pollution (2008). *New Materials in the Environment: The Case of Nanotechnology. Surrey: Office of Public Section Information.* http://www.rcep.org.uk/novel materials/Novel Materials report.pdf (last accessed Mar 26, 2009).

Royal Society and Royal Academy of Engineering (2004). *Nanoscience and Nanotechnologies: Opportunities and Uncertainties,* London: Royal Society.

Sargent, J. F. (2008). *Nanotechnology and U.S. Competitiveness: Issues and Options*, Congressional Research Service Report for Congress. http://fas.org/sgp/crs/misc/RL34493.pdf (last accessed Mar 26, 2009).

Savage, N., Diallo, M., Duncan, J., Street, A. and Sustich, R. (2009). *Nanotechnology Applications for Clean Water*, Norwich, NY: William Andrew.

Scheufele, D.A., Corley, E.A., Dunwoody, S., Shih, T.J., Hillback, E., and Guston, D.H . (2007). Scientists Worry about some risks more than the public, *Nature Nanotechnology*, 2, pp. 732–734.

Sobsey, M. D. Stauber, C.E., Casanova, L.M., Brown, J.M., and Elliott, M.A. (2008). Point of use household drinking water filtration: A practical, effective solution for providing sustained access to Safe Drinking Water in the Developing World, *Environ. Sci. Technol.*, 42, pp. 4261–4267.

Soil Association (2008, January 17). Soil Association First Organisation in the World to Ban Nanoparticles—Potentially Toxic Beauty Products That Get Right under Your Skin (press release). http://www.soilassociation.org/web/sa/saweb.nsf/848d689047cb466780256a6b00298980/42308d944a3088a6802573d100351790!OpenDocument (last accessed Mar 26, 2009).

Strathmann, T., Werth, C. J. and Shapley, J. R. (2009). Heterogeneous catalytic reduction for water purification: Nanoscale effects on catalytic activity, selectivity and sustainability. In Savage, N., Diallo, M., Duncan, J., Street, A. and Sustich, R. *Nanotechnology Applications for Clean Water*, Norwich, NY: William Andrew, pp. 269–280.

Sustich, R. C., Shannon, M. and Pianfetti, B. (2009). Introduction: Water purification in the twenty-first century—Challenges and opportunities. In Savage, N., Diallo, M., Duncan, J., Street, A. and Sustich, R. (eds) *Nanotechnology Applications for Clean Water*, Norwich, NY: William Andrew, pp. xxxi-xl.

Tahaikt, M.; Haddou, A., Habbani, R.E., Amor, Z., Elhannouni, F., Taky, M., Kharif, M., Boughriba, A., Hafsi, M. and Elmidaoui, A. (2008). Comparison of the performance of three commercial membranes in fluoride removal by nanofiltration. continuous operations, *Desalination*, 225, pp. 209–219.

Theron, T., Walker, J. A. and Cloete, T. E. (2008). Nanotechnology and Water Treatment: Applications and Emerging Opportunities, *Critical Reviews in Microbiology*, 34, pp. 43–69.

UN Department of Economic and Social Affairs, Division for Sustainable Development (2004). Protection of the Quality and Supply of Freshwater Sources: Application of Integrated Approaches to the Development, Management and Uses of Water Resources. http://www.un.org/esa/sustdev/documents/agenda21/english/agenda21 chapter18.htm (last accessed Mar 26, 2009).

Uskokovic, V. (2007). Nanotechnologies. What we don't know, *Technology in Society*, 29, pp. 43-61.

World Business Council for Sustainable Development (2008). *The Waterbox: Royal Dutch Shell.* http://www.wbcsd.org/Plugins/DocSearch/details.asp?Doc TypeId=24&ObjectId=MzAyNDk. (last accessed Mar 26, 2009).

Wiesner, M.R.; Lowrycarnegie, G.V., Alvarez, P., Dionysiou, D., og Biswas P (2007). Er Nanomaterialer Farlige? *Global Aekologi*, 14(2), pp. 16-19.

Wickson, F. (2008). Narratives of nature and nanotechnology, *Nature Nanotechnology*, 3, pp. 313–315.

Wildson, J. and Willis, R. (2004). *See-through Science: Why Public Engagement Needs to Move Upstream, London:* Demos.

Wilsdon, J. Wynne, B. and Stilgoe, J. (2005). *The Public Value of Science, or How to Ensure that Science Really Matters*, London: Demos.

Wynne, B. (2006). Public engagement as a means of restoring public trust in science. Hitting the notes but Missing the music, *Community Genetics*, 327-T2, pp. 1-10.

Xue, X. Fu, J.F., Zhu, W.F. and Guo, X.C. (2008). Separation of ultrafrine TiO2 from aqueous suspension and its reuse using cross-flow ultrafiltration (CFU), *Desalination*, 225, pp. 29-40.

Chapter 9

Nanotechnology in Foods: Understanding Public Response to its Risks and Benefits

Lynn Frewer & Arnout Fischer

This chapter focuses on the importance of understanding public responses for successful introduction of nanotechnology in the agri-food sector. Historically, innovations in food production technology have provided important benefits to society in terms of food security and quality. Societal acceptance of these innovations, and the technology underpinning these changes, has been equivocal. Potential determinants of societal responses to agri-food technology can be understood from a psychological perspective (for example, in relation to food choice and risk-benefit perception). Research in this area is of direct relevance to understanding potential societal reactions to nanotechnology in a potentially controversial area of application, the agri-food sector. Understanding the psychology of risk perception will also provide information of direct relevance to developing a communication strategy with consumers.

Nano Meets Macro - Social Perspectives on Nanoscale Sciences and Technologies
by K L Kjølberg & F Wickson

www.panstanford.com
978-981-4267-05-2

1. The Importance of Understanding Public Perceptions

Many diverse applications of nanotechnology are currently emerging, ranging from those within the medical and pharmaceutical sectors, the development of new materials, personal care products, to applications in agriculture and food (from here on referred to as the agri-food sector). Whilst all areas of application have the potential to be associated with societal controversy, it is application of nanotechnology in the agri-food sector which is under consideration in the current chapter.

Given recent societal responses to the introduction of agri-food technologies (genetic modification being a prominent example, at least in Europe), the use of nanotechnology in the agri-food sector has the potential to be the focus of controversy. Food consumption implies the introduction of foreign particles into the body, whether on a voluntary or involuntary basis. Some concerns specifically related to the consumption of "nanoparticles" may arise. For example, the potential for nanoparticles to pass through membranes previously impermeable to non-nano sized particles (e.g. the blood brain barrier) may represent a focus of health-related concern. Related to this is the potential for bioaccumulation of nanoparticles in specific body tissues, which may also represent an important concern. This is likely to be particularly problematic if these health risks are perceived to be involuntary in terms of consumer exposure. Additionally, environmental concerns, especially in relation to packaging and cosmetics, may arise.

At the time of writing, societal attitudes to nanotechnology remain uncrystallised compared to other food production technologies (Fischer *et al.* in press). However, as the public is increasingly exposed to information about, and actual products of, nanotechnology, public opinions may change, as for example was the case following the introduction of genetically modified organisms (Frewer *et al.* 2002), and is currently occurring with other food-related technologies such as nutrigenomics (Pin *et al.* submitted).

Agri-food nanotechnology would not be developed unless the developers anticipate clear commercial and societal benefits. In the area of agri-food, various potential benefits can be identified among applications under development. For example, nanotechnology could improve food availability by reducing spoilage and losses, improve nutritional content by implementing targeted delivery of nutrients, and contribute to food production by targeted pesticide delivery or disease detection in animal production systems. Improved food quality (for example, extended shelf life or improved taste, appearance and other hedonic qualities) and safer food (for example, prevention or timely detection of microbial or toxicological contamination) or more sustainable production (for example, reduced requirements for pesticides or irrigation, or improvements in animal health) may also be achieved. In spite of these potential benefits, successful introduction and commercialisation of nanotechnology is contingent on the societal response to the technology and its various applications.

The relationship between technological innovation and societal responses in general has a long and complex history. A case in point is represented by the example of citizen protest and mobilisation against nuclear energy which has been observed since the 1950's (Flynn & Bellaby 2007; Gupta *et al.* in preparation). However, there does seem to be domain specific societal sensitivity towards the use of technology in food production (for example, see Frewer *et al.* 1997) as compared to, for example, medical or pharmaceutical developments. People's attitudes to different applications of nanotechnology need to be understood in terms of the underlying determinants of attitudes and their consequences in terms of positive or negative public response, if the introduction of nanotechnology in the agri-food sector is to be successful.

2. A Brief History of Food Technologies and Societal Responses

The domestication of plants and animals represents an early example of 'technology' applied to food production, an activity initiated in prehistoric times (Diamond 2002). The ability to grow food crops, as opposed to gathering existing wild varieties from the environment, resulted in societal changes. For example, the nomadic lifestyles extant in hunter-gatherer societies evolved into static, permanent communities developed around immovable buildings and structures (Diamond 2002). Subsequently, the members of these developing agrarian societies expanded their diets to include complex preparation of otherwise inedible or indigestible foods and food ingredients, (for example by cooking or fermentation processes), as well as developing and applying food preservation techniques, in order to develop and maintain food security (Hillman & Davies 1990). The implementation of more efficient production processes was further facilitated by the introduction of agricultural tools and implements. The introduction of these new technologies also had 'transformative' impacts in societal terms. For example, the introduction of the mouldboard plough in medieval Europe (Gimpel 1992) allowed the European population to recover from the period of scarcity and population reduction after the collapse of the Roman Empire. One consequence was repopulation of Europe during the middle ages, which arguably resulted in the critical urban population mass with enough time available to develop the intellectual inputs needed for the occurrence of the renaissance. Other technological changes have, over the last millennium, resulted in improvements in food availability, safety and quality, and the nutritional content of foods. For example, the introduction of mechanized farming in the 19th century increased the rate and amount of food production. At the same time, the need for agricultural labour reduced, resulting in a population shift from rural to urban environments in order to avert rural unemployment. Agricultural workers concerned about changes in traditional employment, and their traditional ways of living, protested about the new 'steam' technologies which they perceived to be the cause of those changes (Sale 1996).

The agricultural revolution continued when chemical fertilizers and pesticides were developed. The widespread international application of these intensive farming methods occurred in the middle and latter half of the 20th century: the 'green revolution' (Evenson & Gollin 2003). Opponents of pesticide use (for example, environmental NGOs) have claimed that indiscriminate use of pesticides results in accumulation of toxins in the ecosystem. The publication of the book "Silent spring" by Rachel Carson in 1962 signalled broader societal disquiet regarding pesticide usage, and contributed to the development of the environmentalist movement in the latter half of the 20th century, as it highlighted the potential negative environmental impact of agrichemicals. Other societally controversial technologies applied in the agri-food sector have included the introduction of irradiated foods (for example, see Behrens *et al.* 2009; Bruhn 1995), and novel nutrients such as cholesterol lowering phytosterol, which can be potentially associated with health benefits but can also have negative impacts if used incorrectly (Fransen *et al.* 2007). Debates about food technologies continue at the present time, with environmental NGOs questioning, for example, the environmental impact of genetically modified crops, as well as their unknown effects on animal and human health (Frewer *et al.* 2004). Indeed, NGOs have recently started to voice some concern about the unchecked application of nanotechnology (see e.g. Chapter 17).

Understanding potential societal responses to food technologies is important from a policy and regulatory perspective (for example when considering how regulatory frameworks might incorporate societal concerns). Industry and food producers are also concerned about consumer responses to, and choice for the products of novel technologies, as this may influence profitability, consumer trust in the activities, and motives of (competing) industry actors.

3. The Psychology of Food Choice and Food Technology Acceptance

Consumer food choices involve decision–making associated with nutritional requirements. Food choices are also associated with cultural elements, for example because of regional and temporal variation in what is considered to be a “healthy”, “risky”, or “high quality” food. Indeed, food routinely consumed in one culture or society might be deemed a subject of cultural taboo, or even illegal to consume, in others. For example, consumption of snake, pig, cow, dog or cat meat is not uncommon in some cultures, whilst being unacceptable in others. In many parts of the world insects are considered part of the diet, but in others are unacceptable. A high frequency of lactose intolerance in certain populations means that milk products, regarded as a delicacy in some parts of the world, are not consumed because they are indigestible. Cultural variations in rules and practices about how food is produced, prepared, presented and eaten are extensive (Rozin 2007). Cultural specificity in what constitutes acceptable food choices is also a potential influence on risk and benefit perceptions associated with different kinds of foods and food production technologies.

Many food products are still produced using traditional methods developed in specific regions. The European Union facilitates protection of traditional products like Feta cheese, Parma ham and champagne (Van Rijswijk *et al.* 2008; Frewer & Van Rijswijk 2008). Perceived "naturalness" of both product and production process in relation to food is particularly valued by consumers (see for example, Frewer *et al.* 1997). Foods made with novel technologies are potentially perceived to be "unnatural" or "untraditional", and therefore less likely to be accepted.

The question arises as to what sort of benefit associated with foods produced using novel technologies is sufficient for consumers to accept food products (Frewer *et al.* 2003). The acceptability of novel applications of nanotechnology in the agri-food sector will be dependent on a complex interaction between consumer evaluations of the perceived benefits and risks associated with novel products in general, together with the risks and benefits specifically associated with nanotechnology (Frewer *et al.* 2004). Based on the societal debate about the introduction of genetic modification in the agri-food sector, it might be assumed that discussions regarding its introduction will include not only risk perceptions associated with health and environmental impact assessment, but also include broader issues associated with social impact (for example, differential impact on the rural economy in regions of the world utilising different agricultural systems, or issues associated with social equity in different farmer groups) (see Cope *et al.* in press)

3.1 A brief history of risk psychology

Recognition of the importance of risk perception, as well as technical assessment, and how these influence risk communication, has become an integral part of risk analysis in the last 40 years (starting with the landmark publication by Chauncey Starr, 1969). Much of the research on societal acceptance of technologies and products relates to risk perception. Relevant psychological factors that determine people's responses to a particular hazard were recognised as the underlying reason that lay responses to risk often differ from expert assessments (Fischhoff *et al.* 1978; Slovic, 1987; 1992; 1993; Slovic *et al.* 1982). An important finding was that factors that are not explicitly addressed as part of technical risk estimates may influence peoples' perception of a given risk. These factors include the extent to which a risk is perceived to be unnatural, potentially catastrophic, or to which an individual perceives their own exposure to be involuntary. These results have been confirmed in empirical research including those pertinent to the area of food choice (Fife-Schaw & Rowe 2000). There is extensive literature regarding consumer perceptions of risk associated with food and food production technologies (for example, see Bredahl 2001; Frewer & Fischer 2005; Hansen *et al.* 2003; Miles & Frewer 2001; Siegrist 2000; Townsend 2006; Sparks & Shepherd 1994; Verbeke 2001). Research in this area has demonstrated that, for example, perceptions of unnaturalness, or compromises in important societal values such as environmental protection or animal welfare, may be influential determinants of consumer acceptance of its products. Other psychological factors may influence consumer acceptance of emerging agri-food technologies. For example, some consumers exhibit a fear of new products, technologies or ingredients used in food products ("neophobia"), and this has been found to be an important determinant of their reactions to novel foods in particular (e.g. Bredahl 2001; Pliner & Hobden 1992; Tuorila *et al.* 2001). The value of food neophobia may be protective insomuch as consumers avoid the potentially toxic effects of novel foods (Rozin & Vollmecke 1986). Emotional responses may also contribute to the rejection of novel foods (Rozin *et al.* 2000). Disgust, for example, is a strong emotion that can prevent consumption or acceptance of foods which are contaminated, for example through chemical contact or biological decay, or where this potential contamination is unknown to the person consuming them (Oatley & Johnson-Laird 1987). Disgust may be invoked by events, processes or associations that have little or no direct relation to actual food contamination (Scherer 1997), but nonetheless may result in negative consumer responses to products made with novel technologies.

Another issue of relevance to discussion of consumer psychology is that of "optimistic bias" (Weinstein 1989; Miles & Scaife 2003), where people perceive that their personal risks are less than those experienced by comparative others. Many 'lifestyle' situations, for example, consumption of foods over which an individual has a high perceived level of personal control, but which may also be associated with a negative impact on health, may result in consumers underestimating their personal risks from a particular hazard. Additionally, consumers may react negatively to the introduction of specific food technologies such as food irradiation, genetic modification of foods, and nanotechnology applied to food production, overestimating the health and environmental risks in comparison with the estimates provided by experts.

In this context, *expert* responses to lay risk perceptions have been contextualised by the observation that consumers accept exposure to some risks which may result in a high level of harm in human health terms, whilst at the same expressing higher levels of concern about those which are technically 'smaller'. An example is provided by expert concern about continued consumption of 'unhealthy' foods such as high caloric foods (cookies and chocolate), or foods containing high levels of saturated fats (certain types of meat), despite public health campaigns advising healthier consumption behaviour. At the same time, public concerns about the presence of pesticide residues on foods produced using conventional agricultural techniques have been dismissed by some expert communities as irrelevant. This has led experts to assume that the public were in some way 'deficient' in their views, a deficit which could be corrected by provision of technical risk information (Hilgartner 1990). Contemporary thinking regards this model as obsolete, although research has not yet delivered the solution to the issue of how best to incorporate the views of society into science policy and regulation, and scientific application (Rowe & Frewer 2000; 2004; Rowe *et al.* 2004).

4. Communicating about Food Technologies

For governments working on regulation, scientists developing applications, and businesses aiming to market nanotechnology applications in food, it is essential to develop an effective communication strategy to inform the public, stakeholders and other interested end-users about risks and (potentially) the benefits of novel products. In any case, adequate communication about nanotechnology is important if consumers are to be provided with the informational tools which enable them to make informed decisions about whether or not to consume products developed using nanotechnology. When developing a communication strategy, it is essential to take consumer psychology as the starting point to develop content and form of the message. In consumer psychology, it has been found that the attitude towards an event or object is likely to subsequently determine peoples behaviours associated with the event or object. For example, if people have very negative attitudes towards nanotechnology applied to food production, it is less likely that they will buy the products made using nanotechnology, or containing nanoparticles. Conversely, positive attitudes are likely to result in consumers buying products. It is therefore of interest to understand under what circumstances consumers might change their attitudes.

Attitude change models have been developed to explain the circumstances which motivate people to change their attitudes following presentation of *persuasive information* (information intended to change an attitude in a particular direction) about a given topic. An important assumption in this line of research is that information needs to be processed in an 'in-depth' and thoughtful way if attitude change is to result. In other words, people must carefully consider and think about the information, and only then will a long term change in attitude be achieved (Petty & Cacioppo 1986). 'In depth' processing requires cognitive (mental) effort. That thinking requires effort is illustrated by the experience of feeling tired after being involved in an intellectually challenging debate about science or politics, or after studying for an exam. This 'in-depth' information processing mode is known as 'systematic' or 'elaborative' information processing (Chen & Chaiken

1999). To initiate elaborate information processing, the individual receiving the information needs to be motivated to think carefully about the topic which is the focus of communication. In other words the individual needs to be willing to 'expend' cognitive effort to process the information. In addition, the individual also needs to be able to access appropriate 'cognitive resources' to engage in processing activity. To extend the example provided, an individual who is already very tired may not be willing or able to think about complex information in detail. Another example relates to the co-presence of competing stimuli. For example, if an individual is sitting in a bar with loud music, it may not be possible for them to focus their attention on processing complex information. Both motivation to process information and cognitive resources vary between individuals, and may also vary according to the content of thc information provided, and the contextual situation in which the information is received by an individual (Cacioppo & Petty 1982). Other psychological factors, such as habit (where an individual continues to behave in the way that they always do, independent of novel information about an activity) or unrealistic optimism (where information is perceived to be irrelevant to the individual but directed towards someone else instead[1]) means that the individual receiving the information may not process the message in an elaborative way that will result in attitude change, (Fischer & de Vries 2008), or they may ignore it altogether.

Experts in a particular area may be surprised that non-experts are less motivated to attend to information relevant to their area of expertise than they are themselves. Experts may overestimate the motivation of other people to process information about the expert's own area of interest, leading to the development of information which does not reflect the target audiences concerns and interests. If individuals are not highly motivated to expend cognitive effort on information about a particular topic, or have only limited cognitive resources available (e.g. because of fatigue or information overload) they are unlikely to process the

---

[1] For example, an individual may perceive information about excessive alcohol use to be directed towards a particularly vulnerable comparative person, perhaps someone suffering from alcohol rclated illness, rather than the self

information in an 'in depth' way. Instead, they apply decision-rules known as *heuristics* or other mental 'shortcuts' in reasoning. Heuristic behaviour regulation relies on 'rules of thumb' such as 'food produced using any technology is unnatural' that will swiftly lead to a decision not to consume the associated food products, as unnatural is seen as risky (e.g. Slovic, 1987). Heuristic processing results in fewer cognitive resources being required by the individual to process a particular piece of information. However, if heuristic processing is applied, attitude changes are less predictable and stable compared to situations where elaborate or systematic processing has been applied.

Recent research has suggested that elaborate and heuristic processing may occur simultaneously (Petty & Wegener 1999); and ideally communication should be consistent in its heuristic and systematic content. For example, the content of the message needs to address the concerns of the message receiver (which will increase their motivation to process the information), as well as provide information amenable to systematic or elaborative processing. People may be more motivated to process information about health and safety of ingested nanomaterials, or potential for environmental impact of nanoparticles, if the message is accompanied by a heuristic cue, for example if the information is derived from a highly trusted source.

Information processing theories may explain why attempts to 'educate' the public about the technical risks and benefits of a particular technology, based on 'logical and rational' arguments relating to risk assessment alone, may be unsuccessful. Lay people are unwilling and unmotivated to process complex information which is not relevant to their concerns (Fazio & Towles-Schwein 1999), particularly if this information is perceived to be from an untrustworthy source (for example, see: Frewer *et al.* 1997; Cvetkovic *et al.* 2002). Instead they may resort to "heuristic" processing, and fail to process the information contained in the messages. As a consequence, providing members of the public with technical information about the safety of a new technology and its products is unlikely to persuade them to accept the technology.

Furthermore, consumers may be differentially motivated to search for information regarding risks and benefits of emerging technologies (Fischer *et al.* 2006; Kornelis *et al.* 2007). Demographic and psychological factors may account for profound differences observed between different consumers regarding their responses to emerging technologies and associated applications, as well as other risk issues. In particular, targeted information provision is required, which meets the needs of different groups of consumers, (for example, in terms of their concerns about technologies). Peoples' responses to risk (and potentially benefit) information may also vary according to predictable individual differences (Fischer & Frewer 2008), for example, women may be more risk-adverse than men.

In summary, the development of an understanding of consumer risk – benefit perceptions associated with nanotechnology, and the development of relevant and balanced information about both the risks and benefits, possibly targeted to individual differences in information needs of different population groups, is likely to be an important factor influencing the commercialisation of nanotechnology in the agri-food (and potentially other) consumer sectors.

5. Consumer Acceptance of Nanotechnology in the Agri-food Sector

Based on the introduction of other emerging food technologies, consumer and societal acceptance of agri-food nanotechnology will be dependent on the delivery of tangible benefits to individual consumers and society more generally. For novel foods to be accepted, consumers must perceive that any potential benefits outweigh potential risks or negative effects (for example, potential for negative impact on the environment, human and animal health, or ethical concerns such as animal welfare or social equity).

This is complicated by the potential for the introduction of new products and technologies to introduce new potential hazards to the food chain. These might include allergic reactions to novel proteins (Kuiper *et al.* 2002; Van Putten *et al.* 2006) or unexpected negative environmental impacts, such as the negative effects of the non-toxic chlorofluorocarbons (CFCs) on the ozone layer (Mulder 2005). Even among experts, uncertainty associated with risk-benefit judgments may exist in the context of consumer protection and environmental impacts of new technologies (Marvin *et al.* 2009). Whether the potential benefits of nanotechnology are sufficient to result in consumer acceptance of food products with specific benefits is a topic worthy of further research. It is also important to note that people are unlikely to utilise risk and benefit information equally in decision-making, as people generally pay greater attention, and assign more weight, to risk information compared to benefit information (Kahneman & Tversky 1979). Furthermore, risk and benefit perceptions are often inversely correlated (Alhakami & Slovic 1994), indicating that people may not evaluate risk and benefit independently from each other. Rather, increased risk perception may be associated with reduced perceptions of benefit associated with a particular event or object, and *vice versa.* In general, research on *benefit* perception, and how consumers make trade-offs between food-related *risks* and *benefits* has less frequently been a focus of research (Frewer *et al.* 1997; Verbeke *et al.* 2008; Fischer & Frewer in press). Further research is required to examine the relationship between perceived risk and benefit, and to understand the implications these have for the development of effective communication about (for example) emerging technology acceptance. This issue is of particular relevance in the case of nanotechnology applied in the agri-food sector as public attitudes have not yet crystallised, and are likely to be informed by whatever information becomes available, within the framework of the constraints of cognitive information processing described earlier.

In summary, consumer perceptions associated with the introduction of novel food technologies are characterised by a range of specific perceptions (i.e. associated with the technology under consideration), as well as those which can be generalised to technological innovation (see e.g. Frewer 2003; Frewer *et al.* 2004; Siegrist *et al.* 2008). It is arguable that the agri-food sector may be potentially vulnerable to consumer concerns associated with the introduction of novel technologies, and there is no reason to assume that nanotechnology will not raise similar societal concerns unless lessons regarding effective development, communication and commercialisation strategies are learned from historical precedents.

6. Conclusion

While nanotechnology applied within the agri-food sector is still a technology "under development", there has already been extensive comparison with other recent technological developments, in particular the introduction of genetically modified foods. The traditional emphasis on risk communication (or indeed communication about the absence of risk) may be less relevant to consumer decision-making, as it has become increasingly evident that consumers are making decisions about the acceptability of specific foods and production technologies based on a complex interaction of perceptions of risk and benefit associated with specific food choices. Understanding consumer psychology is essential if understanding and predicting peoples' responses to nanotechnology applied in the agri-food sector is to be anywhere near accurate. An effective research and development strategy must simultaneously address emerging societal needs and societal preferences for novel foods and production processes. Ensuring public and stakeholder participation in the debate about the strategic development and commercialisation of nanotechnology is important if societal responses are to be addressed in an adequate way relevant to the development of societally acceptable governance systems (Rowe & Frewer 2005). As part of this, regulators, scientists and industry need to ensure that societal participation leads to visible changes in policy practices if these are indicated.

Questions for Reflection:

1. What can we learn from previous public responses to the introduction of new technologies in food production?
2. What is your attitude towards the use of new technologies in food production? What is this attitude based on?

3. You have created a new commercial application of nanotechnology in food. How would you approach the task of communicating with the public about this? How would your communication strategy be different if you worked for an anti-nanotechnology movement?

Bibliography

Alhakami, A. S. and Slovic, P. (1994). A psychological study of the inverse relationship between perceived risk and perceived benefit, *Risk Analysis,* 14(6), pp. 1085-1096.

Behrens, J. H., Barcellos, M. N., Frewer, L. J., Nunes, T. P. and Landgraf, M. (2009). Brazilian Consumer views on food irradiation, *Innovative Food Science and Emerging Technologies*, 10(3), pp. 383-389.

Bredahl, L. (2001). Determinants of consumer attitudes and purchase intentions with regard to genetically modified food - Results of a cross-national survey, *Journal of Consumer Policy,* 24(1), p. 23.

Bruhn, C. M. (1995). Consumer attitudes and market response to irradiated food, *Journal of Food Protection,* 58, pp. 175-181.

Cacioppo, J. T. and Petty, R. E. (1982). The need for cognition, *Journal of Personality and Social Psychology,* 42(1), pp. 116-131.

Carson, C. (1962). *The Silent Spring,* Boston, MA: Houghton Mifflin.

Chen, S. and Chaiken, S. (1999). The heuristic-systematic model in its broader context. In Chaiken, S. and Trope, Y. (eds) *Dual Process Theories in Social Psychology,* New York, NY: Guilford Press, pp. 73-96.

Cope, S., Frewer, L.J., Dreyer, M. and Renn, O. Potential methods and approaches to assess social impacts associated with food safety issues, *Food Control, in press.*

Cvetkovich, G., Siegrist, M., Murray, R. and Tragesser, S. (2002). New information and social trust: Asymmetry and perseverance of attributions about hazard managers, *Risk Analysis,* 22(2), pp. 359-367.

Diamond, J. (2002). Evolution, consequences and future of plant and animal domestication, *Nature,* 418(6898), pp. 700-707.

Evenson, R. E. and Gollin, D. (2003). Assessing the impact of the Green Revolution, 1960 to 2000, *Science,* 300(5620), pp. 758-762.

Fazio, R. H. and Towles-Schwein, T. (1999). The MODE model of Attitude-Behavior Processes. In Chaiken, S. and Trope, Y. (eds) *Dual-Process Theories in Social Psychology,* New York, NY: Guilford Press, pp. 97-116.

Fife-Schaw, C. and Rowe, G. (2000). Extending the application of the psychometric approach for assessing public perceptions of food risks: some methodological considerations, *Journal of Risk Research,* 3(2), pp. 167-179.

Fischer, A. R. H. and De Vries, P. W. (2008). Everyday behaviour and everyday risk: An exploration how people respond to frequently encountered risks, *Health Risk and Society,* 10(4), pp. 385-397.

Fischer, A. R. H. and Frewer, L. J. (2008). Food safety practices in the domestic kitchen; Demographic, Personality and Experiential Determinants, *Journal of Applied Social Psychology,* 38(11), pp. 2859-2884.

Fischer, A. R. H. and Frewer, L. J. (in press) Product familiarity and the perception of food related risks and benefits, *Food Quality and Preference.*

Fischer, A. R. H., Van Dijk, H., De Jonge, J. and Frewer, L. J. (in press). The impact of information on attitudinal ambivalence: The case of nanotechnology in food production.

Fischhoff, B., Slovic, P. and Lichtenstein, S. (1978). How safe is safe enough? A psychometric study of attitudes towards technological risks and benefits, *Policy Sciences,* 9, pp. 127-152.

Flynn, R. and Bellaby, P. (eds) (2007). *Risk and the Public Acceptability of New Technologies,* Hampshire, UK: Pelgrave.

Fransen H. P., De Jong, N., Wolfs M., Verhagen H., Verschuren W. M., Lütjohann D., Von Bergmann, K., Plat, J. and Mensink R. P. (2007). Customary use of plant sterol and plant stanol enriched margarine is associated with changes in serum plant sterol and stanol concentrations in humans, *Journal of Nutrition,* 137, pp. 1301-1306.

Frewer, L. J. and Fischer, A. R. H. (2005). Consumer perceptions of risks from food. In Lelieveld, H. L. M. , Mostert, M. A. and Holah, J. T. (eds) *Handbook of hygiene control in the food industry*, Cambridge, UK: Woodhead, pp. 103-119.

Frewer, L. J. and Van Rijswijk, W. (2008). Consumer perceptions of food quality and safety and their relation to traceability, *British Food Journal,* 110(10), pp. 1034-1046.

Frewer, L. J., Howard, C. and Shepherd, R. (1997). Public concerns in the United Kingdom about general and specific applications of genetic engineering: Risk, benefit, and ethics, *Science, Technology, and Human Values,* 22(1), pp. 98-124.

Frewer, L. J., Howard, C., Hedderley, D. and Shepherd, R. (1997). Consumer attitudes towards different food-processing technologies used in cheese production — The influence of consumer benefit, *Food Quality and Preference,* 8(4), pp. 271-280.

Frewer, L. J., Miles, S. and Marsh, R. (2002). The media and genetically modified foods: Evidence in support of social amplification of risk, *Risk Analysis,* 22(4), pp. 701-711.

Frewer, L. J., Lassen, J., Kettlitz, B., Scholderer, J., Beekman, V. and Berdal, K. G. (2004). Societal aspects of genetically modified foods, *Food and Chemical Toxicology,* 42(7), pp. 1181-1193.

Frewer, L. J., Scholderer, J. and Lambert, N. (2003). Consumer acceptance of functional foods: Issues for the future, *British Food Journal,* 105(10), pp. 714-731.

Gimpel, J. (1992). *The Medieval Machine: The Industrial Revolution of the Middle Ages*, London: Pimlico.

Gupta, N., Fischer, A. R. H. and Frewer, L. J. A review on determinants of societal response to transformative technologies, in Preparation.

Hansen, J., Holm, L., Frewer, L. J., Robinson, P. and Sandoe, P. (2003). Beyond the knowledge deficit: Recent research into lay and expert attitudes to food risks, *Appetite,* 41(2), pp. 111-121.

Hilgartner, S. (1990). The dominant view of popularisation: Conceptual problems, political issues, *Social Studies of Science,* 20, pp. 519-539.

Hillman, G. C. and Davies, M. S. (1990). Measured domestication rates in wild wheats and barley under primitive cultivation, and their archaeological implications, *Journal of World Prehistory,* 4(2), pp. 157-222.

Huffman, W. E., Rousu, M., Shogren, J. F. and Tegene, A. (2004). Consumer's resistance to genetically modified foods: The role of information in an uncertain environment, *Journal of Agricultural and Food Industrial Organization,* 2(2), article 8.

Kahneman, D. and Tversky, A. (1979). Prospect theory: An analysis of decision under risk, *Econometrica,* 47(2), pp. 263-292.

Kornelis, M., De Jonge, J., Frewer, L. J. and Dagevos, H. (2007). Consumer selection of food-safety information sources, *Risk Analysis,* 27(2), pp. 327-335.

Kuiper, H. A., Noteborn, H. P. J. M., Kok, E. J. and Kleter, G. A. (2002). Safety aspects of novel foods, *Food Research International,* 35, pp. 267-271.

Marvin, H. J. P., Kleter, G.A., Frewer, L. J., Cope, S., Wentholt, M. and Rowe, G. (2009). A working procedure for identifying emerging risk at an early stage, *Food Control,* 20(4), pp. 343-356.

Miles, S. and Frewer, L. J. (2001). Investigating specific concerns about different food hazards, *Food Quality and Preference,* 12(1), pp. 47-61.

Miles, S. and Scaife, V. (2003). Optimistic bias and food. *Nutrition Research Reviews,* 16(1), pp. 3-19.

Mulder, K. F. (2005). Innovation by disaster: The ozone catastrophe as experiment of forced innovation, *International Journal of Environment and Sustainable Development,* 4(1), pp. 88-103.

Oatley, K. and Johnson-Laird, P. N. (1987). Towards a cognitive theory of emotions, *Cognition and Emotion,* 1(1), pp. 29-50.

Petty, R. E. and Cacioppo, J. T. (1986). *Communication and Persuasion: Central and Peripheral Routes to Attitude Cchange*, New York, NY: Springer-Verlag.

Petty, R. E. and Wegener, D. T. (1999). The elaboration likelihood model: Current status and controversies. In Chaiken, S. and Trope, Y. (eds) *Dual Pprocess Theories in Ssocial Ppsychology*, New York, NY: Guilford Press, pp. 37-72.

Pin, R., Frewer, L.J. and Seydel, E. Intentions to adopt nutrigenomics: personalized nutrition and functional foods, submitted.

Pliner, P. and Hobden, K. (1992). Development of a scale to measure the trait of food neophobia in humans, *Appetite,* 19(2), pp. 105-120.

Rowe, G. and Frewer, L. J. (2000). Public participation methods: A framework for evaluation, *Science Technology and Human Values,* 25(1), pp. 3-29.

Rowe, G. and Frewer, L. J. (2004). Evaluating public participation exercises: A research agenda, *Science Technology and Human Values,* 29(4), pp. 512-556.

Rowe, G. and Frewer, L. J. (2005). A typology of public engagement mechanisms, *Science Technology and Human Values,* 30(2), pp. 251-290.

Rowe, G., Marsh, R. and Frewer, L. J. (2004). Evaluation of a deliberative conference, *Science Technology and Human Values,* 29(1), pp. 88-121.

Rozin, P. (2007). Food choice: An introduction. In Frewer, L. J. and Van Trijp, H. (eds) *Understanding Consumers of Ffood Products,* Cambridge: Woodhead, pp. 3-29.

Rozin, P. and Vollmecke, T. A. (1986). Food likes and dislikes, *Annual Review of Nutrition,* 6, pp. 433-456.

Rozin, P., Haidt, J. and McCauley, C. R. (2000). Disgust. In Lewis, M. and Haviland-Jones, J. M. (eds) *Handbook of Emotions* (2nd ed.), New York, NY: Guilford, pp. 637-653.

Sale, K. (1996). *Rebels Against the Future: The Luddites and Their War on the Industrial Revolution,Llessons for the Computer Age*, New York: Perseus.

Scherer, K. R. (1997). The role of culture in emotion-antecedent appraisal, *Journal of Personality and Social Psychology,* 73(5), pp. 902-922.

Şiegrist, M. (2000). The influence of trust and perceptions of risks and benefits on the acceptance of gene technology, *Risk Analysis,* 20(2), pp. 195-204.

Siegrist, M., Stampfli, N., Kastenholz, H. and Keller, C. (2008). Perceived risks and perceived benefits of different nanotechnology foods and nanotechnology food packaging, *Appetite,* 51(2), pp. 283-290.

Slovic, P. (1987). Perception of risk, *Science,* 236(4799), pp. 280-285.

Slovic, P. (1992). Perceptions of risk: Reflections on the psychometric paradigm. In Goldring, D. and Krimsky, S. (eds) *Theories of Risk,* New York, USA: Praeger.

Slovic, P. (1993). Perceived risk, trust, and democracy, *Risk Analysis,* 13(6), pp. 675-682.

Slovic, P., Fischhoff, B. and Lichtenstein, S. (1982). Why study risk perception? *Risk Analysis,* 2(2), pp. 83-93.

Sparks, P. and Shepherd, R. (1994). Public perceptions of food-related hazards: Individual and social dimensions, *Food Quality and Preference,* 5(3), pp. 185-194.

Starr, C. (1969). Social benefit versus technological risk, *Science,* 165(899), pp. 1232-1238.

Townsend, E. (2006). Affective influences on risk perceptions of, and attitudes toward, genetically modified Food, *Journal of Risk Research,* 9(2), pp. 125-139.

Tuorila, H., Lähteenmäki, L., Pohjalainen, L. and Lotti, L. (2001). Food neophobia among the Finns and related responses to familiar and unfamiliar foods, *Food Quality And Preference,* 12, pp. 29-37.

Van Putten, M. C., Frewer, L. J., Gilissen, L. J. W., Gremmen, B., Peijnenburg, A. A. C. M. and Wichers, H. J. (2006). Novel foods and food allergies: A review of the issues, *Trends in Food Science and Technology,* 17(6), pp. 289-299.

Van Rijswijk, W., Frewer, L. J., Menozzi, D. and Faioli, G. (2008). Consumer perceptions of traceability: A cross-national comparison of the associated benefits, *Food Quality and Preference,* 19(5), pp. 452-464.

Verbeke, W. (2001). Beliefs, attitude and behaviour towards fresh meat revisited after the Belgian dioxin crisis, *Food Quality and Preference,* 12(8), pp. 489-498.

Verbeke, W., Sioen, I., Pieniak, Z., Van Camp, J. and De Henauw, S. (2005). Consumer perception versus scientific evidence about health benefits and safety risks from fish consumption, *Public Health Nutrition,* 8(4), pp. 422-429.

Verbeke, W., Vanhonacker, F., Frewer, L. J., Sioen, I., De Henauw, S. and Van Camp, J. (2008). Communicating Risks and Benefits from Fish Consumption: Impact on Belgian Consumers' Perception and Intention to Eat Fish, *Risk Analysis,* 28(4), pp. 951-967.

Weinstein, N. D. (1989). Optimistic biases about personal risks, *Science,* 246(4935), pp. 1232-1233.

Chapter 10

My Room

Lesley L. Smith

I was never coming out of my dark room.

"Tyler, come to dinner right now" mom yelled from the kitchen. So much for that plan. My virtual bedroom was just starting to look cool, I couldn't stop now. I'd added glass-doors that opened onto stylized gardens, skylights, and a specially sprung wooden floor. Instead of a religious shrine, I'd added a media center with multiple screens and computer access ports.

"I'm busy with v-world" I said, and I really didn't want to go into the light.

"Now, Tyler!" dad said. He sounded mad.

"I'll come if you turn out the kitchen lights" I yelled back at him. This was met with silence. A few moments later, dad was at my door, rattling the knob.

"We've had just about enough of you, young man. Come out of there right now!" I couldn't bear to look at either one of them. The things I'd seen today would scar me for life.

"I'm not hungry" I said.

"This isn't a debate, young man," dad said "you come out here or I'm coming in to get you." I wouldn't put it past dad to break down my door. No doubt the kids at school would get a kick out of that too.

"Okay, okay." I rushed to the door before he could knock it down. "I'm coming." Stepping into the hall, I kept my head down.

"What's wrong with you?" he asked, grabbing my chin and lifting my head up. Ugh, more dirt for the kids at school. No matter what I did I was doomed. I almost darted straight back into my refuge.

Nano Meets Macro - Social Perspectives on Nanoscale Sciences and Technologies
by K L Kjølberg & F Wickson

www.panstanford.com
978-981-4267-05-2

"Where did you get that black eye?" dad asked.

"I dunno." I was no rat, and if I told, tomorrow would be worse than ever. I couldn't believe that bully Logan kicked my ass so easily. In front of Brianna—that was the worst! I never got my ass kicked in v-world.

Dad pointed toward the kitchen.

"What's wrong with you, Tyler?" he said as we walked to the table.

"Nothing," I mumbled, sitting down. "Can't we turn out the lights?" They ignored that request. At least mom's spaghetti looked and smelled good. I reached for the bowl. Mom grabbed my hand.

"Not so fast, young man. We're going to say grace." Oh, no. I'd get teased about that at school for sure. Why me? My life was hell. I kept my head down while they prayed. Finally they finished. I filled my plate and started ladling the food into my mouth. Things had changed. If I had to be out here in the light, it wasn't going to be for very long.

"What's going on Tyler?" mom asked.

Before I could say 'nothing' again, the kitchen computer pinged and an ad started, "Enjoy spaghetti? Then you'll love Homecookin's all new extra spicy marinara sauce." The Homecookin' jingle started.

"Huh" dad said. "That's a funny coincidence. The ads during dinner always seem related to what we're eating." I kept shoveling in the food. The computer pinged again and we heard mom's boss' voice:

"Ashley Morgan will be docked seven minutes pay for the extra time she took at lunch." Mom's brow wrinkled.

"How did they know?"

Duh, my parents were total idiots. At least I was just about done eating and hadn't done anything super embarrassing. Could it be I would escape dinner unscathed? Then we heard my principal's voice.

"Tyler Morgan report to detention tomorrow for the fight you were in after school." 'Crap.'

"Ah, ha," dad said. "I knew you were in a fight. What did we tell you about fighting?" He scowled as he looked for a cigarette.

"Good thing your principal found out about it and told us," mom said. She was an idiot.

"Good thing?!" I exploded. "Only an idiot wouldn't find out about it." The computer pinged.

"Smoking within city limits is illegal. Evan Morgan you have been fined one hundred new dollars."

"What the hell?" Dad had just inhaled his first lungful and he paused, savoring it, before letting it out. He stared at the computer. "How do they always figure it out?" I jumped up.

"You guys are idiots!" Mom and dad just looked at me.

"Don't you know what's going on here?" I demanded "Why do you think I wanted to turn out the lights? The nano-cams are everywhere. You can't escape them! They're probably crawling all over us right now. They see everything! And now the hackers have opened the stream online. The kids at school are already watching!" Mom closed her mouth and her face turned red.

"Everything? Surely not. You're just being dramatic Tyler." Her head would downright explode if she knew what I'd been forced to watch after school today.

"You're mistaken, young man," dad said. "The surveillance cams just watch criminals and airports."

"Get a clue, dad! They are everywhere!" I choked back a sob. "They even see when you have sex! It's all over the web, everything everybody does. For God's sake turn out the lights!"

I ran back to my room and slammed the door. There was no way I'd let the kids at school see me cry. In my place of solace I could cry all I wanted and no one would know. As I sat there feeling sorry for myself, I couldn't help noticing the band of light under my door went dark. I fell asleep listening to my parents arguing.

After school the next night I went straight to my room and worked on my v-gardens. Among the plants I inserted a bunch of planets and stars of different sizes. I did peek at my dad via the surveillance cams and he was at the computer in the kitchen checking out someone else. When my mom came home I found out who dad had been surveilling: mom. They had a huge fight about it. I wouldn't come out to eat and they didn't make me. I don't think they even ate dinner.

The next morning, they forced me to go to school but that night they had the biggest fight ever. All their stupid noise made it hard to concentrate on v-world. I'd been meaning to add some neat re-sizing algorithms so I could shrink down and go to the planets in my v-garden, but instead I'd surveilled dad surveilling mom. When I caught a glimpse of her kissing some other guy and both of them turning red in the face, I quickly switched away.

When she got home, mom went crazy:

"I gave you something to watch, you perv."

"How could you?" dad screamed. It went on and on. I turned up the volume on my headphones but still didn't make any progress in v-world. I'd thought that was as bad as it could get, but I was wrong. It turned out that that man with mom was her client and she got fired because of what they did. My parents screamed at each other every night. Until they stopped screaming.

One night mom stood outside my door and tried to get me to let her in, but I wouldn't. She started talking and crying through the door, something about how she and dad had had problems for a while and she was leaving, but I turned up my music. I was mad at her for being such an emotional loser. Mom moved out and I never saw her again. There was no way I was going to surveill her - who knew what the hell she'd do. A couple of months later, dad forced his way into my room, crying,

"She's gone. She killed herself." He sank down on my bed. "I can't believe it." He rocked back and forth, crying, and repeating "I can't believe it. I can't believe it." The light from the hall shone on us, so I was careful not to touch him. I didn't want to look pathetic, too, pathetic like dad, and super-pathetic like mom. What kind of loser offs herself right in front of everyone? I hated her.

Eventually dad pulled himself together enough to stop crying.

"Are you okay, Tyler?" He reached for me and I jerked back.

"I'm fine" I said.

"It's okay to be sad and even to cry because you miss your mom" he said. I didn't miss her; I hated her!

"Whatever."

Dad looked at me for a few minutes.

"What happened wasn't your fault, Tyler" he said

"I know" It was her stupid fault, her and her stupid client and my stupid dad. It was all their faults!

"I'm here to talk, whenever you're ready" dad said.

"Whatever" I shrugged.

Dad forced me to go to the funeral, saying he'd take away my computers if I didn't. He didn't understand anything. I had to surveill people to know if they were surveilling me. So, I went to the funeral. It was stupid with everyone wailing and crying like my dad. Why would they do that? Didn't they know people were watching? I was careful not to move or show weakness, and I definitely didn't cry. I just stared at the floor at the colored light from those stained glass windows. Dad had to poke me to get me to stand up when we were supposed to. When Brianna came over after the service and said she was sorry I just shrugged and kept looking at the floor. Having a stupid mother who offed herself was mortifying. I wasn't going to show any weakness. When I glanced up at Brianna though, she was showing weakness. Her chin was quivering and her face was white. A middle-aged woman clutched her hand. I looked away before I lost my cool too.

If there was one thing my mom's situation taught me, it was that r-people can't be trusted. So I wised up. It was better to interact with v-people.

Dad was never the same after the funeral. He'd sit in his room alone crying most nights. He still kept the light on though - I couldn't understand that. Once in a while, I'd surveill him and he'd be saying things like '...all my fault. Why didn't I give her another chance?' He was so pathetic I couldn't stand it for long.

One night, the fridge was empty and I went over to his room to tell him. That was a mistake. From the hall I could tell from the smell that he was not okay. I never went in his room. I definitely never surveilled what he did to make that smell. I texted the authorities and they came and took him away. Some lady, she said she was a social worker, tried to get me to come out of my room. She was crazy. Grownups really didn't get it. I refused.

It turned out I wasn't the only kid whose parents offed themselves and there weren't enough foster-whatevers for all of us. So I became a ward of the state. All that meant was I got to stay in my room almost all the time. Somebody delivered meals as long as I went to v-school and got decent grades. During all my boring schooling, I built up an awesome v-persona, a boxer/wrestler/karate champ, and won lots of tournaments. That was my passion. I practiced my moves almost all the time so I could accumulate skills and v-body enhancements. I almost never had to leave my room. It was sweet.

I was celebrating my latest tourney win in the Chatsubo, a favorite haunt after martial arts contests - gloating over the losers was part of the fun -, when a woman approached me and my v-buddies. It was odd enough to see a woman in the Chatsubo, but her v-persona looked human-normal. I hadn't seen a human-normal woman in v-world in, I didn't know how long.

"Hey, Tyler, is that you?" Her voice took me back to a day in church spent staring at the colored floor. "I caught the end of the tournament and thought I recognized you. Your v-face is almost the same as your r-face." My v-persona was b-u-f-f. I grinned. I was glad she recognized me after knowing the scrawny wimp I'd been in r-world.

"Brianna?"

She nodded, her long brown hair falling in front of her face.

"Can we talk alone?"

My buddies whooped it up and slapped me on the back.

"Sure" I said. Once we were out of earshot of my buddies, I asked "So, what's up? How are you?" She pulled a lock of hair behind her ear.

"Not so good. I'm sick of v-world. Don't you ever want to go out into the real world and interact with a real person?"

"No." I shook my head. "Why would I want to do that?" My mind skittered towards that night in the hall with that bad smell. "R-world sucks. V-world rocks!"

"But you still live in the neighborhood, don't you?" Brianna asked with an edge in her voice.

"Yeah," I said slowly. What was she getting at?

"Tyler, I need to see a real human being. Come on. Don't you?" Her chin quivered just like I remembered.

"No." R-world and the people in it had never done anything for me. My r-parents in particular were total losers. Good riddance. I was better off without them. Totally.

"But...are you just going to sit in the dark alone in front of your computer until you die?" It beat getting beat up or interacting with losers like my mom and dad. I shrugged.

"You're crazy." Her eyes filled.

"Whatever" I said.

"Fine." She glared at me and stomped away. It was just as well. I didn't need her or anybody else.

I never heard from Brianna again. That was the way I wanted it. Totally.

Questions for Reflection:

1. How would your life change if everything you did could be seen by anyone at any time?
2. To what extent do you have a 'v-world persona' (e.g. through facebook, myspace, blogs, online games, personal/professional homepages, google hits etc) and to what extent has it been consciously created?
3. Surveillance cameras are already widespread. Where do you think the line should be drawn in their use and application?

Periodic Table

Teresa Majerus

The work "Periodic table" is based on a TEM microscopic picture showing early γ'-precipitations in the superalloy Waspaloy (Patrick Majerus, Masters thesis, 1999). The creative process started with an idea of a garden that only slowly reveals its secrets, through smooth transitions from the microscopic to an imaginary real world. The link to science goes via chemistry where all the known elements have their well-defined place within the different groups of elements. The background was painted using acrylic wet-in-wet. The branches follow the shadows of the original nanostructure. The γ' precipitations were then drawn as fruits of the trees, representing at the same time chemical elements, their symbol being indicated on each. Every group of elements (e.g. nonmetals, noble gases) is further painted in a different color. We finally see a nano-chemical garden.

Teresa Majerus grew up and studied physics in Poland. She earned a PhD in natural sciences in Jülich, Germany. Since 2005 she has lived and worked in Luxembourg. Her scientific background in natural sciences does not contradict but complements her growing love for art. Her way to artistic cognition was guided by her passion for understanding the world, and searching for its beauty. She mainly paints with acrylics, where quick decision-making is needed, but also enjoys watercolors. Teresa's artworks mostly start with discovering or rediscovering natural elements from unexpected views. Regardless of the instrument chosen or medium employed, it is always the subject that defines the styles and techniques of her work. When she paints, her goal is to stimulate emotions. Every person who enjoys her work is a reward for her efforts.

Question for Reflection:

Do you believe that many stories can be hidden in microscopic pictures?

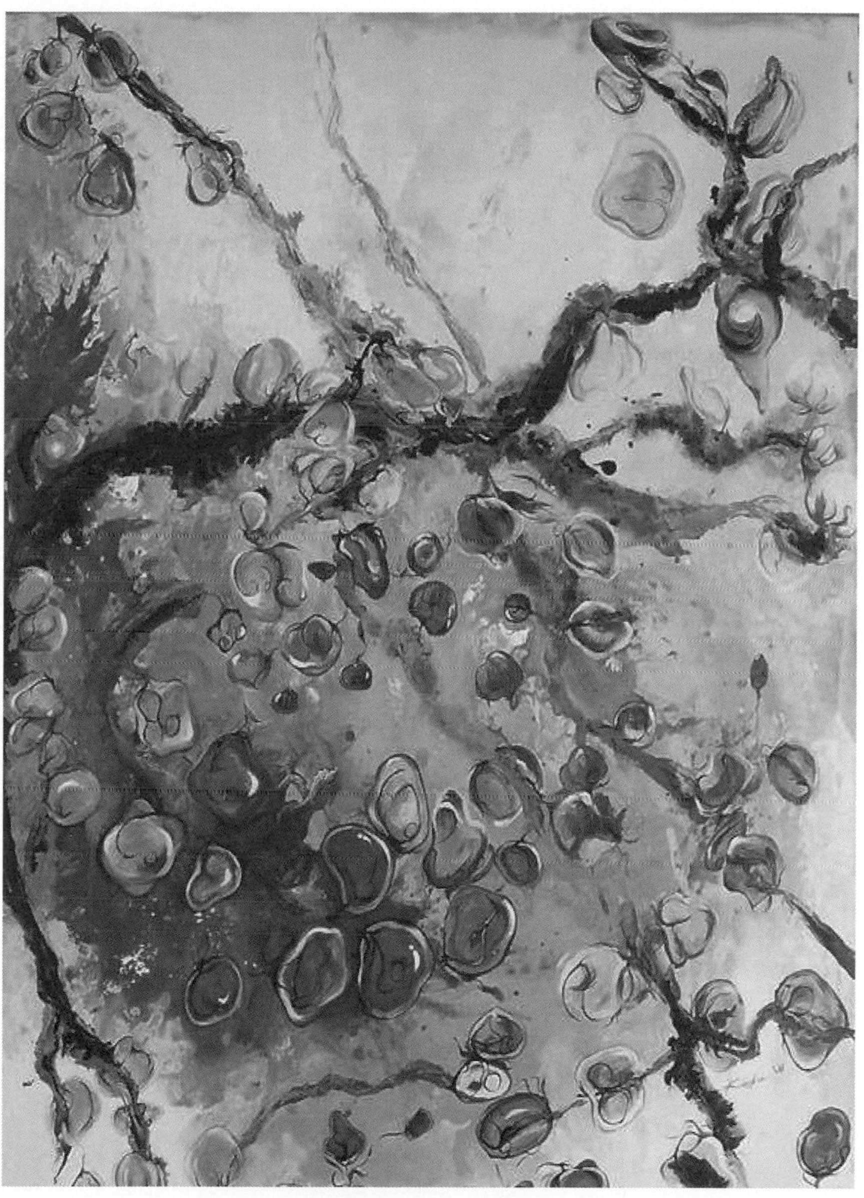

For colour reference turn to page 550.

Brainbots

Tim Fonseca

This image depicts artificial nerve cells inhabiting the spaces between biological brain cells – to improve memory, allow total virtual reality immersion, telepathic communication, the internet inside your head, etc.

In the second half of the 1990s I was introduced to the emerging technology community. I was excited and very enthusiastic about creating computer graphic artwork for these futurists. However, within a few years I began to become disenchanted, and even revolted, by the community's ultra-extremist, teeth-gnashing insistence for a materialistic answer to questions of meaning and origins: Where do we come from? Why are we here? Where are we going? One of many faith-based beliefs of this emerging technology movement is that human consciousness, in its entirety, can be digitized and downloaded into machines. Human beings are nothing but zeros and ones - our minds, our thoughts, and our emotions are nothing but mathematical algorithms. It reduces existence to mere 'Computationalism'. All living entities are merely biological mechanisms, with no spiritual dimension, soul, or psyche. Until the emerging technology community can rid itself of its radical materialistic reductionism, and learn to work side by side with religious, spiritual, and depth psychology communities, I fear it will lose force, credibility and respect. It has already lost my respect and my choice to contribute to it with my graphic illustrations. Technology is helpful, but it is not everything, and it is not the final solution to all human inquiry.

Question for Reflection:

How does reading this text affect the way you feel about the image?

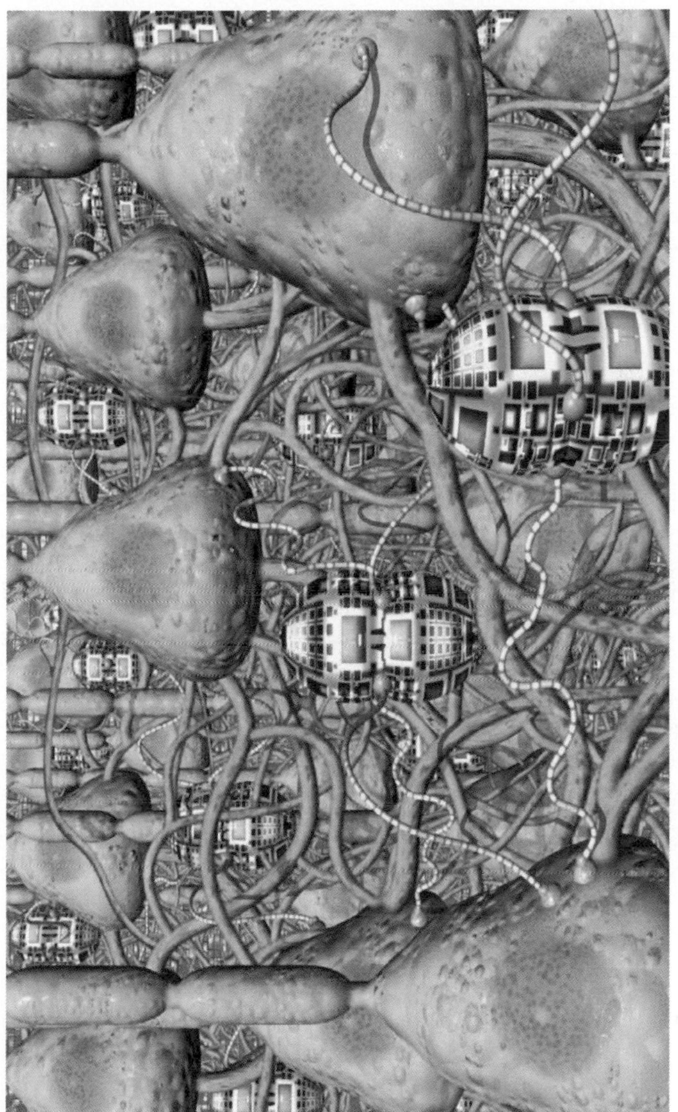

For colour reference turn to page 551.

Nano Meets Macro:

In the Big Questions

This section presents perspectives on the interface between nanoscale sciences and technologies and 'the big questions' relating to metaphysics, epistemology and ethics. In other words, what might be thought of as more philosophical issues around things such as the nature of reality, what it means to be human, the relationship between living things and machines, the role of humans in social and biological environments and so on. This node collects works that explore how actual developments and/or future visions of nanoST interact with these types of questions. Central concepts for this node include: personhood, human/machine relations, human/nature relations, the natural vs the artificial, morality, and ethical norms. All the contributions in this section have a specific focus on the 'big question' of how we do/should relate to technology, nature and each other.

In concrete terms, this section consists of three academic chapters, one short fiction story and three artworks. In chapter 11, 'Inside, Outside: Nanobionics and human bodily experience', Renée Kyle and Susan Dodds take a closer look at nanobionics being developed to overcome sensory deficits and neurological disorders. Drawing on phenomenology and feminist theory, they argue for the importance of the body as the locus of experience and engagement with the world and suggest that altering bodies through nanobionic devices may therefore change how the world is experienced and how the self is perceived. Kyle and Dodds also question whether the location of the device (inside or outside the body) could alter our moral response. In their conclusion, they consider the relationship between nanobionics and transhumanism and suggest that drawing a line between therapy and enhancement is particularly difficult because what it means to be human is constantly changing.

In chapter 12, 'Enhancing Material Nature', Alfred Nordmann suggests that the vision of enhancing the human body through the incorporation of technological developments is just a small and far distant possibility in a general and far more significant programme: that of enhancing material nature. This chapter provides a detailed exploration and analysis of the particular notion of 'with nature beyond nature', which Nordmann sees as permeating visions of nanoST. Through this chapter, Nordmann explores the dual concept of nature at work in this notion and points to how it dissolves traditional boundaries and categories. He particularly focuses on the significance of the way in which this notion simultaneously presents nature as technological and technology as natural. In this way, the chapter explores the understandings of nature and technology informing and driving the development of nanoST and asks what the apparent uncritical acceptance of a programme of enhancing material nature tells us about our hopes for the future and our perceived role in the world.

Chapter 13 is entitled 'The Nano Control-Freak: Multifaceted strategies for taming nature'. Here Arianna Ferrari examines the idea of control in relation to nanoST. Following Bensuade-Vincent, she describes two cultures at work in nanoST: that of the engineer and of the chemist. She relates these two cultures to the difference between top-down and bottom-up strategies for nanotechnology, and to the famous debate on this between K. Eric Drexler and Richard Smalley. Ferrari argues that across these two cultures, we see different notions of control; the engineering approach seeking control through mastery and the chemistry approach seeking control through a construction process allowing for and embracing surprise. Ferrari, however, argues that what the two approaches importantly share, is a vision of nature as a machine, an instrumental resource, and a 'plastic engineer' able to be reshaped from the outside and continuously reshaping itself from within. In her conclusion, Ferrari points to some of the ways in which these ideas can be seen to shape social and ethical debates and suggests that there is a relationship between the perception of material plasticity and an idea of social plasticity in these debates.

The short story 'It's Perfect and I Want to Leave' takes place in an imagined future in which many currently proposed uses of nanoST have become embedded in daily life. In this story, Lupin Willis uses a conversation between characters from different generations to explore what it is that we value in the way we live our lives, and how this can shift through time – particularly through the uptake of new technologies. Exploring big questions about what we mean by 'progress' and 'perfection' and how this underlies 'human/technology/nature relations', she encourages us to consider the desirability of different technological trajectories.

In her artwork 'The Metamorphosis of Forms', Giuliana Cunéaz asks the big question of the nature of reality through highlighting the way in which she sees recurring patterns across different levels. Seeing the cosmos reflected within the microcosm, she presents the world of the nanoscale as filled with flowers and forests. Through the ambiguity of her artwork, she challenges the boundaries we create between concepts such as the conscious/unconscious and the visible/invisible.

In the artwork 'The Crucifixion of Nemesis', Lukas Sivak depicts the persecution of the Greek goddess of limit and moderation. In his written presentation, Sivak reminds us that without limit nothing would exist and suggests that our modern insistence on raging against the imposition of limit, represents a crucifixion of Nemesis that can only result in our own eventual pain and suffering. Through this artwork, Sivak stimulates us to reflect upon the hubris of the endless human pursuit for greater control and the extent to which nanoST contributes to the crucifixion of Nemesis.

Jack Mason uses his artwork 'Entanglio' to question how we gain knowledge of a nanoscale reality. In doing so, he is touching on cross-nodal themes such as the production of imagery of the nanoscale and the representation of this scale in a public sphere. Mason suggests that in his art he works like a conduit helping to rearrange immortal atoms into new and particular forms of expression. In this way, the artwork engages big questions around differential experiences of time and the role of mortal beings in manipulating fundamental constants of matter.

While this section contains a number of important perspectives on nanoST in relation to 'Big Questions', there are some fields of interest that we were unable to include. This includes perspectives on whether nanoST raises any fundamentally new ethical questions or simply engages old debates in new ways; how nanoST may be thought through in relation to particular ethical theories or frameworks; and the role of moral norms vs moral imagination in approaching ethical questions for nanoST.

NanoST, like all new fields of science and technology, requires consideration in terms of its potential social and ethical ramifications. As this particular field works with fundamental elements of matter, it can't help but also engage a number of fundamental philosophical questions and deeply ingrained social categories. All of the contributions in this section present a perspective on how nanoST impact and/or reconfigure basic concepts, such as those of nature, technology, humanity, as well as alter our perceptions and experiences of the relations between them. In engaging 'the big questions', these perspectives allow us to see the potential relevance and importance of nanoST as existing not only in their power to reshape our physical world, but also our mental world of concepts, categories and understandings. In working at this level of fundamental questions, it makes sense that contributions in this node can also often have relevance for work in other nodes for interest. For example, when thinking about how to govern nanoST, it may be wise to consider the role of ethics in governance practices as well as the desire and/or ability to govern things such as perceptions of reality/nature/self in addition to processes causing physical harm. For the node interested in how nanoST are presented and understood in the public sphere, it can be worthwhile to look to how fundamental concepts such as nature, humanity and the role of technology in our lives both shape and are shaped by particular representations. Finally, it can be important to consider how our fundamental concepts, categories and approaches to understanding the world are both employed and challenged in the making of nanoST itself.

Chapter 11

Inside, Outside: Nanobionics and Human Bodily Experience

Renée Kyle & Susan Dodds

Developments in nanobionics promise to provide people with a range of impairments (including physical disabilities and neural conditions) with new treatments or functional prosthetic devices. In this chapter we wish to explore some of the ways in which individuals offered these new technologies may experience them in relation to their self-understanding and engagement with the world, and some of the ways in which the availability of these nanotechnologies may shape societal understanding of people who have impairments and the value of medical nanobionics. Drawing on theoretical work in phenomenology, we ask whether those medical nanotechnologies that are external or 'wearable' may differ in their ethical and social significance from those that are internal or implantable; how use of implantable devices may shape our self-understandings; whether the developments in these technologies may yield new modes of human perception and sensory experience; and whether these amount to 'transhuman' transcendence of human limitations or more familiar extension and gradualist change in the human condition.

Nano Meets Macro - Social Perspectives on Nanoscale Sciences and Technologies
by K L Kjølberg & F Wickson

www.panstanford.com
978-981-4267-05-2

1. Introduction

When many people discuss bionic devices, visions of human-machine hybrids familiar to science fiction often present themselves in their minds: like the title character in the US television series "The Six Million Dollar Man" (in which an American military officer who is seriously wounded is turned into a 'super-agent' courtesy of a range of bio-mechanical implants that enhance vision, hearing, speed and endurance). Largely encouraged by science fiction, the image of cyborgs – also known as post-humans – symbolize a new understanding of what it means to be human. Whilst there is a significant literature on what developments in science and technology may mean for humanity as a race (for example Haraway 1991), there is little discussion on the impact of such technologies on a person's bodily experience, knowledge and self-understanding. In this chapter we will ask, and answer, the following question: How might nanobionics and nano-wearables shape human perceptions, experiences and understandings of self and the world?

Our objective is to demonstrate that the body plays a significant role in a person's engagement with the world, and that changes to a body and its capacities – through implantable or wearable technology – will impact the structure and meaning of that person's experience of, and engagement with, the world.

2. Nanotechnology and Bionics

In developed countries around the world, research on applications of nanotechnology in health is progressing rapidly. The emerging field of nanobionics – the merging of biology, mechanics and electronics using recent advancements in nanotechnology and neuroscience – shows enormous potential in improving the health and wellbeing of people who have specific sensory deficits (loss of sight or hearing) or neurological disorders (nerve damage, epilepsy). Materials that can conduct electricity, such as carbon nanotubes and organic conducting polymers, are increasingly being integrated into the design and development of bionic devices to achieve better communication between living nerve cells and bionic devices (Wallace & Spinks, 2007). This means that these nanotechnologies and nanostructures may improve the ability of recipients of bionic devices to get the intended benefits (eg improved hearing or sight, or better nerve function) of their devices. It is possible that some future developments in medical bionics using advances in nanotechnology and neuroscience may allow for the creation of devices that will allow people to have sensory experience, perception or capacities beyond the normal human range. These future developments would constitute enhancement technologies.

The cochlear implant – commonly known as the bionic ear – is perhaps the most successful bionic device. The multiple channel cochlear implant (Clark *et al.* 1978) was developed to assist deaf people to understand speech and participate in conversation. Used by thousands of people around the world who are profoundly deaf or severely hearing impaired, the cochlear implant circumvents deafness by bypassing damaged portions of the inner ear and directly stimulating auditory nerves. The cochlear implant uses an external microphone, speech processor and a transmitter to send signals to the device implanted in the cochlea, which then stimulates the auditory nerves, sending a signal to the brain which is interpreted as meaningful sound. The implant does not allow a deaf person to actually hear, but it does provide them with a simulated sensation of speech. The cochlear implant is neither a wholly internal nor wholly external bionic device: while the receiver/stimulator is implanted beneath the person's skull, in the cochlea, there are also external components (microphone, speech processor etc). Currently, some developments are underway to implant all components beneath the skin (Zenner & Leysieffer 1998; Miller & Sammeth 2008, p. 331).

Although the multiple channel cochlear implant originated some time ago and is not a 'nanotechnology' as such, improvements to the interface between an individual's nerves and the implant's electrodes has been facilitated by nanoscience. For example, by coating the implant's electrodes with a nanoscaled coating of polypyrrole (a plastic polymer with electro-conducting properties that can be activated within the body), the electrodes are able to deliver neurotrophins ('helper' molecules that enable a neuron to network with its neighbours) to the inner ear while also delivering an electric current through nanoscaled electronics

(Richardson *et al.* 2007). This indicates that such coatings can rescue neurons from dying after deafness and also protect them from further damage that may occur when the implant is inserted.

Advances in these smart plastics are also enabling researchers to develop other kinds of bionic devices. At the Bionic Ear Institute (BEI) in Australia (as well as other labs), polymers are being used to create intelligent scaffolds that can deliver neurotrophic factors to aid the functional repair of damaged nerves, including damaged spinal cord nerves (BEI 2007). Gaining insight into how nerves regenerate with these intelligent polymer scaffolds may lead to the development of a 'bionic spine' (BEI 2007). The upshot of this technology is that the bionic spine may enable those people with limited mobility to regain movement and function in their legs.

In a similar vein, several research groups around the world are researching the development of a bionic implant with the intention of restoring reading vision to people suffering from a range of eye diseases or injuries (for example Ellis-Behnke *et al.* 2006). Like the cochlear implant, the bionic eye is comprised of both internal (electronic chip) and external (video camera, processor) components. It consists of an array of nanoscaled electrodes implanted on the surface of the retina to stimulate the optic nerve in the eye. These electrodes act as a substitute for the photoreceptors that have been badly damaged or have degenerated and no longer function effectively. The visual image is then captured on a miniature video camera and is processed, then transmitted wirelessly to the bionic eye through an ultra-high frequency radio, after processing to select for the target image (for example, human faces, or large objects, depending on whether the aim is to assist the ability to recognise other people or to assist in negotiating the physical environment). By electrically stimulating the appropriate electrodes, the optic nerve is activated and it transmits the signal to the visual cortex of the brain. Figure 1 provides a visual representation of how the bionic eye will work.

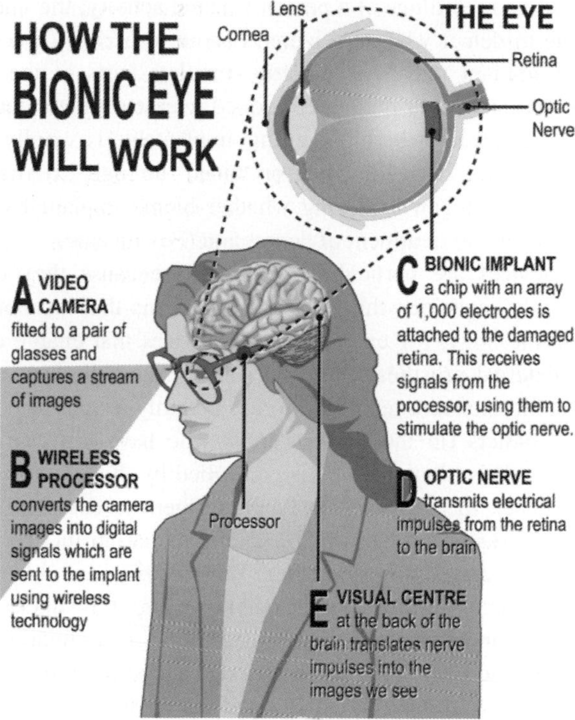

Figure 1: The bionic eye (BEI 2008).

Advances in nanoscience are not only enabling scientists to develop devices for those with sensory disabilities. Individuals with neurological conditions, such as Parkinson's disease, epilepsy and stroke may also benefit from medical applications using nanoscience. A neurological condition that is characterised by abnormal electrical activity in the brain, epilepsy, produces seizures in individuals that range from mild to severe, and they can significantly impact a person's quality of life. A multidisciplinary research team in Australia is investigating how directly stimulating the brain using an electronic implantable employing nanoscience may allow the recognition, and control, of seizures (BEI

2007). Based on recordings of a person's neural activity, the implantable may be able to detect when a seizure is about to occur. To control the seizure, the implant may then directly stimulate parts of the person's central nervous system or deliver controlled release of therapeutic drugs. Unlike the bionic ear or bionic eye, this implantable is wholly internal, although the device is manipulated or 'tuned' through external means. Researchers have begun exploring whether bionic implantable devices may also assist in the treatment of depression (Fox no date)

Bionic devices are particularly interesting because they change a person's bodily capacities through artificial means that are both within and outside the body, for example, those devices that enable cognitive control of external prostheses. In the United States the Department of Defense commissioned a team to develop a highly advanced bionic arm for use by soldiers (in the first instance) who have lost one in battle (Adee 2008). This bionic device is controlled by thought and provides sensory feedback to the recipient, enabling them to complete complex motor-neural tasks, such as pulling a credit card from their pocket and stacking cups by controlling grip force. These devices are not commonly available outside of military contexts, however, it is likely that the application of nanotechnologies into non-military rehabilitation settings will follow. At the same time, the knowledge and applications developed for rehabilitation purposes may also have further military purposes, for example, the bio-mechanical premises of the bionic arm may also enable military scientists to develop highly sensitive devices that will allow personnel to disarm explosives from safer distances.

It is clear that the range of bionic devices can be adapted beyond their original purpose, as in the case of the bionic arm being extended to use in disarming explosives, or the potential to develop brain stimulators that override normal psychological function (memory, sleep or emotional responses). For this reason, some critics are concerned about the risks associated with the 'dual use' of research intended for health. The literature on dual use emphasizes the social and ethical obligations of researchers to ensure that their research (developed with peaceful intention) does not fall into the hands of governments or terrorists who may misuse the research for malevolent purposes (Miller & Selgelid 2007).

2.1 Inside/outside: Implanted and wearable bionics

Technologies such as the bionic arm and bionic eye are designed with the intention of redressing certain physical or sensory disabilities using implanted devices. There are, however, a number of other devices incorporating nanotechnology that are external or 'wearable' that can provide aid in physical recovery, monitor bodily states or serve as assistive devices for people with disabilities. For example, the 'intelligent knee sleeve' (Munro *et al.* 2008), developed for use by people who have suffered knee damage due to sporting injuries, is a biofeedback device that monitors the wearer's knee joint motion. As the wearer's knee bends, the electrical resistance of the fabric sensor changes, and when the knee joint reaches the correct angle for safe landing, the signal processor beeps, so that the wearer knows when their knee ligaments are under potentially damaging stress and can learn through practice to jump, tackle and run in a way that avoids risk of damage. Currently being trialled by the Australian Football League to help its players avoid knee injuries, the knee sleeve has the potential to prevent cruciate ligament injuries sustained by a range of sportspeople, potentially saving significant dollars in health care costs.

Other wearables that incorporate nano-polymer technologies are used as artificial skin for improved wound care, or patches to monitor bodily states (eg blood pressure or glucose levels) and may be combined with drug delivery devices. In these cases, the nanotechnology uses the skin and changes to the skin surface as a means of introducing drugs into the body and as a source of information about what's going on in a person's body (Lu *et al.* 2008).

As this discussion of the range of nanotechnologies used in bionics suggests, some devices are primarily external to the body and others are more fully integrated in sensory organs, brain or neural system. Most current and future applications of nanotechnology as they relate to health are not easily categorised as either wholly internal or external to the human body; rather, they exist on a continuum. One question that arises in assessing the ethical significance of bionics is whether the location of the device (on the skin, or under the skin) affects our moral responses. This question is made more complicated because implanted devices frequently have an external component, and wearable devices permeate the skin to either sense body states or deliver therapeutic agents or stimulation.

3. Phenomenology

For those interested in our lived bodily experience (i.e. how we understand, perceive and come to know the world through our movement in it), bionic devices are particularly interesting. This is because they may alter a person's bodily capacities – and hence their perception of the world – through artificial means that are both within and outside the body. To understand why this is philosophically significant, we need to explore how people's bodily movement with the world and their experience of the world shapes them and the ways they understand and conceptualise themselves and the world. One of the fields of philosophy concerned with this question is phenomenology. With its origins in the 19th and 20th century works of Edmund Husserl, Martin Heidegger, and Maurice Merleau-Ponty amongst others, phenomenology is concerned with our being-in-the-world, with the experience of being a subject in the world (i.e. a thinker, perceiver, experiencer), and how being a perceiving subject who makes meaning of her experiences differs from being an object in the world. As such, it can provide us with an important and unique insight into how nanotechnologies that are both inside and outside the body may change a person's engagement with and experience of the world by changing their bodily capacities, perceptions and experiences.

In the *Phenomenology of Perception* (1962), Merleau-Ponty articulates a theory of embodied subjectivity. Embodied subjectivity concerns what it means to be a person or self, given that we have human bodies, and that we perceive the world, and gain knowledge through our bodily engagement as humans with the world (sight, touch, hearing, taste) and the movement of our bodies through the world (that is negotiation of the physical environment, exploring spaces and so on). Further, we are subjects because we are able to direct our attention to different aspects of the world to control our focus on different bodily experience. In this theory, human consciousness, the world and the human body are intertwined and mutually engaged. Meaning is constructed through bodily involvement with the world, experiences and shared meanings. Our conceptualizations and understandings of the world can be traced through bodily capacities. The body is unique in that it both shapes and is shaped by our experience in the world, so it is both an object that we sense and the subject of sensory experience. In other words to be an embodied human is to be both 'sensible' – something that can be perceived by other objects in the world – and 'sensing', a perceiver of experience. It does these things simultaneously.

> *Our body, to the extent that it moves itself about, that is, to the extent that it is inseparable from a view of the world and is that view itself brought into existence, is the condition of possibility, not only of the geometrical synthesis, but of all expressive operations and all acquired views which constitute the cultural world.* (Merleau-Ponty 1962, p. 388).

The body is the locus of our *being in the world* (Merleau-Ponty 1962, p. 78-81).

A person's understanding of her self in relation to the world requires her to have a view of her body as both part of what is her, but also part of the world which provides her with the point of contact between self and world. In this way, embodied subjectivity involves both, what Gallagher (2005) terms body image (roughly, a person's mental image of their body) and body schema (how their body moves through the world). Body image, then, involves the person's perceptual experience of her body; her conceptual understanding of the body in general; and her emotional attitude towards her body (Gallagher 2005, p. 25). It is, in short, how she sees her body. Body schema, on the other hand, does not involve perceptions or beliefs; it is a system of sensory-motor capacities that function without awareness or the necessity of perceptual monitoring (Gallagher 2005, p. 26). Human cognition or understanding of bodily experience relies on a scaffold of meaning-making experiences and responses that have developed through experience of bodily perception and the world. Knowledge comes from the engagement of the body in purposeful movement in the world and reflective experience. As such, phenomenologists view human being as inherently involving striving to make sense of our selves and the world around us through our bodily engagement with the world. Therefore, changing our sensory perception of the world or disruptions in the relationships among our experience, perception and shared conceptualisations of significant features of the world may change not only what we know but also how we make sense of the world and ourselves.

3.1 Phenomenology and bodily difference

Taking the phenomenological view that the body is the locus of our experience and engagement with the world, feminist theorists would seek to explain how bodily differences, and social responses to these bodily differences, shape the ways that individuals understand themselves, each other and engage with the world. According to Iris Marion Young (2005), patriarchal culture profoundly influences the way women and girls think of, and use, their bodies. Women, in Western cultures, have not been encouraged, traditionally, to use their full body capacities, with their bodies presented as existing for others (as objects for sexual pleasure, or for the reproduction of the species) and not as sites of action and agency. The effects of these expectations about how women can or should use their bodies are often internalized and become manifest in how women use their bodies in engagement with the world. For example, women tend not to fully extend their limbs when engaging in athletic activities such as throwing or meeting the motion of a ball, because of oppressive social attitudes towards the capabilities of women's bodies and how much space such bodies should occupy. "… we are physically inhibited, confined, positioned and objectified. As lived bodies we are not open and unambiguous transcendences that move out to master a world that belongs to us" (Young 2005, p. 42-43).

Young argues that a person's body particulars shape her bodily comportment, spatiality, and performance of tasks (Young 2005). For example, to become a gymnast a person needs strong leg muscles and good lower body flexibility, but if she lacks bodily co-ordination, her capacity to successfully engage in gymnastics is limited. To become a fully integrated self, a person has to accept the material and cultural reality of her physical existence; that she not only has a body, but she has *this* body, a body which has specific limitations and possibilities and that the significance of these are shaped by social and cultural values. Young also draws attention to the ways in which societies 'scale' bodies, that is bodies that have certain socially valued characteristics (often associated with race, class, gender, sexuality and ability) are ranked as being more desirable than those that lack those characteristics.

Young's understanding of the ways in which individuals' bodily differences shape and are shaped by social structures extends the phenomenological account by clearly acknowledging how social relations, culture, history and social structures intersect with individual subjectivity and physical reality. In relation to people with physical disabilities this is most evident.

> *In so far as it makes sense to say that people with disabilities are a social group, for example, despite their vast bodily differences, this is in virtue of social structures that normalize certain functions in the tools, built environment, and expectations of many people* (Young 2000, p. 98).

When a person is physically disabled, her possibilities for bodily action, gesture, movement are restricted. For some people with disabilities, "[t]he world is experienced as overtly obstructive, surprisingly non-accommodating. Actions are sensed as effortful, where hitherto they had been effortless" (Toombs 1995, p. 15). It is also the case that disabled bodies are understood through social encounters, structures and relationships that may further restrict the subjectivity of the person who has the physical impairment. (Seymour 1998, p. 51). The visibility of a disabled body and its restricted capacities increase the vulnerability of a person to the scrutiny and appraisal of others, and may further socially disadvantage people in these circumstances (Seymour 1998, p. 54).

So, why is phenomenology of interest to discussion of nanobionics? At face value, bionics promises to extend the embodied possibilities of people with sensory, neurological or other impairments, and so would appear to enhance human embodied subjectivity: a good thing. However, the story may be more complicated. In the final section of this chapter we will look at some of the ways in which bionic devices can both enhance and threaten our sense of agency because of the social and embodied aspects of human subjectivity.

4. Implications of Phenomenology for Understanding the Moral Significance of Bionics

In this section, we will explore the implications of bionics for human embodied subjectivity, informed by the phenomenological account provided above. We make three kinds of claims. The first concerns the balance of the potential benefits to people with illness or disability from bionics when compared with the limited resources available to those for whom nanobionics are not available (whether due to inability to afford them, or conditions for which bionic devices are not developed or appropriate). Second, we consider the impact of nanobionics on a person's sense of agency or embodied subjectivity (her body image and body schema), suggesting that the ability of nanobionics to improve a person's life may depend (in part) on the degree to which the device is experienced as integrated into the person's bodily capacities. Third, we consider the significance of nanobionics for a trans- or post-humanist world, one in which limits on human capacities are indeterminate and more directly open to individual choice and control.

4.1 Disability and disadvantage

The use of nanotechnologies in medical bionics can provide people with sensory, neurological or other impairments with new opportunities for bodily engagement with the world, and hence can enhance subjectivity. Technologies that assist in communicating with others, improving mobility, and increasing personal autonomy and control over one's body and life (amongst other things) can play an important, and positive, part in experiences of the lived body. Their use may mean less dependency on others for personal care and more spontaneity in physical movements, issues which greatly affect, and generate stress in, some individuals with physical disabilities (Seymour 1998). The increased bodily possibilities that nanobionics offer may allow some people to feel like they can transcend their relative impairment, which is indeed valuable.

At the same time, however, the use of technologies as a solution to a complex social, physical and individual problem may displace a significant social concern: how people with disabilities are regarded, disvalued and marginalised. Being disabled is not simply a matter of being physically impaired; rather, a person's understanding of her disability is shaped by a complex interaction of one's bodily particulars and social attitudes towards that disability. For example, a person who uses a wheelchair for mobility may not consider herself disabled until she is unable to access a building because it does not provide ramps. At that point it is not clear whether the limitation should be understood as bodily or engineered. Our assessment of what is normal, disabled or enhanced in relation to human capacities is, inevitably, informed by cultural expectations and values attached to activities.

> *[P]romises to eliminate the biological causes of disability assume that some culturally neutral, biomedical definition of disability can be agreed upon, and they obscure the fact that disability is socially constructed from physical and mental difference.* (Wendell 1996, p. 84).

Implantable devices may 'invisibilise' disability by creating a "technological fix" and may exacerbate the relative disadvantage of those who do not have access to the technologies.

4.2 Subjectivity: Alienation and integration

Nanobionics can also more directly affect embodied subjectivity because of the ways in which the technologies are experienced (or the ways in which experiences are perceived as a product of the technology or the person's own bodily processes). Nanotechnologies that provide responses to chronic conditions like sensory impairment or neurological disorders like epilepsy, or psychological conditions like depression, may be alienating or integrating of a person's subjectivity, depending on whether the person does (or can come to) experience the technology as simply an extension of her body schema and can incorporate the technology into her body image or not.

Where the technology is implanted – hence less visible or cumbersome and more readily integrated into a positive body image – and is (perhaps after a period of practice) experienced as functioning seamlessly with a person's body schema, these technologies support a person's sense of agency and integration in the world. However, these technologies do not always work this smoothly and are not as readily integrated, hence they may lead to an ambivalent agency: the person feels both that she can achieve her goals through her bodily control, and feels at the same time that she cannot do so confidently or without awareness of the mediation of the technology (which may be exacerbated by social responses to the visible aspects of the technology, or her perception of the device as 'not quite' simulating the 'real thing').

In the case of implants that affect brain processes, the perception by the agent that her thoughts or brain are being interfered with may undermine her subjectivity and autonomy (her actions, reactions, emotions may appear to be the product of the technology, not genuinely 'hers'). However, for someone who has experienced significant epileptic seizures, or major depression, for example, the condition being treated may also be experienced as external, alienating and threatening to her understanding of herself as an agent – for example if she is unable to pursue her preferred job because her uncontrolled seizures make doing the job impossible. A person may feel degrees of control, integration and identification with the condition or the technology. As such, her subjectivity may be (to greater or lesser degree) ambivalent and her bodily intentionality (her confidence in her ability to engage her body with the world, the bodily 'I can') inhibited (Young 2005, p. 38). It would be a matter of judgement by the individual whether, all things considered, the technological response improved her embodied subjectivity: her sense of herself as able to engage her intentionality effectively with and in the world.

It is not surprising, for example that there is a substantial group of people with profound hearing impairment who have cochlear implants, but who turn the devices off or decide not to use the device some or all of the time (Wheeler *et al.* 2007). For some people, even after they have used the device for a period and have improved their brain's ability to discern meaningful speech from the processed stimuli, the device is experienced as an impediment to subjectivity, rather than an aid. While the technology is improving, the auditory experience of those who use cochlear implants is still a distant second best to normal hearing and requires a level of concentration and focus that can be draining and often presents a barrage of noise from which the individual has to struggle to pick out the intended communication (Wheeler *et al.* 2007). Just as a person may give up on attempts at holding a discussion in a noisy restaurant or subway station, so may the deaf person elect to do without the intrusion of using the cochlear implant and some may decide that they would prefer not to pursue using the device at all.

Wearables that may be more readily removed can be less challenging to subjectivity, because a person can control directly the use of the device.

4.3 Transhumanism and extending the limits of human capacities

The third aspect of nanobionics that we want to consider is the implications of the development of these technologies for the future of what we understand as human-ness. The evidence from phenomenology suggests that the self is mutable (that is, there is a significant capacity for our idea of who we are and what we can do to change over time, inresponse to social and environmental conditions) and cognitive science that the human brain is plastic (that is, neural structures are able to change in response to stimuli and the environment). In light of that evidence, it would be a mistake to assume that what it is to be human is unchanging and fixed. Therefore, some of the concerns raised by those who view nanobionics as the first step down the slippery slope to a world in which human nature has been left behind, and our capacities and potential have become wholly within the control of humans and their choices, are misplaced.

Some authors seek to defend a strong distinction between 'therapy' (treatments aimed at restoring normal human function or protecting health) and 'enhancement' (interventions that extend the range of human capacities beyond what is currently understood as "normal" whether understood as a statistical norm or some idealised notion of human flourishing) (Juengst 1997). This distinction has been useful in promoting technologies that redress health problems while stopping short of those thought to be mis-used when directed towards improving on nature or normal biological function. However, it may prove difficult to use this distinction consistently, given that what we now take to be normal or acceptable interventions in health (such as antisepsis, vaccination, immunisation, transplantation) reflect significant enhancements to human capacities over what were once assumed to be the limits of the human condition. What is considered to be normal healthcare and therapy is in part a product of social, historical, and technological conditions: we accept as normal what we know and can understand.

If the therapy/ enhancement distinction is not tenable, then does that commit those who support the development of nanobionics to transhumanism: or the unlimited enhancement of human capacity? 'Transhumanism' or 'posthumanism' is the view that human capacities and potential can be disarticulated from natural or normal biological endowment. Transhumanists look forward to a future in which individuals or parents can choose their (or their children's) mental and physical capacities, limited only by the state of the technology.

> *Ultimately, it is possible that such enhancements may make us, or our descendants, 'posthuman', beings who may have indefinite health-spans, much greater intellectual faculties than any current human being – and perhaps entirely new sensibilities or modalities – as well as the ability to control their own emotions.*(Bostrom 2005, p. 203)

However, the phenomenological approach discussed above suggests that there are greater limits to the plasticity of subjectivity and neural capacity than the transhuman optimists suggest. The roles of body image and body schema in our understanding of ourselves as agents in the world and the relationships among these and our social environment point to the view that our ability to incorporate bodily change in our self-understanding will be influenced by our ability to experience these as within the scope of our current (and emerging) body image, schema and social relations. While change may occur in a way that enhances subjectivity, this is more likely where gradual or incremental changes are slowly integrated into the individual's understanding of her self, as opposed to radical or extreme changes which may be disruptive, alienating or simply incomprehensible to a sense of subjective agency. For the changes to effectively extend subjectivity, they will need a degree of internal and external supports through social relations, neural conditioning, and/or an integrated body schema.

5. Conclusion

Nanobionics show promising potential in improving the health and well-being of individuals suffering from a range of illnesses, diseases and disabilities. However, as we have argued, these devices also present a risk to individuals, societies and the future of human subjectivity. We have attempted to draw on work in phenomenology to demonstrate the significance of the body, perception and social relations for subjectivity and agency. On this view bionic implants may contribute to increased potential for subjective control and support bodily engagement with the world, enhancing the subjectivity of people whose bodily engagement was previously limited by physical or neural impairment. At the same time, nanobionics (whether implanted or worn) may contribute to inhibited intentionality, limited engagement and social and self-alienation. Finally, we have shown that the possibilities for transhumanism may be less disturbing than previously thought because of the ways in which social, neural and subjective change relies on gradualist and integrative steps.[1]

Questions for Reflection:

1. Does radical technological alteration to our perceptual apparatus change what we can know and who we can be?
2. How could the incorporation of technological devices into a person's body, change what it means to be an 'embodied subject'?
3. Can you draw a line between therapy and enhancement? To what extent does the existence of this line rely on a stabile and universal concept of what is 'normal'?

[1] The research for this paper was supported by Australian Research Council funding of the Australian Centre for Excellence in Electromaterials Science (ACES) CE0561616.

Bibliography

Adee, S. (2008). A "Manhattan Project" for the Next Generation of Bionic Arms, IEEE Spectrum Online. http://www.spectrum.ieee.org/mar08/6069 (last accessed Apr 1, 2009).

Bionic Ear Institute (2008). *22nd Annual Report 2007-2008. http://www.bionicear.org/bei/annualreport.html* (last accessed Apr 1, 2009).

Bostrom, N. (2005). In defence of posthuman dignity, *Bioethics*, 19(3), pp. 202-214.

Clark G. M., Tong Y. C., Bailey Q. R., Black R. C., Martin L. F., Millar J. B., O'Loughlin B. J., Patrick J. F. and Pyman B. C. (1978). A multiple-electrode cochlear implant, *Journal of the Oto-Laryngological Soc. of Australia,* 4, pp. 208-212.

Ellis-Behnke, R. G., Liang, Y. X., You, S. W., Tay, D. K. C., Zhang, S., So, K. F. and Schneider, G. E. (2006). Nano neuro knitting: Peptide nanofiber scaffold for brain repair and axon regeneration with functional return of vision, *PNAS*, 103(13), pp. 5054-5059.

Fox, R. Submission to the Review of the National Innovation System, St Vincent's Hospital, Melbourne A Centre for Medical Bioengineering Appendix B. http://www.innovation.gov.au/innovationreview/Documents/334St_%20 Vincent's _%20Melbourne%20(R).pdf (last accessed Apr 1, 2009).

Gallagher, S. (2005). *How the Body Shapes the Mind,* New York: Oxford University Press.

Haraway, D. (1991). *Simians, Cyborgs and Women: The Reinvention of Nature,* London: Free Association Press.

Heidegger, M. (1962). *Being and Time*, J Macquarie (trans.), New York: Harper and Row.

Husserl, E. (1962). *Ideas: General Introduction to Pure Phenomenology*, WRB Gibson (trans.), London: Collier Books.

Juengst, E. T. (1997). Can enhancement be distinguished from prevention in genetic medicine? *Journal of Med. and Phil*, 22(2), pp. 125-142.

Lu, J., Do, I., Drzal, L. T., Worden, R. M. and Lee, I. (2008). Nanometal-decorated exfoliated graphite nanoplatelet based glucose biosensors with high sensitivity and fast response, *ACS Nano*, 2(9), pp. 1825-1832.

Merleau-Ponty, M. (1962). *Phenomenology of Perception*, C Smith (trans.), London: Routledge & Kegan Paul.

Miller, D. A. and Sammeth, C. A. (2008). Middle ear implantable hearing devices. In Valente, M., Hosford-Dunn, H. and Roeser, R. J. (eds) *Audiology: Treatment* (second edition), New York: Thieme, pp. 324-343.

Miller, S., and Selgelid, M. (2007). Ethical and philosophical consideration of the dual use dilemma in the biological sciences, *Sci. and Eng. Ethics*, 13, pp. 523-580.

Munro, B. J., Campbell, T. E., Wallace, G. G. and Steele, J. R. (2008). The intelligent knee sleeve: A wearable biofeedback device, *Sensors and Actuators B: Chemical*, 131(2), pp. 541-547.

Richardson, R. T., Thompson, B., Moulton, S., Newbold, C., Lum, M. G., Cameron, A., Wallace, G., Kapsa, R., Clark, G.and O'Leary, S. (2007). The effect of polypyrrole with incorporated neurotrophin-3 on the promotion of neurite outgrowth from auditory neurons, *Biomaterials*, 28(3), pp. 513-523.

Seymour, W. (1998). *Remaking the Body: Rehabilitation and Change*, Sydney: Allen & Unwin.

Toombs, S. K. (1995). The lived experience of disability, *Human Studies*, 18, pp. 9-23.

Wallace, G. G. and Spinks, G. (2007). Conducting polymers – Bridging the bionic interface, *Soft Matter*, 3, pp. 665-671.

Wendell, S. (1996). *The Rejected Body: Feminist Philosophical Reflections on Disability*, London: Routledge.

Wheeler, A., Archbold, S., Gregory, S., and Skipp, A. (2007). Cochlear implants: The young people's perspective, *Journal of Deaf Studies and Deaf Education*, 12(3): pp. 303-316.

Young, I. M. (2000). *Inclusion and Democracy*, New York: Oxford University Press.

Young, I. M. (2005). *Throwing Like a Girl and Other Essays,* London: Oxford University Press.

Zenner, H. P. and Leysieffer, H. (1998). Totally implantable hearing device for sensorineural hearing loss, *Lancet* 352, p. 1751.

Chapter 12

Enhancing Material Nature

Alfred Nordmann

Nanotechnologies and the idea of their convergence with other technosciences are rooted in the notion that they are capable of mobilizing nature in order to go beyond nature. At times this notion declares itself with surprising clarity and at others it is merely implied. Often it is rhetorical accompaniment to conventional research, and sometimes it pronounces far-flung possibilities. As a dream (or nightmare) of reason (Dupuy 2007, p. 242), this visionary theme deserves close examination. The main purpose of this examination is to appreciate its queerness – in other words, the way in which this notion transgresses traditional categories and expectations, the way it tweaks language and stretches received concepts. The aim is not to question the significance of the notion, but to create a certain critical distance to it. Such distance is needed not only in regard to its technical feasibility, which ought not simply to be assumed by an all too obliging ethical discourse (Nordmann 2007a). Critical questions also need to be raised about the desires and hopes that are brought into focus by nanotechnological aspirations to enhance material nature. This is because quite independently of their fulfillment, these desires and hopes are producing effects even now which are far from insignificant. Especially

Nano Meets Macro - Social Perspectives on Nanoscale Sciences and Technologies
by K L Kjølberg & F Wickson

www.panstanford.com
978-981-4267-05-2

one of these effects will come to the fore in the following pages: in our thinking about nature we are challenged by nanotechnologies to adopt a performance-orieated engineering attitude rather than to think, for example, the kind of materialism and reductionism that is associated with genetics.[1]

1. Introduction

The notion that we might recruit nature to surpass nature is not new. It has a long history, at least in alchemy and other magical sciences, in romantic philosophy of nature (*Naturphilosophie*), but also in theories of self-organization and, not least, in biomimetics or bionics. If there is anything new about it in the context of nanotechnologies, then it is its innocuous appearance and the way in which it is taken for granted. In fact, it assumes a kind of spectacular prominence only in the debates about so-called transhumanism and its goal of technologically enhancing human nature, which is to be enabled, somehow, by nanotechnologies and their convergence with other emerging technologies. In contrast, the idea of mobilizing nature to go beyond nature goes just about unnoticed where it proves to be fundamental, namely with regard to the project of technologically enhancing or surpassing material nature. This basic idea therefore needs to be brought to light before it can be properly contextualized historically and appreciated philosophically.

---

[1] This is a revised and expanded version of "Mit der Natur über die Natur hinaus?" which appeared in K. Köchy, M. Norwig, G. Hofmeister (eds.) Nanobiotechnologien: Philosophische, anthropologische und ethische Fragen, Munich, 2009, pp. 131-147. It benefited from discussions at the EthicSchool on Ethics of Converging Technologies, 21-26 September 2008, Romrod/Alsfeld, Germany (a Specific Support Action funded by the European Union under Framework Programme 6, Science and Society programme, contract number 036745). I would like to thank its participants and contributors, especially Bernadette Bensaude-Vincent and Astrid Schwarz, also Thomas Vogt and Michael Stöltzner at the University of South Carolina. Finally, Kathleen Cross, Reinhard Heil, Daniel Quanz, and Travis Rieder offered valuable advice on editorial questions and the translation.

The first of these tasks and just bringing the idea to light can be absorbing enough. It is not simply a matter of providing exhaustive documentary evidence of the fact that among the founding myths of "nanotechnology" in the singular is the expectation of recruiting nature to surpass nature.[2] Conceptually, it requires a different approach: how can one make any sense at all of the programmatic notion to venture "with nature beyond nature!" and how, then, might nanotechnological research and its associated discourse render it meaningful?

2. Mobilizing Physical Knowledge to Leave behind Biological Givenness

On first sight, the program to go 'with nature beyond nature!' sounds too dubious to be considered meaningful at all, and perhaps this is why it often remains silently in the background. One way of assigning meaning to it would be to allocate different indices to the word 'nature' which occurs twice in the phrase: recruiting $nature_1$ to go beyond $nature_2$, where $nature_1$ would be nature as it is conceived by physicists, based on laws and regularities, and $nature_2$ would encompass all the conditions of life that we encounter in the world and that are the subject-matter of evolutionary biology and ecology. In this view, then, nanotechnologies are seen as using the principles, rules and laws of nature as elaborated by physics to surpass nature, our natural environment, or conditions of life in the evolutionary or ecological sense.

In this dual use of the concept of nature, though, there resides a certain disingenuousness that plagues nanotechnologies, biomimetics, synthetic biology, and nanobiotechnology in equal measure. The first, physicalist concept of nature is thin: it holds any particular form of existence to be contingent or accidental, especially the existence of humans or of the earth as a habitable place. From that point of view,

[2]Indeed it would not be easy to produce such evidence systematically and in a methodologically sound manner without incurring the accusation of selectivity and over-interpretation. But see Bensaude-Vincent & Hessenbruch 2004; Dupuy 2009; Hayles 2004; Milburn 2008; Schwarz 2004; Toumey 2008.

these forms of existence are the lawful consequence of some initial conditions that happened to obtain at some time and place. Accordingly, those who go down to the level of molecules and adopt the perspective of how molecules "see" one another (to cite a popular metaphor) do not see certain categorical differences and thus do not see what is usually called 'life.' In contrast to this, the biological concept of nature is thick.[3] Essential to this concept are the highly specific conditions of life that have emerged in the course of evolution. Generally, those who call for sustainability or worry about environmental and resource problems, or about the dignity of particular entities in relation to universal laws, are not arguing at the nanoscale but in respect to human scale and conditions of life. 'With $nature_1$ beyond $nature_2$' would mean to surpass, overhaul, or recreate the contingently given highly specific conditions of life with reference to the abstract regularities of nature's principles. Accordingly, some of the most pointed formulations of this idea refer to a second creation story or to the fact that from now on we are in a position to take evolution into our own hands.[4]

The aforementioned disingenuousness arises from the fact that this construction of "with $nature_1$ beyond $nature_2$" enables an opportunistic line of reasoning: nanotechnologies can legitimize themselves as being in agreement with nature even when, judged against the standards of

---

[3]The differentiation between "thick" and "thin" refers back to what was first characterized by Gilbert Ryle as "thick" and "thin description" and was then taken up by Clifford Geertz for ethnography. The physicists' concept of nature offers a comparatively thin description that does not appreciate our rather tenuous dependence on historically evolved features of the world that are contingently given and lack physical necessity. See Ryle 1968 and Geertz 1973.

[4]Nobel laureate Horst Störmer is quoted in one of the founding documents of nanotechnology as saying: "Nanotechnology has given us the tools … to play with the ultimate toy box of nature – atoms and molecules. Everything is made from it … The possibilities to create new things appear limitless" (NSTC 1999, p. 2). Gerd Binnig also received the Nobel Prize (for the development of the scanning tunnelling microscope) and expresses himself even more explicitly: "At this time we humans are witnesses and shapers of a second genesis, a fundamentally new evolution of material structures that we are as of yet not even able to name properly" (Binnig 2004). Binnig develops this idea more extensively in his book *Aus dem Nichts*: "We have to become familiar with the idea that there is nothing inferior about dead matter. All the wonders of the world are contained, for example, in a stone, as all the laws of nature (and thus all the possibilities that can emerge from them) are reflected in it" (Binnig 1992).

sustainability or nature conservation, it acts against nature and the protection of those conditions of life that make human existence possible. On this account, nanotechnologies are not actually bound to $nature_2$-as-given and yet do not violate in any way the natural order as denoted by $nature_1$ which is inviolable since it is constituted by the laws of nature.[5]

In order to construct a seamless transition between the lawfulness of inanimate nature and the construction of specific conditions of life, one speaks of "nature's own nanotechnology" not only at "higher" levels when clams build their shells or when proteins and whole organisms are constructed from DNA and RNA interactions, but also in relation to perfectly ordinary biochemical 'lock and key' causalities at the nano level (e.g. Davies 2007, p. 4). In either case, the fact that nature uses technology to achieve its ends signals that it can be used to construct different things, including the ones that are, in fact, being constructed by nature now.

Likewise, only the physicalist conception of "nature's own nanotechnology" can explain the use of the word "incidentally" in the following statement by chemist Roald Hoffmann: "Nanotechnology is the way of ingeniously controlling the building of small and large structures, with intricate properties; it is the way of the future, a way of precise, controlled building, with incidentally, environmental benignness built in by design" (NSTC 1999, p. 4). Thus, right from the start and quite effortlessly it appears that the ecological benignness of nanotechnologies is unquestionable, simply because it cannot do otherwise but to follow the principles of nature and obey the laws of nature. Ecological problems—so it has been said rather pointedly elsewhere—will take care of themselves once we have nanotechnologies,[6] because anyone who is guided by the fundamental

---

[5]Bionics has for some time now been suspected of using this sleight of hand: "Even bionics has no direct, unmediated, value-free access to nature. Instead it chooses a technically mediated and technically induced access to nature in order to create a bridge from life understood in technical terms to technologies optimized for life: not from life to technology, but from technology to life!"(Schmidt 2002).

[6] This was said in the introductory presentation by the Wuppertal Institute at a citizens' conference organized by the European "NanoDialogue" project on 7 October 2006 in the German Museum in Munich.

ways of nature will supposedly reach their goal directly and efficiently, without producing any waste and without squandering any resources. Indeed, anyone who has understood the basic principles of nature$_1$ may be capable of judging and relativising nature$_2$-as-given and of seeing that inefficiencies and redundancies have crept in over the course of evolutionary history. Nano researchers Frans Kampers and Bernhard Roelen, for example, state that cows are extremely inefficient meat producers and that a sustainable, green nanotechnology would manufacture meat more efficiently in the laboratory (Kampers 2007; Löhe 2009). Freeman Dyson goes further with his opinion that nano and biotechnologies – and especially synthetic biology – can clean up, correct, and straighten out the tortuous course of evolutionary history *in toto*. Green biotechnology need no longer frantically concern itself with the diversity of species because there is no longer any need for species once any number and variety of phenotypes can be engineered genetically.[7]

Whether or not all this represents a perversion of biomimetics, of the idea of sustainability or of "green technology", is a question I shall leave open. What is abundantly clear, however, is that in this view it is nature$_1$ – nature conceived of purely in terms of physics – that opens up spaces of possibility and horizons of expectation for nanotechnologies. Anything that does not contradict the laws of nature counts as technically feasible and permissible. Just as technoscientific research is loath to restrict itself to nature$_2$ as its technological role model, as its regulatory idea, moral instance or point of reference, it is quite reluctant to distinguish between what is physically possible and what is technically possible. In this way it gives itself carte blanche – in line with the motto "shaping the world atom by atom" – to create, in the name of physicalist nature$_1$, what might turn out to be a totally new, totally different nature$_2$, or at least to technologically optimize the latter and to surpass it even in the name of green technology. 'With nature$_1$ beyond nature$_2$,' then, signifies an intensification, improvement, or *enhancement* of nature

[7]Freeman Dyson makes a single exception for the human species: it can and probably will be preserved, if only because it is humans, after all, who are synthesising other organisms using biotechnology (Dyson 2007)

within which we exist as living creatures and to which we owe the fact that this planet is habitable for us at all.

3. Technological Processes for the Self-Enhancement of Nature

There may be various other possibilities for assigning nanotechnological meaning to the seemingly paradoxical notion of venturing "with nature beyond nature." However, only one other makes sense to me.[8] It represents both a more subtle and a more naïve approach that puts to the test traditional concepts of nature and technology in terms of their relationship to one another.

This second reading does not interpret the word 'nature' in a deliberately ambiguous way that draws on different scientific conceptualizations, namely at first on that of physics and then on that of biology. Rather, the concept of nature is framed technologically from the very start: nature is an engineer and employs technical processes of self-formation, and thereby surpasses itself all the time. In this view, nature is a technical system, a collection of processes and properties which, like all technical systems, is capable of extrapolation, optimization, and enhancement – not in the sense of biological evolution with its gradualistic mechanism of selection and adaptation, but in the sense of algorithms and procedures that can be mobilized for the purpose of enhancing nature as a system. In this case, the two-fold occurrence of the word "nature" refers to the two manifestations of a single process. To put it in anachronistic terms by using old language for a new way of thinking, the motto of this technological programme is: 'With *natura naturans* beyond *natura naturata*.' *Natura naturans* encompasses the dynamic and creative principles that produce or realize any particular

---

[8]It should be noted here that the point is to assign to the formulation a "nanotechnological" meaning. In principle, of course, another candidate would be the "autopoietic" variant from the 19th century tradition of philosophy of nature which is clearly distinct from the technological appropriation that is developed in the following. Far from considering a technical process, 19th century *Naturphilosophie* starting with Immanuel Kant and ending, perhaps, with American philosopher Charles Sanders Peirce viewed all of nature as a kind of organism with its dynamic, self-organizing life-processes.

state of nature (the *natura naturata*) where each manifested state is merely contingent, thus accidental, questionable, incomplete, and imperfect.

Now, when it is said that each given *natura naturata* is intensified, enhanced, surpassed or improved by *natura naturans*, this involves far more than noting that as a matter fact there is always a next state of nature after the present one. And it also means more than claiming that the next state is differently configured and may represent a higher level of complexity and organization, while preserving and processing all that went before within a single ongoing process of formation.[9] Beyond all that, this technological conception of nature refers to a radical liberation, a 'setting free' of nature by technology such that the improvement of nature and the increase of its efficiency are only side-effects. This places the notion of a technical enhancement of human and material nature into a tradition of thought that appears to be at odds with a technological conception of nature. According to this tradition, "to philosophize about nature means to heave it out of the dead mechanism to which it seems predisposed, to quicken it with freedom and to set it into its own free development" (Schelling 2004, p. 14).

It may indeed seem well-nigh blasphemous to interpret as a technological concept of nature something that is associated principally with an idealist-romantic, dynamic conception of nature. But this is exactly what nanotechnology does, as do the discourses of ethics and present day philosophy of nature: they secularize and vulgarize philosophy of nature by absorbing it into the idiom of engineering.[10] And herein lies the main thrust of this second mode of interpreting the notion "with nature beyond nature". If there is anything new and different about

---

[9]Only G. Khushf interprets the development of nanotechnologies in the context of convergence as a process of intellectual and moral self-formation. He therefore rejects the distinction of a *naturphilosophische* and technological interpretation of "with *natura naturans* beyond *natura naturata*." See Khushf 2007.

[10]See the previous two notes: Along the way and as if in passing, this "vulgarized" *Naturphilosophie* obliterates the distinction between organism and technical artifact. Schelling's dynamic conception of self-organizing nature has precursors, of course, especially in Spinoza, and it has descendants such as Charles Sanders Peirce, whose philosophy of nature prepares the ground for technoscience. On this, see Nordmann forthcoming.

nanotechnologies and anything that we don't already know from well-worn technological utopianism and hype, it is this incidental, if not entirely unreflected appropriation by a technological programme of a theme from philosophy of nature and an ideal for natural science. The lofty and provocative idea of 'self organization' accordingly reappears quite matter-of-factly in the notion of 'bottom-up engineering'.[11] So, much of what is new and different here is simply the casual confidence with which this appropriation is undertaken. It occurs in the absence of metaphysical or epistemological reflections on the limits of knowledge and control, on the methodological problems with, perhaps outright untenability of the technological programme. As technical concepts of 21st century technoscience, the concepts *natura naturans* and *natura naturata* are anachronistic – out of time and out of place. In the following, therefore, I want to illuminate the queerness of this technical appropriation of a conception of nature that sees it as a kind of self-organizing dynamism.

3.1 Enhancements

What, then, is that nanotechnological programme of liberating nature, and setting it free? This is currently much discussed especially in regard to human enhancement and the discourse of surpassing human nature. Supposedly, the convergence of nanotechnologies with other technosciences will realize human potential far beyond its current development (Roco & Bainbridge 2002). Ordinarily, these transhumanist techno-phantasms are not considered in the context of the 19th centuy theme of enhancement as a purpose of nature. Instead, they are usually discussed as a program for the technological perfection of a technically deficient human being. However, the narrower construal fails to acknowledge the particularity of a discourse that is informed especially by nanotechnological programs and achievements.

---

[11]To be sure, even this "novelty" has been prepared, for example, by John von Neumann's theory of automata.

The term "converging technologies" is often taken to denote the integration at the nanoscale of biotechnologies and information technologies with the insights and objectives of the cognitive sciences.[12] The idea is that the natural principles of self-organization at work on the nanoscale should feed into the technical design of a new human being, into improving his or her physical and intellectual performance (Healey & Rayner 2009). Only in alliance with nature's self-organising, bottom-up strategies can this enhancement take place. The lofty ambition of this programme which is to involve a renewal of the sciences and therefore promises a 'new Renaissance' (Roco & Bainbridge 2002, p. 1ff.) is suggestive of Ernst Bloch's term *Allianztechnik* which draws on Schelling's philosophy of nature and pronounces that human emancipation depends on the liberation of nature and vice versa (Bloch 1995, p. 803-817). Humanity will realize its full potential only by technologically allying itself with nature's dynamics of self-organization.[13] Here, then, the project of the nanotechnologically enabled convergence can no longer be considered merely as part of an anthropological trajectory that views the history of technology as an ongoing compensation of human deficiency (Gehlen 1965).[14] Instead, it is apparent that the project aims to enter into an alliance or 'unparalleled entanglement' that 'really builds humans into nature' (Bloch 1995, p. 817). And this implies a re-contextualization of the currently all-too intense debate about technologies for improving human performance: human enhancement now appears as merely a special limiting case at the far horizon of the more general and more significant programme of enhancing material nature.

---

[12]This is not the place to discuss the various conceptions of "convergence" that were developed in the US, Canada, and Europe. Here, I limit myself to "NBIC-convergence" because it is most clearly indebted to the nanotechnological enhancement theme.

[13]G. Khushf makes this connection to Bloch explicit, at least in conversation. See Nordmann 2007b.

[14]I refer to this as an anthropological trajectory because it found its most lucid expression in Arnold Gehlen's philosophical anthropology that goes back to Johann Gottfried Herder and Ernst Kapp and has since been taken up, at least implicitly, by John Harris.

3.2 New neighbours for free atoms

The term 'enhancement' offers a fitting technological expression for the appropriation of a notion of nature that is permanently surpassing, if not transcending itself.[15] However, with regard to material nature this does not tell us as of yet how the idea of nature liberated by nanotechnologies concretely manifests itself. A first pointer in this direction leads back to one of the origins of nanotechnology (Nordmann 2009).

As far as concrete applications of nanotechnologies are concerned, it is currently nanoparticles that are being discussed first and foremost. While they occur naturally and have been manufactured for many years first by artisans and then on an industrial scale, their technologically promising particularity has only been appreciated since the early 1980s. Accordingly, there is nothing novel about nanoparticles themselves. But nevertheless, they were reborn in a very real way to become interesting new nanoparticles through the way in which they were characterized by materials researcher Herbert Gleiter.

Materials properties of solids had normally been determined from the point of view of solid state physics. It derives material properties from knowledge of the crystalline structure on the assumption that nearly all atoms in a solid body are integrated into crystal lattices and that comparatively few atoms are located on or near the surfaces or at boundary layers that have been introduced into the solid. The atoms' neighbourhood relationships are fixed within the crystal lattice in which they are bound up. As far back as 1972 Herbert Gleiter and Bruce Chalmers inquired how to envisage "departure from the perfect crystal." What interested them most in this was the question of what happens when "an atom moves to a position in which its nearest-neighbor configuration is changed and its departure from equilibrium interatomic spacing is outside the linear Hookean range" (Gleiter & Chalmers 1972, p. 2).

---

[15]What is noteworthy here is the contrast between the technical metaphor of "human enhancement" and the organismic metaphor of "human flourishing" which derives from "flowering." These metaphors frame the improvement of the human condition in very different ethical and metaphysical perspectives.

Gleiter presented the corresponding technological vision along with his first experimental findings in 1981: when a body is so small that more than 50% of its atoms are localized in the surface area, and when this body is compacted with other bodies like this to form a material, new degrees of freedom and unforeseeable material properties emerge from the resulting neighbourhood configurations on the numerous boundary surfaces that now dominate the compacted material (Gleiter 1981). Liberated from their crystal lattices, atoms enter into new neighbourhood relationships and when these comparatively unconstrained and unpredictable atoms dominate the material, they open up new technical possibilities. The structural properties of the new material thus obtained (the crystalline structures from which the nanoparticles have been dislodged, so to speak) are irrelevant to the discovery and technical use of these possibilities; by contrast, the unpredictable properties that only become observable through the creation of the new material are crucial.

Here, then, the liberation of material nature means quite literally the 'liberation' of atoms from the 'dominance' of the crystal lattices, where there is a free play of properties that is not tied to structure and where a space of new material possibilities opens up that represents at the least a technical expansion of nature.

3.3 Licensed properties

This reconstruction of the discovery of nanoparticles sounds exaggerated —a poetic interpretation of a rather mundane development, namely the pursuit by materials science of ever smaller grain sizes with ever larger surface areas. However, the language of dominance on the one hand, surprising behaviour on the other comes from the researchers themselves, as does their sense of euphoria about unpredicted novel properties and the associated technical possibilities. Even if these researchers did not for the first time encounter particles of this size, Gleiter's approach signaled, quite literally, a kind of nanoparticle renaissance.

Euphoria about an old and familiar material becoming reborn as a fundamentally different kind of entity can be found in places other than research laboratories. It is part of a larger, mythical process of supposed "dematerialization" (Bensaude-Vincent 2004) that extends to what is known as the knowledge economy.

One small case-study may serve to illustrate this. It concerns the company Evonik Degussa, which has made its new business model fit a general conception of nanotechnology and nanotechnology fit its new business model.[16] Under the heading "What is nanotechnology?" it announces quite casually: "Substances with structures in the nanometer range often have completely new properties. These can be used to create innovative applications and improved products" (Evonik 2009).[17] The story behind this innocuous formulation is that of a material called Aerosil®, which Degussa has been producing since the 1940s and which literally changed its nature or state of being when it was reborn as a nanostructured material.

Aerosil® is a powder that has all kinds of uses and occurs in many technological applications. As long as it was conceived as a material substance, Degussa's contribution consisted in manufacturing it on an industrial scale and selling it in large quantities. Its buyers used it as a component that added a desired quality to their end-product. With the advent of nanotechnologies, however, Aerosil® ceased being a manufactured powder. It was transmuted into a collection of properties that owe not so much to their structural substrate and material nature but to a nanostructured surface.[18] Moreover, this collection of properties represents actual and potential solutions to present and future technical problems. Accordingly, Aerosil® is no longer marketed principally as a powder or bulk product but for its potential functionalities as an innovative solution—for example in dispersions that meet buyers' needs.

---

[16]Evonik Degussa takes part in many public debates and political initiatives on nanotechnologies (see, for example, the explicit statement on dealing responsibly with nanotechnology at www.degussa-nano.com/nano/en/dialogue/positions/leitlinienanote Chnolo gie/ (accessed: 14.4.09).

[17]EvonikDegussa GmbH, "What is nanotechnology?", not dated, accessible at www.degussa-nano.com/nano/en/nanotechnology/ (accessed: 14.4.09).

[18]It is no accident that the word "transmutation" is used here. There are frequent 'to nanotechnology as a kind of new alchemy (NSTC 1999, p. 4), and there is an alchemical background also to 19th century philosophy of nature (Magee 2001; Liedtke 2003).

Over the course of 50 years, the material had become commonplace and routine, but the advent of nanotechnology turned it into a bundle of surprisingly attractive possibilities, and research efforts are now devoted to discovering novel uses and market-opportunities which will attract developers.[19] Instead of selling a bulk product, Evonik Degussa is moving towards licensing its knowledge of these properties, their potential functionality, and its skill in handling them. Unlike the material itself, this knowledge does not change owners when it is sold: like the buyer of a software package, the buyer of Aerosil® uses this knowledge without becoming its sole owner, and the nanoparticles themselves are something akin to the DVD on which the software is delivered.

This transition from control through the industrial reproduction of a defined structure to the creative diversification of useful properties amounts to an uprooting and reorientation of these properties. They are no longer conceived primarily as dependent on structure and thus on their nature, but in regard to how they can be functionalized and thus in regard to varied human interests and uses. Now that nano research has devoted itself to discovering surprising properties at a scale where there are no lawfully predictable structure-property relations, Evonik Degussa is celebrating this liberating separation of functional properties from causally determinate structure.[20]

---

[19]Evonik Degussa's business model also involves coming up with these potential applications themselves and stimulating interest among buyers. On this, see the company history of "Degussa Advanced Nanomaterials" at www.advanced-nano.com (accessed: 14.4.2009). This start-up firm within the Evonik Degussa company structure offers "Solutions for You" in the areas of coatings, cosmetics, electronics, catalysis, adhesives. Under the first of these headings it advertises that "We offer joint developments based on our experience in nanomaterials in order to provide you with a tailor-made solution."

[20]This agrees with a conception of the knowledge economy that refers to the sale of licenses (non-tradable goods like knowledge) instead of classical products (tradable goods like powders). – To be sure, cynics will maintain that the story of Degussa Evonik is not a story about nanotechnology and the changing nature of a nanostructured material at all but instead a story of opportunistic marketing. But there is no contradiction here. The marketing strategy brought about the liberation of the material from its defined substantial nature and the shift to the free play of novel properties as potential functionalities—and *vice versa.*

3.4 From matter to material to function

The release of properties described here is not, of course, a liberation as envisioned by Schelling, Bloch or Khushf, namely a setting-free of self-organising processes that emancipate humans and nature simultaneously. Instead, what is going on is the technical appropriation of a quite simple idea: freed from a context of nature which ties them down to their causal structure, the properties enter a context of human use that relies on their relatively more flexible capacity of being adapted to technical purposes. Liberation here, then, means first and foremost the opening up of innovative potential, the discovery of new markets and the disclosure of technical possibilities.

The studies of Bernadette Bensaude-Vincent serve to place this transition in a historical and philosophical context (Bensaude-Vincent 1992; 2001; 2006; Bensaude-Vincent & Hessenbruch 2004). Chemical substances were initially defined largely by their phenomenological manifestation and their local origins. Each had its own natural history until the end of the 18th century when in the context of Antoine Lavoisier's reform of chemistry and its nomenclature they became standardized and comparable samples of chemically composed matter. They became subject to common measures and universal treatment through a novel use of scales in closed laboratories that were governed by the principle of conservation of matter. These samples of matter were further transformed by the materials science of the 20th century. Now, matter became material. At first, this meant that naturally occurring substances were examined as to their suitability to serve as, for example, building materials. In the course of its development, though, materials science has itself undergone a transition. Whereas it first inquired about the functions that the materials known to us can fulfill, it soon began the more ambitious project of actively developing materials capable of fulfilling specific desirable functions. Bensaude-Vincent shows that this approach involved a 'de-substantialisation' of matter: the substance or material object is no longer the recognized starting point or point of reference for research but is rather seen as a constraint that needs to be overcome. This development continued until the relationship between structure and function became marginal and until process and function came to be considered independently of underlying structure. In evidence of this Bensaude-Vincent quotes Ahmed Zewail, who received the Nobel Prize in chemistry in 1999 and heralded a new era of chemistry. According to him, functions are now directly accessible through the interpretation of their dynamics, without having to trace them back to structures or a corresponding set of causal events. So, if 'with nature beyond nature' refers to the realization of as yet unrealized possibilities of nature, then this story of emancipation deals first and foremost with the discovery of an unlimited space of possibilities in which poor old pathetic dead matter is replaced by new, 'intelligent' designer materials, much as was the case when plastic was introduced (Barthes 1957).

3.5 Technological use of self-organization

The foregoing discussion has been restricted to the area of new materials, an area that, while important for nano research, is still rather limited in view of the larger ambitions associated with nanotechnologies. In contrast to this, the programmes of nanobiotechnology and synthetic biology open a huge gulf between accomplished research and envisioned application. Especially in synthetic biology the intent is to merge technology and nature and thus to go beyond nature with technical means. One perhaps extreme vision speaks of plants with black silicon leaves whose purpose is to gather and store energy (Dyson 2007). Such envisioned syntheses express the aspiration of a technological convergence (VDI 2003) that is enabled by nanotechnologies' indifferent manipulation of all molecular structure: by surpassing the distinction between the organic and the inorganic, between the natural and the artificial, molecular recombinations are to enable the fusion of plants and solar panels. This programme of technological convergence is dedicated by no means to human enhancement, in particular, but includes, for another example, visions of using information technology to penetrate and enhance all ecosystems (Banfield 2002).

3.6 Material and device

The anachronistic queerness of the notion 'with nature beyond nature', interpreted in a technical sense, becomes ever more apparent when one includes in the analysis also the rhetoric of nanotechnological programmes. In regard to this rhetoric it is immediately apparent that materials research is used to illustrate the immediate economic success of nanotechnological research but that the materials themselves are far from being seen as the real thing as of yet. The nanotechnologies that are yet to come are thought to be technical systems, devices and machines. Nanotechnological ambition demands more than taking materials out of their context of nature and placing them into a context of use with its greater degrees of freedom where properties can serve a variety of functions. For nature to surpass itself by technical means and for technology to become natural, the agency of self-propelled nature would have to link up with technical functionality. This merely regulative or heuristic idea has not been realized as yet but is anticipated in the marketing rhetoric of nanotechnology, especially in attempts to ascribe to as yet dead and dumb matter that it is now animated and has become smart 'self'-acting material.

The expectation that nature might become technical and technology natural through their alliance in the production of self-acting systems finds expression, for example, in talk of 'selective surfaces', 'intelligent materials', 'self-propelled' or 'autonomous' motion, 'self-organized structures', and most prominently of 'self-cleaning glass', much of it is hardly more than 'dirt resistant.' Where an activity of cleaning and self-maintenance is invoked, most self-cleaning glass is actually an entirely passive structure. It is the rain that provides the action of rinsing dirt from a finely prickly surface that prevents dirt from sticking too well in the first place. Other kinds of self-cleaning glass claim a 'dual-action process.' The sun's ultraviolet light provokes the release of oxygen ions from a coating on the glass, these ions break down the dirt on the glass (this is the first action) and the rain then washes it off (the second action). If this glass has a more proper claim to agency, this is because 'the greatest feature of Pilkington Activ™ Self-Cleaning Glass is that it relies on the forces of nature to keep your windows looking beautiful' and with this, indeed, 'an impossible dream' has become reality (Pilkington 2009). The glass appears to have incorporated within itself what appears to be a natural propensity or affordance that may well continue in perpetuity (compare Harré 2003). If it still sounds like an exaggeration to speak here of a perfect and transcendent fusion of nature and technology and the transformation of glass from inert and dumb translucency to active and smart self-maintenance, the reason for this is only that the glass in question does not quite live up yet to the expectations associated with nanotechnological systems.

3.7 Uncontrollably controlled agency

Nowhere is the ascription of the 'self' so enigmatic and ambiguous than in the concepts of self-assembly and self-organization, whose many and varied uses in the context of nanotechnological research and whose philosophical interpretations cannot be analyzed here. At times these are nothing other than a fancy way of stating what happens in any chemical reaction; sometimes, though, it is also used to describe a technology that is no longer constructed but grown by mobilizing powers of self-organization.

The fact that this ambiguity is preserved, that these and further connotations are implied in terms such as "bottom-up engineering," and that there is no sense of urgency to clarify these terms, testifies to the 'ontological indifference' identified by Peter Galison in nanotechnological research (Galison 2006). Traditionally, scientific understanding is not indifferent at all to what exists and what does not, and to what is the precise meaning of theoretical concepts. In contrast, the striking indifference regarding nanotechnological appeals to "self-organization," "self-assembly," or "bottom-up engineering" can be viewed as a resultant of two forces that are acting together. There is on the one hand a powerful idea that was articulated by Bernadette Bensaude-Vincent, namely the implicit notion of the presumed totipotency of technoscience and converging technologies (Bensaude-Vincent 2009). The notion of "totipotency" came to the fore in discussions of adult and embryonic stemcells where the latter are said to be more powerful and significant because they can become any cell whatsoever. It is their totipotency that makes embryonic stemcells attractive objects of biomedical desire since it expresses a kind of organic and biological power that is projected in a general way by nanotechnologies on the free play of functions, on molecular recombination, and unbounded technological potential in the material world (Schwarz 2004, 2009). The notion of "self-organization" serves as a cipher for this totipotency. There is on the other hand the queerness that owes to the technical appropriation of a concept from philosophy of nature: whatever might be meant by “self-organization” philosophically or scientifically, the technical meaning of the term is not in question because nature, as a kind of super engineer, always already shows its human imitators how ‘bottom-up engineering’ is done. To note that nature organizes itself is merely another way of saying that it has constructed everything we see—including ourselves.

At this point it becomes clear how nanotechnologies adopt nature as their role model—but only after they conceptualize nature as nothing other than an engineer. This shift in meaning allows visionaries of self-organization to describe technical control over natural and technical processes at the nanoscale as a mere illusion without renouncing the idea of control:

The problem is the illusion of control—what we want to do is reverse engineer. We harness self-assembly in a non-linear way to get what we want. To do this at the nanoscale will be a big breakthrough because we can then start to control things, put them in compartments and let them evolve. We don’t need the illusion of control. We let the system select what it needs according to its local environment. We can’t be an engineer at that level if we want to use bottom up. Nature takes this approach and it works very well.[21]

---

[21] Statement made by a nano researcher at a scenario workshop run as a part of the EU project DEEPEN, see Macnaghten/Kearnes 2007.

This on first sight rather confused notion of control is another sign of ontological indifference: converging on the business of creating and recreating a world, nanotechnologies and the various technosciences put themselves at the mercy of a self-organizing and thereby presumably self-controlling nature. This enables nanotechnological researchers to distance themselves from false notions of precision control at the nanoscale and at the same time to wait and see what interesting and useful properties they might discover there.

4. Conclusion

'With nature beyond nature' could mean that there are evolutionary processes that lead to emergent phenomena, to a 'higher' level of organization or further development of nature. This is a notoriously difficult conception to grasp. In the technoscientific practice of nanotechnological research it becomes even more exasperating precisely because a grand conceit has become a matter of casual indifference. According to this conceit, nature reveals itself as a nano engineer and thus vanishes inside a generalized notion of technical construction, while technology entrusts itself to nature by seeking to absorb the dynamics of self-organization into the design process.

I here described this unassuming indifference to the grand conceit as queer because it violates received categorical distinctions without paying the price of justifying itself theoretically. It neither reduces spirit and culture to matter and nature, nor does it celebrate the scandalous creation of hybrids or monsters. Instead, it simply dissolves the received categorical difference of science and technology, nature and culture, organism and artefact, the natural and the artificial into the idiom of engineering.[22]

---

[22] One of the few explicit statements to that effect can be found in Roco & Bainbridge 2002, p.11: "Some partisans for independence of biology, psychology, and the social sciences have argued against 'reductionism,' asserting that their fields had discovered autonomous truths that should not be reduced to the laws of other sciences. But such a discipline-centric outlook is self-defeating, because as this report makes clear, through recognizing their connections with each other, all the sciences can progress more

Questions for Reflection:

1. This text follows the common assumption that there is a sharp distinction between philosophy and basic science on the one hand, and engineering and technology on the other hand. Where does this assumption come from and how might it be justified?
2. Some argue that technology always goes beyond the limits of nature. Others maintain that technology can only work within the limits of nature. Given this contrast, what do you think the nanotechnological ambition to go 'with nature beyond nature!' means?
3. How do you feel about the program to enhance material nature vs. the program to enhance human performance? What do you think is the relationship between the two?

---

effectively. A trend towards unifying knowledge by combining natural sciences, social sciences, and humanities using cause-and-effect explanation has already begun, and it should be reflected in the coherence of science and engineering trends and in the integration of R&D funding programs."

Bibliography

Banfield, J. (2002). Making sense of the world. convergent technologies for environmental science. In Roco, M. C. and Bainbridge, W. S. (eds) *Converging Technologies for Improving Human Performance*, Arlington: National Science Foundation, pp. 260-264.

Barthes, R. (1991 [1957]). Plastic. In Barthes, R., *Mythologies*, New York: The Noonday Press, pp. 97-99.

Bensaude-Vincent, B. (1992). The balance: Between chemistry and politics. In *The Eighteenth Century*, 33(2), pp. 217-237

Bensaude-Vincent, B. (2001), The Construction of a discipline: Materials science in the USA. In *Historical Studies in the Physical and Biological Sciences*, 31(2), pp. 223-248.

Bensaude-Vincent, B. (2004), *Se libérer de la matière? Fantasmes autour des nouvelles technologies*, Paris: INRA editions.

Bensaude-Vincent, B. (2006). Materials as machines, paper at workshop "Science in the Context of Application" at the Centre for Interdisciplinary Research (ZiF) Bielefeld, October 2006.

Bensaude-Vincent, B. (2009). *Les vertiges de la technoscience: Façonner le monde atome par atome*, Paris: La découverte.

Bensaude-Vincent, B. and Hessenbruch, A. (2004). Materials Science: A field about to explode? *Nature Materials*, 3(6), pp. 345-346.

Binnig, G. (1992). *Aus dem Nichts: Über die Kreativität von Natur und Mensch*, Munich: Piper.

Binnig, G. (2004). Foreword. In Boeing, N., *Alles Nano?! Die Technik des 21. Jahrhunderts*, Berlin: Rowohlt, pp. 7-9.

Bloch, E. (1995 [1954]). *The Principle of Hope*, Cambridge, MA: MIT Press.

Davies, G. (2007). Introduction. In Davies, A. G. and Thompson, J. M. T. (eds) *Advances in Nanoengineering: Electronics, Materials and Assembly*, London: Imperial College Press, pp. 1-7.

Dupuy, J. P. (2007). Some pitfalls in the philosophical foundations of nanoethics, *Journal of Medicine and Philosophy*, 32(3), pp. 237-261.

Dupuy, J. P. (2009). *On the Origins of Cognitive Science: The Mechanization of the Mind*, Cambridge: MIT Press.

Dyson, F. (2007). Our biotech future, *New York Review of Books*, 54(12), pp. 4-7. http://www.nybooks.com/articles/20370 (last accessed Apr 14, 2009).

Evonik Degussa (2009). What is nanotechnology? www.degussanano.com/nano/en/nanotechnology/ (last accessed: Apr 14, 2009).

Galison, P. (2006). The Pyramid and the Ring, lecture given at a conference of the "Gesellchaft für analytische Philosophie" (GAP, Society for Analytical Philosophy), Berlin, September 2006.

Geertz, C. (1973). Thick description: toward an interpretive theory of culture. In Geertz, C., *The Interpretation of Cultures*, New York: Basic, pp. 3-30.

Gehlen, A. (1965). Anthropologische Ansicht der Technik. In Freyer, H., Papalekas. J. and Weippert, G. (eds) *Technik im technischen Zeitalter: Stellungnahmen zur geschichtlichen Situation*, Düsseldorf: Schilling, pp. 101-118.

Gleiter, H. (1981). Materials with ultra-fine grain sizes. In Hansen, N., Leffers, T. and Lilholt, H. (eds) *Proceedings of the Second Risø International Symposium on Metallurgy and Materials Science*, Roskilde: Risø National Laboratory, pp. 15-22.

Gleiter, H. and Chalmers, B. (1972). *High-Angle Grain Boundaries*, Oxford: Pergamon Press.

Harré, R. (2003). The materiality of instruments in a metaphysics of experiments. In Hadder, H. (ed) *The Philosophy of Scientific Experimentation*, Pittsburgh: University of Pittsburgh Press.

Hayles, N. K. (ed) (2004). *Nanoculture: Implications of the New Technoscience*. Bristol: Intellect Books.

Healey, P. and Rayner, S. (eds) (2009). *Unnatural Selection: The Challenges of Engineering Tomorrow's People*, London: Earthscan.

Kampers, F. (2007). Interview in the *BBC Focus Magazine*, No. 175/April 2007, p. 42.

Khushf, G. (2007). An ethic for enhancing human performance through integrative Technologies. In Bainbridge, W. S. and Roco, M. C. (eds) *Managing Nano-Bio-Cogno Innovations: Converging Technologies in Society*, Dordrecht: Springer, pp. 255-278.

Liedtke, R. (2003). *Das romantische Paradigma der Chemie*, Paderborn: Mentis.

Löhe, F. (2009). Das Steak aus dem Reagenzglas, *Frankfurter Rundschau*, No. 124, May 30/31, p. 35.

Macnaghten, P. and Kearnes, M. (2007). DEEPEN Deliverable 4: Working Paper - Scenario Planning and Draft Design of Focus Groups, Durham, April 2007, unpublished.

Magee, G.A. (2001). *Hegel and the Hermetic Tradition*, Ithaca: Cornell University Press.

Milburn, C. (2008). *Nanovision: Engineering the Future,* Durham: Duke University Press.

National Science and Technology Council (NSTC) (1999). *Nanotechnology. Shaping the World Atom by Atom*, Washington: National Science and Technology Council/Committee on Technology.

Nordmann, A. (2007a). If and then: A critique of speculative nanoethics, *Nano-Ethics*, 1(1), pp. 31-46

Nordmann, A. (2007b). Renaissance der Allianztechnik? Neue Technologien für alte Utopien. In Sitter-Liver, B. (ed) *Utopie heute. Zur aktuellen Bedeutung, Funktion und Kritik des utopischen Denkens und Vorstellens*, Vol. 1, Fribourg/Stuttgart: Academic Press/Kohlhammer, pp. 261-278.

Nordmann, A. (2009). Invisible foundations: Herbert Gleiter and the contribution of materials science, *Perspectives on Science*. 17:2, pp. 123-143.

Nordmann, A. The hypothesis of reality and the reality of hypotheses. In Heidelberger, M. and Schiemann, G. (eds) *The Significance of the Hypothetical in the Natural Sciences*, Berlin: deGruyter, forthcoming.

Pilkington (2009). *Pilkington Activ™ Self-Cleaning Glass*. http://www.pilkingtonselfcleaningglass.co.uk/ (last accessed March 21, 2009).

Roco, M. C. and Bainbridge, W. S. (eds) (2002). *Converging Technologies for Improving Human Performance: Nanotechnology, Biotechnology, Information Technology and Cognitive Science*, Arlington: National Science Foundation.

Ryle, G. (1971 [1968]). The thinking of thoughts. What is 'Le Penseur' doing? In Ryle, G. *Collected Papers*, Vol. 2, London: Hutchinson, pp. 480-496.

Schelling, F. W. J. (2004). *First Outline of a System of the Philosophy of Nature Ideas for a Philosophy of Nature*, Albany: State University of New York Press.

Schmidt, J. C. (2002). Wissenschaftsphilosophische Perspektiven der Bionik, *Thema Forschung. Das Wissenschaftsmagazin der Technischen Universität Darmstadt*, 2, p. 19.

Schwarz, A. E. (2004). Shrinking the 'ecological footprint' with nanotechnoscience? In Baird, D., Nordmann, A. and Schummer, J. (eds) *Discovering the Nanoscale*, Amsterdam: IOS Press, pp. 203-208.

Schwarz, A. E. (2009). Green dreams of reason: Visions of excess and control. In Nordmann, A., Malsch, I., Quesada, M. A. and Simons, E. *Ethics of Converging Technologies* (e-learning DVD developed in the EU funded EthicSchool project), Nijmegen: Radboud University.

Toumey, C. (2008). Reading Feynman into nanotechnology: A text for a new science, *Techné*, 12:3, pp. 133-168.

VDI Technologiezentrum (ed) (2003). Ansätze zur technischen Nutzung der Selbstorganisation Düsseldorf; Verein Deutscher Ingenieure.

Chapter 13

The Nano Control-Freak: Multifaceted Strategies for Taming Nature

Arianna Ferrari

This chapter will aim to explore the idea of the control of nature in emerging nano(bio)technology and its ethical, social and political dimensions. After a brief introduction of the topic and an explanation of the initial idea for the analysis (the link between epistemic visions of the control over nature and societal dimensions of nano(bio)technology), I will analyse two different conceptualizations of control. Here I will rely on the two cultures present in these technologies - the engineer's and the chemist's cultures, following the analysis initiated by Bensaude-Vincent (2004). I will further explore the possibility of whether there are common features beyond these two epistemic cultures, i.e. what is the common vision of nature embedded in nano(bio)technologies. I will define this common vision as the idea of nature as a plastic engineer. I will also show how this vision is linked to control. Then I will try to explain the way in which these epistemic visions relate to the societal images of these technologies as well as shape important characteristics of the political debate.

Nano Meets Macro - Social Perspectives on Nanoscale Sciences and Technologies
by K L Kjølberg & F Wickson

www.panstanford.com
978-981-4267-05-2

1. Nanotechnology and Control

Because the increase of knowledge through scientific discoveries has often been attached to an intention of intervening and transforming the world, it could be argued that the idea of control has—at least since Bacon[1]—always characterized science as human enterprise. The question of whether nanotechnology,[2] in its conceptualization(s) of nature, will offer some new elements or move on a line of continuity with ideas and concepts of the past is still under discussion and cannot be clarified in this chapter. The purpose here is to explore how nano(bio)technology is particularly connected with the idea of control and how it reinterprets it in light of the new knowledge gained at the nanoscale, as well as the knowledge it is aspiring to gain.

Nanotechnology is commonly described as having two different approaches: top-down and bottom-up. On the one hand, nanotechnology is permeated by the aim of miniaturization of technology (its triumph often seen in micro-lithography, which is the most advanced approach in this field). In fact, in almost every introduction to nanotechnology there is reference made to the so-called 'Moore's law' which describes the progressive miniaturization trend of microelectronics[3] (see for example Goodsell 2004; Crandall 1996; Wood *et al.* 2003; Royal Society 2004). On the other hand, nanotechnology seems to also offer a new way of transforming materials, i.e. the direct manipulation of atoms and molecules for the construction of new devices which are expected to be

---

[1] Francis Bacon (1561-1626) was a Renaissance author. He wrote extensively about science and technology and he stressed the role of knowledge as a form of liberation for human beings from prejudices and as a form of gaining power. One of the goals of science and technology according to Bacon was also to establish dominion over nature. In his utopian novel "The New Atlantis" (1627) he envisioned a mythical land, Bensalem, where among others the State sponsored a scientific institution, the Solomon's House, in which scientific experiments were conducted. The goal of these experiments was to apply the collected knowledge for the betterment of society.

[2] Here I refer to nanotechnology using the singular form to focus on the role of nano as a unified program of research. When I use the plural form I refer to the bunch of different disciplines and projects in the nano area.

[3] Gordon Moore is one of the inventors of the integrated circuit. In one article in 1965 he formulated a correlation, predicting that every two years it would be possible to duplicate the number of transistors on an integrated circuit.

capable of creating themselves in autonomous ways: this is the bottom-up approach, also known as self-assembly. Both of these two approaches inform the work of nanotechnologists. Although they refer to different 'directions' of construction, they both have the idea in common of the manipulation of reality at the nanoscale. Therefore, the central and particularly challenging idea of nanotechnology is the possibility of controlling matter at the atomic scale, an idea which was already launched by Richard Feynman. Although he didn't mention the word 'nanotechnology' in his famous speech in 1959 (where he said that the principles of physics do not speak against the possibility of manoeuvring things atom by atom (Feynman (1960)), Feynman is considered one of the fathers of nanotechnologies. The ideas presented in his speech have been reworked in later works, in particular in K. Eric Drexler's vision of *molecular manufacturing,* as well as in the launch of the National Science Foundation's scientific program on nano, entailed in the report "Nanotechnology: Shaping the world atom by atom" (National Science and Technology Council 1999).

Although human beings have always wanted to exercise control over natural objects, in modern times, the notion of control appears dissociated from considerations of the meaning and value of our activities and social arrangements (Lacey 1999). Indeed, it is characterized by intense efforts to expand and implement this capability of control because these efforts are seen as fundamentally important in order to meet human needs and wants. The furtherance of the modern value of control over natural objects is dependent on the expansion of understanding achieved through what Lacey (1999) calls "materialist strategies", i.e. a materialist metaphysics.[4] This is because these strategies and control results are mutually dependent, share interests and conditions and show intertwined causal dynamics and a strong social appeal and force. It would be too easy and, therefore, wrong to explain the connection between materialism and control of nature by saying that

---

[4] Materialism is the philosophical theory that considers physical matter as the only reality: everything, including thought, feeling, mind, and will, can be explained in terms of matter and physical phenomena. Metaphysics is the branch of philosophy that studies the essential properties related to things. Materialist metaphysics therefore means that materialism is the underlying philosophical principle to explain reality.

the aim of science is the control of nature, as if materialism were the point of departure and thus capable of providing a complete explanation of the affirmation of control. As a matter of fact, "materialist understanding does not explain, for example, the social forces and the sources of the social values that have come to be woven into social forms along with the modern values of control" (Lacey 1999, p. 136). On the contrary, the real question regards how the values of control came to be deeply rooted in modern society and in its institutions and how these values interrelate with the most accepted and, therefore, important metaphysics of science and technology, that is, the materialistic one. Problems and difficulties are therefore not only located at the level of material understanding, since it represents a remarkable contribution to knowledge, but the fact that materialistic conception intertwines with and reinforces the search for control over nature and has done so mostly since the beginning of modern science in Western societies. This phenomenon has been endorsed by the success gained through technology.

Here it seems that the relationship between control and materialist strategies is reinforced in nanotechnology by the idea that the nanoscale is viewed as the basic level of materiality, in other words, the one in which the fundamental building blocks of matter are to be found. However, as we will see, nanotechnology is also connected with interesting strategies of imitation of autonomous forces of nature and complex phenomena so that it seems to introduce new elements whose relationship with the notion of control is worth investigation.

2. The Multifaceted View of Control in Nanotechnology: Mastery and Surprise

In 2004, Bernadette Bensaude-Vincent described two different cultures in the work of nanotechnologists: the engineer's and the chemist's. The engineer's approach is exemplified by the works of Drexler (1986; 1992; 2006), who is himself an engineer, and Ray Kurzweil (1999; 2000). It is an approach in which nature and living systems are perceived as blind mechanisms without intentionality, acting under the control of a program. The idea here is that by substituting nature's program with a nanotechnological one, we can achieve control at the molecular and atomic scale. Drexler has argued for the construction of molecular assemblers which are mechanisms for guiding chemical reactions by positioning reactive molecular tools that are themselves made of organic materials, such as carbon, oxygen, nitrogen, and hydrogen. Under the examples of 'true nanoscience and nanotechnology' Kurzweil (2005) has listed the construction of *nanobots*, which will be used in a variety of therapeutic and diagnostic functions in order to expand our experiences and capabilities (providing fully immersive virtual reality and "experience beamers" for flowing our sensory experiences and emotions) (Kurzweil 2005, p. xliv-xlv).

The chemist's approach starts from a different conceptualization of nature which considers it a marvellous and complex creator and refuses the idea of a direct manipulation of atoms and molecules. In the opinion of the chemist George Whitesides (1995; 2001), for example, the Drexlerian project is destined to fail because nature's mechanisms are too difficult to be reprogrammed and can at best be only imitated. Therefore, in the chemist's approach it is better to imitate the spontaneity and autonomy of nature, aiming at recreating components or tools that spontaneously converge.

The chemist Richard E. Smalley has aimed to show the limits of the approach of the engineer Drexler. The differences between these scientific cultures are well illustrated by the Drexler-Smalley debate (a public debate between the two scientists taking place between 2001-2005). First of all Smalley (2001) observed that Drexler's conception of molecular machines requires a form of self-replication that he believes could take the form of a new parasitic life that could easily run out of control transforming everything into a gray goo[5]. Furthermore, Smalley criticizes scientifically the Drexlerian vision of molecular machines built with atom-by-atom control that can serve as a basis for self-replicators by pointing out two fundamental problems: the problems of the fat and sticky fingers. Since Drexler's molecular machines have to have fingers

---

[5] The hypotheses of "gray goo" was already formulated by Drexler in "Engines of creation" in 1986. Here Smalley was also clearly influenced by Bill Joy's article published in 2000 "Why the future doesn't need us" (Joy 2000).

for positioning the atoms that are themselves made by atoms, it is impossible to imagine room for so many 'fingers' in the cramped space that the machine would have to work (the 'fat fingers' problem) and also how they could exercise atom-by-atom control under molecular attraction forces (the 'sticky fingers' problem) (cf. Bueno 2004). In 2003, Drexler responded to Smalley's article pointing out that he did *not* argue for the manipulation of individual atoms, but rather for the manipulation of reactive molecules and had tried to develop new research models *not* requiring self-replication (Drexler 2003a; 2003b). In a recent publication, written after Smalley's death from cancer in 2005, Drexler has tried to show the fallacies of Smalley's dismissal of Feynman's thesis, by explaining that Smalley's refusal was mainly motivated by fear of self-replicating machines. For Drexler, Smalley's two objections of the fat and sticky fingers are not relevant for atomic manipulation since "chemistry (assembler-based or otherwise) does not require this: as molecules come together and react, their atoms remain bonded to neighbours and need no separate fingers to move them" (Drexler 2006, p. 30).

In this article, Drexler explicitly refers to Feynman's famous speech of 1959 as the beginning of nanotechnology and he defends the idea of the possibility of controlling things at the nanoscale (Drexler 2006, p.29). He regrets the fact that the original Feynman message, which originally inspired the US National Nanotechnology Initiative (NNI) in 2000 (cf. NSF 2001), has been progressively obscured and its basis misunderstood, which has provoked fears among the public and destroyed the real promises.

2.1 The master engineer

The engineer's approach to nanotechnology exemplified by Drexler seems strictly connected to the idea of control of matter as mastery, that is, as direct influence on single molecules or atoms. The Drexlerian conception is very influential, especially in the field of nano(bio)technology. For example, Crandall (1996) sees the promise of nanotechnology in molecular construction and Goodsell (2004) directly refers to Drexler's nanobots when describing the goals of nano(bio)technology (Goodsel 2004). The idea of a precise molecular manufacturing is very present in some of the current scientific work. Kufer and his group (2008) have demonstrated that an atomic force microscope (AFM) can be used as a sort of 'molecular crane' since it is capable of picking up single functional units from a depot area and delivering them to a target area with a precision of around 10nm. Duwez (2008), who was following these results, refers directly to the strategy of molecular manufacturing, especially in the description of three major challenges in current research: to position every single molecule or atom in the right place (the so-called positional assembly); to make the molecule or atom form bonds as required (mechanosynthesis); and to achieve high-throughput fabrication. The examples of the application of this strategy are also interesting: the construction of better computers that can place and wire together millions of logic elements in complex patterns as well as the realization of advanced medical diagnostic devices by placing millions of sensing elements, bio-receptors and components on a single chip (Duwez 2008). Through the construction of precise instruments, scientists materially aim at creating complex devices able to perform new functions, assembling what they see as basic units. An important achievement, considered by some authors even as the representative initial moment in the development of nanotechnology (cf. Nordmann in Boeing 2007), is the positioning of 35 xenon atoms on a nickel surface with atomic precision by Eigler and Schweitzer of IBM Research to spell out 'IBM' (Eigler & Schweizer 1990). Since then, techniques have been refined: examples have been reported of single synthetic polymer chains and biomolecules that have been manipulated in water and other organic solvents through single-molecular force spectroscopy (cf. Clausen-Schaumann *et al.* 2000; Zhang & Zhang 2003), but the idea of the precise positioning of molecules and atoms has remained the same.

Embedded in molecular manufacturing there is the vision of perfect, one-to-one control at the nanoscale. Drexler conceives of everything, both organic and inorganic, as a collection of atoms and molecules and sees a possibility in nanotechnology of understanding the essence, the ontology of nature and of life. It is therefore possible to argue that the idea of control at the atomic and molecular scale is not only conceivable because we currently have nanotechnology at our disposal, but also because the natural components themselves are interpreted as machines (cf. Ferrari 2008b)[6].

The conception of nature as a machine and the possibility of mastering matter is also explicitly referred to by Browne and Feringa (2006) in their review of the advances on molecular machines: Drexler's approach enters not only as a background idea on nature and on the goal of (nano)technology, but also frames the vocabulary used by scientists in their work (who for example speak of 'molecular machines' and 'nanometer-sized factories' and of 'motors'), becoming an expression of what is at stake in these technologies. An exemplification of this approach is also the following quote that represents the beginning of a scientific article on biological nanomotors:

> *By considering how the biological machinery of our cells carries out many different functions with a high level of specificity, we can identify a number of engineering principles that can be used to harness these sophisticated molecular machines for applications outside their usual environments* (Goel & Vogel 2008, p. 465).

2.2 The chemist: ready to give up and be surprised

Different to the previous vision on how nanotechnologies work and what they can achieve is self-assembly. Self-assembly represents a process existing in nature in which a disordered system forms an organized structure or pattern through interactions among the components themselves, without any external direction. The imitation of nature's processes appears very visible in the chemist's culture, which starts from the interpretation of already existing mechanisms and builds upon them.

---

[6] For Dupuy (2008), it is possible to see continuity between cybernetics and nanotechnology concerning the way research is done and in the way nature is conceptualized. The term "cybernetics", coming from the Greek and meaning steersman or pilot, indicates the study devoted to understand and define the functions and processes of systems that have goals. For further reference see Dupuy 2007.

In a recent analysis, Bensaude-Vincent (2006) has distinguished three different strategies for conceptualizing self-assembly which may overlap in practice: hybridisation, biomimetics and integration. In the hybridization strategy, exemplified by the work of molecular biologists and material chemists, structures of the living (in particular of the cell) are used to build nanodevices that assemble without human intervention. In biomimetic's strategy, the old approach of imitating nature appears renewed by the fact that now nature's processes of fabrication are imitated. For these purposes, self-assembly requires reversibility, crucial for the readjustment of parts as well as the fact that reagents contain information (Bensaude-Vincent 2006). Since they are based on the idea that nanodesign has to apply the principles of cell biology and not the ones of conventional machines used for the macroscale (Brownian motion) Richard Jones' *soft machines* can be considered an example of this strategy (see Jones 2004). The third strategy, integration, is based on the connection of the previous two strategies directed toward the crossing of living and non-living systems with the making of artificial devices mimicking natural things.

The approach of imitating natural processes is also taken up in synthetic biology[7] and by the branches of nano(bio)technology that aim to create artificial cells or artificial cell components and is strongly linked with a description of natural components through the metaphor of a machine. Xu and Lavan (2008), for example, have tried to build artificial cells using mathematic models in order to use ion transport with the same efficiency as natural cells. By describing natural processes through the language of engineering sciences (pumps, conductors, channels etc) and at the same time as something capable of spontaneous self-organizing processes (in the view of self-assembly), they argue that artificial cells can be built to use ion transport as effectively as natural

---

[7] Synthetic biology is an area of biological research based on the combination of science and engineering in order to design and build ("synthesize") novel biological functions and systems. Pioneering was the work of Benner and Schultz: in 1989 Benner led a team at ETH Zurich that created DNA containing two artificial genetic "letters" in addition to the four that appear in life as we know it (Gibbs 2004). The research in this field was developed also the mid-to-late 1990s at the Institute for Genomic Research (Rockville, MD, USA), the non-profit research institute founded by Craig Venter in 1992.

cells (Xu & Lavan 2008). In the vision of self-assembly, nature is seen as something marvellous, very complicated and as a source of inspiration to the scientist. In another scientific article we read:

> *The exquisite solutions nature has found to control molecular motion, evident in the fascinating biological linear and rotary motors, has served as a major source of inspiration for scientists to conceptualize, design and build — using a bottom-up approach — entirely synthetic molecular machines. The desire, ultimately, to construct and control molecular machines, fuels one of the great endeavours of contemporary science* (Browne & Feringa 2006, p. 33)

From the philosophical perspective, self-assembly stands for a sort of delegation of human tasks to matter. In referring this idea to the notion of control, it seems that some nanoscientists are ready to give up control if they succeed in constructing the tools or devices they are aiming for. In other words, some scientists accept the wildness and the elements of surprise of the 'new' nanoworld, stemming from quantum mechanisms, because this is the only way for them to enter this world.

3. Nanonature: The Plastic Engineer

Wickson (2008) has pointed out nine narratives of nature in nanotechnology and relies on different conceptualizations of the relationship between these new technologies and the natural world. All these conceptualizations interrelate and simultaneously shape the work of scientists, and so are not to be seen as rigidly separated in the empirical work of research, but rather distinguished for theoretical purposes. This analysis of narratives combines itself very well with the previously described two cultures of nanotechnology and also shows how a bit of each narrative is presented differently in both of these cultures. Although at first sight the chemist's approach, for example, seems more directly connected with the narrative of imitating and using nature, and the engineer's approach with the ones of nanotechnology as natural for its connection with the mechanistic interpretation and with the narrative of transgressing nature for constructing new tools, we can notice that all these different relationships with nature have something in common. There is a sort of instrumental conceptualization of nature lying behind both approaches, in the literal sense of the word, because nature becomes the instrument and the playground for developing new nanotools. This strong orientation toward intervention in the world appears in strict correlation with the 'technoscientific'[8] character of nanotechnology, at least when interpreted as the primacy of 'making' over 'knowledge', i.e. as one way of gaining new knowledge is by first making a new world (cf. Nordmann 2004).

In fact, beyond the differences between the previously described cultural perspectives of the chemist and the engineer, it is possible to trace some common features in the vision of nature embedded in nanotechnologies. On the one hand, nano(bio)technologists very often start with mechanistic conceptions of nature, describing cells and organisms as 'collections of nanomachines' (cf. Brown & Feringa 2006; Goldstein 2006; Schafmeister 2008 see also Ferrari 2008b). On the other hand, they try to imitate nature's autonomous mechanisms by aiming to develop artificial compounds with new functions, in some cases becoming even more efficient than the natural ones. The strategy consists of a combination of reductionism with the acknowledgement of unpredictability inscribed in nature (cf. Kearnes 2006). In other words, we can notice on the one side a reductionist and materialist description of nature in terms of machine through a vocabulary coming from engineering and, on the other side, the recognition of the complexity of nature through expressions of amazement. Nature is then seen as a

---

[8] The concept of "technoscience" is often currently used in the field of STS (science and technology studies) and the philosophy of technology. Although this term was first coined by the Belgian philosopher Gilbert Hottois in the seventies to indicate double internalization process of technology in science (on the one hand technology becomes a "milieu" for science and, on the other hand, it represents the driving force of science) (cf. Hottois 1984). Nowadays the phenomenon of "technoscience" is generally taken as describing the primacy of technology over science. Technoscience differs from previous science in its strong orientation toward the production of tools and devices rather than toward the augmentation of knowledge and the profound understanding of the natural phenomena (cf. Latour 1987).

beautiful and marvellous machine that can be imitated, mechanic but autonomous, a point of reference but independent at the same time. This vision of nature serves as a basis for the construction of nano(bio)devices and synthetic cells or biological compounds that have certain functions. Nature appears here as a *plastic engineer* (Bensaude–Vincent 2009). This conception appears very evident in the following quote from the US National Nanotechnology Initiative-Strategic Plan 2007:

> *A cell includes nanomaterials (proteins, membranes, etc.), nanodevices (motor proteins, ion channels, etc.), and even functional nanosystems (e.g., mitochondria) that are created from individual molecules by hierarchical self-assembly. Biological capabilities can be harnessed directly, borrowed, or taken as inspiration for future nanomanufacturing* (NNI-Research plan 2007, 33).

The conceptualization of nature as a plastic engineer consists in the combination of two fundamentally different traditions, one that relies on nature as a source of spontaneity and complexity and the other a rigid heritage of the engineering sciences which tends to present itself as a new form of systemic approach. In fact, nanotechnology, and more generally converging technologies, are very often presented in the US agenda as the sciences that can interpret the essential message of complexity and systems theory and thereby overcome reductionism. In Roco & Bainbridge (2002, p.x), these disciplines are even referred to as something that will permit a “new Renaissance”, inspired by the old cultural movement and characterized by the fact that ‘scientists’ were active both in the field of art, science and engineering.

For Hunt (2006), however, this reference to systems theory is only instrumental. What he defines as nano technoscience is a science founded on a strong form of determinism and reductionism, very visible in the direct reference to the idea of molecular manufacturing *à la* Drexler and in the general idea that ‘all life is nanotechnology’. Furthermore, because it has struggled between the idea of reductionistic and deterministic control and the emergent unpredictability and indeterminacy, particularly visible in the approach of self-assembly, this core idea of nanotechnoscience relies on a form of inconsistency. The critique of the current nanotechnological research can be done through referring to the problems at the epistemic level, that is, problems and contradictions within the visions of nature.

Nordmann (2005) has seen one core message of nanotechnologies in the suggestion of a form of "noumenal technology"[9] as if, due to their incredible tininess, nanotechnologies are now attached with visions of grasping the ultimate level of reality, the level of "things-in-themselves". This is obviously only an illusion, but a powerful one that becomes very visible in the way in which nanoscientists describe their own creations, which are presented as artifacts indistinguishable from the presence and action of the natural processes that serve as unconsidered background and framework for everyday life. Nordmann (2005; 2008) describes this process as a form of *naturalization of technology* because technology invisibly penetrates the environment, assuming the same function as the one of nature already there, as well as assuming an unpredictable and out-of-control dynamic. In the strategy of self-assembly a search for the uncanny and uncontrollable properties seems predominant, which brings technology back to magic, the place of departure "that we already found untenable when we first thought of controlling, calculating, even mastering it" (Nordmann 2008, p. 184).

However, I think it is still possible to see, beyond the nanotechnological nature as a plastic engineer, a strong orientation toward control, also if we, following Nordmann (2008), stress its closeness to magic. Together with the fact that artificial devices now present autonomous properties, precisely because they are 'constructed', i.e. artificially made by nanotechnologists, even if the plasticity can lead to some unpredictability, they are still under control or at least they rest upon the illusion of control. Perhaps the control here is not done directly

[9] Here Nordmann (2005) explicitly admits to using the Kantian concept of noumena in a broader way than Kant himself. In Kant "noumena" (singular form is "noumenon") refer to things-in-themselves, that is to the real essence of things which is precluded from our sensorial perception, something that is independent of the senses. In Kant's theory of knowledge the human being can always perceive things in the world through his/her "categories of understanding": therefore, knowledgeable things are "phenomena", i.e. noumena filtrated through our categories. These categories include substance, quantity, quality, relation and modality.

by the actions of the scientists but by their orientation, namely by the fact that they want to build something that works and want to act and modify matter. If these scientists were not at least guided by the hope or the illusion of achieving some form of control, they would not even try to build something. In nano(bio)technology and synthetic biology, the very idea of artificiality combines with the struggle for realizing specific goals and constructing instruments capable of specific performances. It is the triumph of 'making' over 'knowing' that implies the search for control as a means for accomplishing specific goals connected with the tools that are built.

This idea seems to be implicitly addressed by one of Drexler's (2006) most recent articles where he responds to the criticisms raised by Smalley (2001) by arguing that molecular assemblers do not need any fingers to move atoms and molecules, but "as molecules come together and react, their atoms remain bonded to neighbours" (Drexler 2006, p. 30). Here Drexler refers to a spontaneous convergence of atoms and molecules, depending on the way in which molecular assemblers are constructed. Apart from the scientific plausibility of this response, it is interesting to notice here that, in the article, Drexler seems to adjust his ideas in order to demonstrate the possibility of controlling matter at the atomic level. Whether through spontaneous convergence or ad-hoc constructed machines with fingers, the core idea of nanotechnology is the controlling of nature in its atomic structure.

The nanoscientists Bath and Turberfield (2007, p. 275) very clearly express this idea of control in self-assembly through referring to the building of machines:

> *If we could cope with the interactions required for a three-dimensional fold we would design more competent machines made, as in nature, from RNA and proteins (...). We make nanomachines from DNA because the simplicity of its structure and interactions allows us to control its assembly* (Bath and Turberfield 2007, p. 275)

Without the illusion of control, scientists would not conceive to construct devices and tools or even to create artificial living components, something that is true for all application areas (cf. Saraniti 2008). In the field of medicine, some scientists are even creating artificial viruses capable of infecting diseased cells with drugs or therapeutic genetic material. Here the artificial virus self-assembles from a nanoribbon to form a double-layered protein sheet that sports short strands of RNA and is capable of inserting RNA into human cancer cells in test tubes. Once inside, the RNA interferes with the cells' malfunctioning DNA with the hope of correcting or treating it (see Yong-Beom *et al.* 2008).

On the one side, we can notice that the emergence of converging technologies, leading to a polarisation of control-oriented technosciences (like molecular biology and biophysics) on the one hand and of complexity- and experience-oriented fields (like ecology or epidemiology) on the other. This could stimulate a growing gap between those scientific fields that have a technological orientation and those developing precautionary practices (Kastenhofer 2007). On the other side, however, since there are many material connections, especially in the US between the NBIC (Nanotechnology, Biotechnology, Information technology and Cognitive science) project and the NNI, and since nature as a plastic engineer interprets precisely this struggle between control and complexity, visions of nature in nano and in the project of convergence could even lead to a rapprochement of different scientific fields. Anyway, the role of these visions still remains under discussion and it is worth further investigation.

The idea of controlling nature and controlling living beings is not new, as it can be traced back to the origin of the molecular biology project where life and its varieties were reformulated as mere variations of the genetic code and its expression. In her study of the experimental approach of molecular biology, Knorr-Cetina (1999) has pointed out that the use of organisms (both cells, organic materials and laboratory animals) as production systems in the laboratory has been motivated and legitimized by a particular conceptualization of life, identified essentially with the reproduction of genetic material. Therefore, the expression 'molecular machines' can be used very well to explain the ontology of the experimental objects used in the laboratory by molecular biologists, including cells, biological materials and experimental animals. The fact that the organisms in the lab are seen as molecular machines (in other words as production sites of specific phenomena the scientist wanted to study under experimental conditions), only becomes possible if organisms in general are described as self-reproducing machines perpetuating their genetic material. Similarly, in nano(bio)technology the cells and molecules existing in nature without being modified, are now themselves described as molecular machines, while acknowledging at the same time their marvellous complexity, perhaps in a more explicit way than in molecular biology. The discovery of the possibility of direct manipulation at the atomic and molecular level combines itself very cleverly with imitation strategies of the complexity of nature.

Moreover, Knorr-Cetina (1999) has also described a "self-reflexive twist" in molecular biology's use of the idea of molecular or biological machines happening at three levels. First, life is conceptualized as a self-reproducing molecular machine. Second, machines are reproduced in the lab in "slow motion", rebuilding them by optimizing some of their parts: the purpose of rebuilding here is to gain an increase of knowledge about the original machines – the natural phenomenon – "by bringing their power and production potential to bear upon themselves, i.e. by using their products to reflect and thereby clarify the original machines" (Knorr-Cetina 1999, p. 156). Third, molecular machines rebuilt in the lab are considered both simulations, i.e. models of real existing processes, not producers of different phenomena that can only happen in the lab.

Bensaude-Vincent (2006) has noticed a fundamental tension in the expression "self-assembly" in nano(bio)technologies: "self" refers to something autonomous that is without an external organizer, i.e. without human intervention; "assembly" is attached to the act of a constructor and it is a word coming from the vocabulary of engineering sciences, thus indicating something artificial. This tension dissolves itself if we consider that nature beyond nanotechnology is presented as something characterized by *plastic machinery*, i.e. as something that integrates the reductionism of engineering sciences and the dynamism of chemistry (Bensaude-Vincent 2009). The engineer is now meant to feel that his enterprise will be crowned by success only to the extent that the system component he has created is *capable of surprising him* (cf. Dupuy 2008). Interestingly we can increasingly find admission of surprise in scientific journals by scientists observing new and unknown properties of materials or living components at the nanoscale (Wiggins *et al.* 2006; Stein 2007; Songmei *et al.* 2008).

The comprehensiveness and the multi-faceted character of the conceptualization of nature in nanotechnologies represent its fascination and its force in providing a wide-ranging explanation of phenomena. This fascination is, however, only explicable if society can accept this kind of vision, and if it is ready to incorporate the fact that natural phenomena are plastic, that we can engineer them and thus can exercise a form of control over them as well as provide the recognition that they are also complex and dynamic.

4. Control in Society: Co-Construction of Nanotechnological Nature and Visions of Society

In this section I will try to combine the two different dimensions of control highlighted already with the idea that the epistemic visions of nature[10] that inform a technology evolve together with visions of the societal role of this technology. On the one hand, the epistemic visions of nature, which determine the essence of nature and, thus, the way in which it is knowable, also influence views of technology, specifically the ways in which man can intervene and modify nature. On the other hand, the direction in which technologies are orientated (i.e. their goals), which relies on the belief of what is possible to achieve through technological development, shapes the ethical and social debate. Their importance can be seen in that they influence what the relevant challenges are that are worth being discussed and, partly implicitly, suggest which kind of problems can be solved through technologies. Therefore, disentangling these images of control is very useful in order to properly understand and to discuss the ethical, social and political implications of nanotechnologies. This kind of analysis can be done for every kind of technology, but it becomes particularly challenging for emerging

[10] By 'epistemic vision of nature' I mean a vision or conceptualization of nature that offers an explanation of how we can analyze nature and learn from it. 'Epistemic' means "related to epistemology", which is the branch of philosophy that studies the nature of knowledge, its presuppositions and foundations, and its extent and validity.

nano(bio)technologies precisely because the epistemic visions of nature are complex and multi-faceted, linked to different visions and projects on how nanodevices have to be built and will work.

In the nanotechnological vision of nature common to both the approach of molecular manufacturing and to the approach of self-assembly, nature is perceived as something that is now reachable in its basic components. Nature as plastic and as an engineer is connected with the idea of *extreme flexibility*: everything can be manipulated and re-shaped according to shared values. In extreme flexibility and ambivalence there also appears the *social use* of this conceptualization of nature. The plasticity of nature can be therefore either oriented toward the promotion of technologies to alleviate ordinary problems and permit a good life or it can be used to transfer the metaphor of recreation and control in society at the individual level, at the level of groups and nations (creating dependency and exploitation) and at the level of the environment. If the material side of the world is fundamentally plastic, the societal part of reality also becomes malleable and this malleability becomes visible in different aspects of the debate around nanotechnologies.

In the following, I will present some initial results of an analysis that is still very much a work in progress. In the current debate, it is possible to identify two moments in which this co-construction of embedded visions of nature and visions on the societal role of technology seem evident. The first refers to the development of the debate on enhancement, while the second considers the way in which goals of technologies are considered in the debate, in particular in so-called ELSI (ethical, legal and social implications) analyses.

4.1 The enhanced engineered control freak

Although the debate on human enhancement didn't originate with nanotechnology, but rather started in relation to changes in the life sciences, especially in medicine (cf. Parens 2005), it soon overcame the strict boundaries of the discussion within the medical sciences, becoming a general topic of converging technologies and also of nano(bio)technology (Roco & Bainbridge 2002; 2006; cf. Ferrari 2008a).

Even if the documents on converging technologies refer to the improvement of human capabilities, it has become clear that the idea of improving all nature represents a powerful narrative of the relationship between emerging technologies nowadays and nature (cf. Wickson 2008): Not only human capabilities, but also the efficiency of plants and the capabilities of animals should be changed (see Harris 2007; Núñez-Mujica 2006; cf. Ferrari 2008a), as well as the characteristics of matter and of entire ecoystems or the environment, as the debate on green nanotechnologies, in particular recent ideas of geoengineering (deliberately manipulating the climate to address global warming) suggests (cf. Forbes/Wolfe Nanotech Report 2006; Varga 2007; Schwarz 2008).

The idea of enhancing nature, rendered possible by this vision of nature as something fundamentally changeable, combines itself very easily with the idea of production - of producing a new nature that now better fits human goals in that it is a direct transfer of human goals into the way nature works. This is very evident in the image of "enhancement evolution", in which the goals of improving human capabilities not only reaches the level of the species, but also even the complex mechanism of evolution, turning the Darwinian theory upside down. The idea here is that through technology it is now possible to guide evolution, whereas in the Darwinian conception, its mechanism is blind (Harris 2007).

Enhancement has also been progressively applied to refer to the societal dimension, along the idea that technologies can also represent tools for shaping societies (Bainbridge 2004). In the article on "The evolution of semantic systems", one of the initiators of the idea of NBIC convergence, Bainbridge (2004), argues that it is even possible to *engineer culture*, i.e. to prepare society for the so-called *memetic engineering*.[11] Continuing with the idea of manipulating every aspect of human life through technology, James Hughes, former President of the

[11] The modern memetics movement dates back to the mid 1980s and was largely influenced by the *selfish gene* theory of Richard Dawkins, who coined the term "meme" to describe a unit of information as analogous to the gene that resides in the human brain and is the mutating replicator in human cultural evolution. For Bainbridge (2004), semantic systems are a set of concepts connected by meaningful relationships and include scientific typologies and ontologies, as well as naturally occurring subcultures.

World Transhumanist Association, has even proposed a sort of *virtue engineering*, a modification of human beings through the means of genetic engineering in order to render them more virtuous and to eradicate bad vices, in particular aggression, thus enhancing the sense of sympathy (Hughes 2003; 2004). Therefore, there seems to be a sort of parallelism or at least linkage between visions of nature as something fundamentally plastic that can be reshaped along engineering principles and society, which in these visions can be also re-designed along an engineering approach. Not coincidentally, both Bainbridge and Hughes speak of "engineering" in their visions of society's modification.

4.2 The society technocontrol freak: the perceptionof the role of nanotechnology in society

Second, attached to the epistemic vision of nature as plastic, a change in the role and expectations and, therefore, in the goals of technology for our society can be noticed. On the one side we can see in the debate about the societal role of nanotechnologies the reproposal of the old idea of technologies solving the most challenging problems of the world. On the other hand, this idea appears combined with the awareness that we as society should be engaged in shaping the nanotechnological development in a 'responsible' direction.

The trend of seeing in technologies great promises of revolution and of fundamental contributions for solving serious problems of the world is not new, since it can at least be traced back to the rhetoric of the "green biotechnological revolution".[12] The case of nanotechnology does however amplify these tendencies. Due to its enabling and ubiquitous character, nanotechnologies are expected to change the approach of manufacturing virtually every product and also the way in which we perceive things and the world. Describing the character of nanotechnology as 'mode 2' (a term introduced in Gibbons *et al.* 1994),

[12] Especially in the field of agriculture emerging biotechnologies have been seen as a means for increasing productivity and thus solving the problem of famine (cf. World Bank 1991).

and/or 'post-academic science' (Vogt *et al.* 2007, p. 330) suggests that nano is directed towards solving compelling problems of our world, like AIDS and climate change. Precisely because nano is expected to permit a new way of manufacturing, it is presented as an open space for possibilities of production and consumption (Nordmann & Schwarz 2008). This fascination becomes very visible not only when considering the multiplicity of products that will be on the market once nanotechnologies have reached their full potential, but also through the fact that these technologies are progressively described as capable themselves of solving the negative aspects of past technologies, such as pollution or the consumption of energy sources (Schwarz 2008). Nanotechnologies, in particular the so-called 'green nanotechnologies', are attached to ideas of reaching a quasi guilt-free and virtually never-ending consumption because through them it will be possible to discover new sources of energy, to stop pollution and even to clean the environment and solve the problem of the scarcity of agricultural resources.

The predominant idea is then that nanotechnologies are now expected to offer incredible new possibilities if developed in the proper way. This representation of nano's potential exercises a powerful seduction that is in some cases perceived as impossible to resist, for the consumers, for the scientists and for the social scientists who cannot resist engaging in the discourse on 'responsible'[13] nanotechnological development. It is an interesting fact that in the scientific journal *Nature Nanotechnology,* a number of articles are dedicated to disentangling the epistemic characteristics and the social implications of nanotechnology. The awareness of the social dimension of science is then increasing among natural scientists together with the acknowledgment of the predominant approach now in STS (science and technologies studies), which is social constructivism. This approach is informed by the idea that science and technologies are social phenomena developing in particular social contexts and that there is a process of coevolution between science and society (cf. Rip 2001). Vogt and colleagues (2007) argue that precisely because nanotechnologies now offer such powerful tools is the reason

---

[13] For an analysis of the rhetoric of 'responsible' development see Ferrari 2009.

that they have to be organized in comprehensive policy programs that must consider the societal implications from the beginning. Therefore, they argue for the introduction of ELSI analyses in research policy as an expression of the understanding that it is no longer possible to argue for the lack of normativity of science or for a strict distinction between pure and applied science (cf. Vogt *et al.* 2007). The idea of complementary ELSI analysis characterizes all research policy on emerging technologies and this becomes very evident when one considers the reports on converging technologies, especially the European one. Whereas the US version of convergence (NBIC) refers only to four scientific domains (nano, bio, info and cogno), the EC commissioned response CTEKS (Converging Technologies for the European Knowledge Society) includes reference to the social sciences and humanities as explicitly a part of the convergence.

To conclude we can notice the emergence of a link between the epistemic visions of nature as a plastic engineer and the versatile role of technology in our society. Nature is plastic in that it can be reshaped along engineering principles. Society is plastic in that it can be reconstructed in a 'scientific' way. In the end the relationship between society and technology also appears as plastic. The idea is that nanotechnologies will render explicit their beneficial potential if we all engage in shaping its development, if we all contribute to rendering it 'responsible'.

5. Conclusion

After having analyzed epistemic visions of nature and the characteristics of the social debate on these technologies, I think that it is possible to notice a mutual transferring of visions of nature between technology and society. On the one hand, nature in nano is described as plastic, something that can be changed and reshaped in every aspect, an idea that has surely long been a trend, and as an engineer, which itself constructs and shapes its components continuously. The conceptualization of nature in nano(bio)technologies, and also increasingly in synthetic biology, incorporates the technoscientific character of the primacy of 'making' over 'knowledge' and it appears permeated by the search for control which, does however, appear in multi-faceted forms (both as mastery or as partial loss) . Precisely because scientists want to reshape the world and construct devices that work, they operate under the guise of control or at least in the illusion of it. The remaining problem in society is whether or not scientists can really achieve the desired control, or if they are only advancing presumptions (that either they themselves are convinced of or that they are trying to sell to society). Obviously, it is not only scientists but also the promoters of these technologies, i.e. policy makers, entrepreneurs and investors, and society as a whole that accepts this illusion of control.

On the other hand, along the conceptualization of nature as a plastic engineer, society also seems to be increasingly presented as a malleable engineer of itself. In the enhancement debate, where this aspect is at its greatest evidence, there are many voices making claims of new kinds of society in which technologies can not only contribute to increasing communication among people and changing ways of doing business, but also with an implicitly more or less strong reference to an engineering of the social dimension (cf. Bainbridge 2004). Social constructivism, as a description of the delicate interplay between society and science, seems to be reinterpreted by some authors to suggest the possibility of constructing the social. If science, technology and society shape themselves continuously, society can also be 'constructed' by science and technologies, not only in the sense that the embedded values in society are influenced by scientific discoveries or new theories, but also in the sense that it is possible to rebuild the social in a similar way to how we can rebuild or change a genome.

The acknowledgment of the mutual shaping of science and society is surely an important and necessary contribution coming from STS studies, but in many cases it appears – especially considering the way in which ELSI analyses are now performed in the case of nanotechnology – as running the risk of distracting from deeper considerations on the role of technology in our society. This doesn't mean that we should go back to a time in which technology was perceived as something evolving detached from the social context and in which scientists were strongly disinterested in the social and political aspects. However, the fact that

ELSI research is now included in research programs and that it is possible to find some publications, some even in scientific journals, should not make us content with this new status quo.

Still, many fundamental questions have not been raised in the debate and have not yet been analyzed. Concerning the question of control as embedded value and its relation with science and technology, it seems important to investigate the cultural processes we have come to accept in society and implement this value as well as the reasons of the force and fascination of this value. Concerning the role of technology, I think it is important to investigate the processes that result in the primacy of science and technology in our time, including on the political and economic level.

Questions for Reflection:

1. What are the contrasts between the culture of the chemist and culture of the engineer in nano(bio)technology? For example, in terms of their visions of nature and control.
2. To what extent do you think that nature, technology and society are 'plastic' and can be actively altered, directed and/or shaped?
3. What role do mastery and surprise play in your life/research?

Bibliography

Bainbridge, W. S. (2004). The evolution of semantic systems. In Roco, M. C., Montemagno, C. D. (eds) *The Coevolution of Human Potential and Converging Technologies*, New York Academy of Sciences (Annals of New York Academy of Sciences, 1013, May 2004, pp. 150-177.

Bath, J. and Turberfield, A .J. (2007). DNA nanomachines, *Nature Nanotechnology,* 2, pp. 275-284.

Bensaude-Vincent, B. (2004). Two cultures of nanotechnology? *HYLE* , 10, (2), pp. 65-82.

Bensaude-Vincent, B. (2006). Self-assembly, Self-organization: A philosophical perspective on converging technologies. Paper prepared for France/Stanford Meeting in Avignon, December 2006, Draft. http://www.u-paris10.fr/servlet/com.univ.collaboratif.utils.LectureFichiergw?ID_FICHIER=6991 (last accessed May 26, 2009)

Bensaude-Vincent, B. (2009). *Les vertiges de la technoscience: Façonner le monde atome par atome*, Paris: Editions La Découverte.

Boeing, N. (2007). Nanotechnologie: Was ist das eigentlich? Ein Gespräch mit Alfred Nordmann, *Die Zeit*, 15.11.2007, 47.

Browne, W. R. and Feringa, W. L. (2006). Making molecular nanomachines work *Nature Nanotechnology*, 1, I1, pp. 25-35.

Bueno, O. (2004). The Drexler-Smalley debate on nanotechnology: Incommensurability at work? *HYLE*, 10, (2), pp. 83-98.

Clausen-Schaumann, H., Seitz, M., Krautbauer, R. and Gaub H.E. (2000). Force spectroscopy with single bio-molecules, *Current Opinions in Chemical Biology*, 4 pp. 524-530.

Crandall, B. C. (ed) (1996). *Nanotechnology: Molecular Speculations on Global Abundance,* Cambridge, MA: MIT Press.

Drexler, E. K. (1986). *Engines of Creation- the Coming Era of Nanotechnology*, USA Anchor Books.

Drexler, E. K. (1992). *Nanosystems: Molecular Machinery, Manufacturing, and Computation*, New York: John Wiley & Sons.

Drexler, E. K. (2003a). Open letter to Richard Smalley, *Chemical & Engineering News,* 81, pp. 38-39.

Drexler, E. K. (2003b). Drexler Counters, *Chemical & Engineering News,* 81, pp. 40-41.

Drexler, E. K. (2006). From Feynman to Funding. In Hunt, G. and Mehta, M. (eds) *Nanotechnology. Risks, Ethics and Law,* UK and USA: Earthscan, pp. 25-34.

Dupré, J. (2003). On human nature. Human affairs, *Journal of the Slovakian Academy of Sciences*, 13, pp. 109-122. http://www.humanaffairs.sk/dupre.pdf (last accessed May 26, 2009).

Dupuy, J. P. (2007). Some pitfalls in the philosophical foundations of nanoethics, *Journal of Medicine and Philosophy,* 32(3) pp. 237-261.

Dupuy, J. P. (2008). *The Mechanization of the Mind: On the Origins of Cognitive Science,* 2nd ed. Princeton: Princeton University Press.

Duwez, A. (2008). Molecular cranes swing into action, *Nature Nanotechnology*, 3, pp. 188-189.

Eigler, D. M. and Schweizer, E. K. (1990). Positioning single atoms with a scanning tunneling microscope, *Nature,* 344, pp. 524 - 526.

ETC Group (2003). The big down: Technologies converging at the nano-scale. www.etcgroup.org/upload/publication/54/02/com8788specialpnanomar-jun05eng.pdf (last accessed May 26, 2009).

ETC Group (2004). Down on the farm. The impact of nano-scale technologies on food and agriculture. http://www.etcgroup.org/en/materials/publications.html?pub_id=80 (last accessed May 26, 2009).

Ferrari, A. (2008a). Is it all about human nature? The ethical challenges of converging technologies beyond a polarized debate, *Innovation. The European Journal of Social Science Research*, 1(21), pp. 1-24.

Ferrari, A. (2008b). "Nanomaschinerie des Lebens": Ethische Herausforderungen der Nanobiotechnologien, In Köchy, K., Norwig, M., Hofmeister, G. (eds.) *Nanobiotechnologien. Philosophische, anthropologische und ethische Fragen*, Karl Alber Verlag, Freiburg, München, pp. 321-350.

Ferrari, A. (2009). Initial and recent developments in the debate on nanoethics: More traditional approaches and the need for a new kind of analysis, In *Nanoethics,* in press.

Feynman, R. (1960). There is plenty of room at the bottom, *Engineering and Science*, February. http://www.zyvex.com/nanotech/feynman.html (last accessed May 26, 2009).

Forbes/Wolfe Nanotech Report (2006). Nanotech Could Give Global Warming a Big Chill, vol. 5, N. 7, July 2006. http://www.qsinano.com/pdf/ForbesWolfe_NanotechReport_July2006.pdf

Gibbs, W. (2004). Synthetic life, *Scientific American*, May, pp. 74-81.

Gibbons, M. Limoges, C., Nowotny, H., Schwartzman, S., Scott, P. and Trow, M. (1994). *The New Production of Knowledge*, London, SAGE Publications.

Goel, A. and Vogel, V. (2008). Harnessing biological motors to engineer systems for nanoscale transport and assembly, *Nature Nanotechnology,* 3(8), pp. 465-475.

Goldstein A. (2006), Potential and dangers of nanotechnology (Part 1). Interview with Arthur Bruzzone in San Francisco unscripted, The Weekly TV Series on The City's Politics, Culture and Personalities, 23.11.06. http://video.google.com/videoplay ?docid=5221560918013409256.

Goodsell, D. S. (2004). *Bionanotechnology: Lessons from nature*, Hoboken, New Jersey: Wiley-Liss Inc.

Gould, K. (2005). The treadmill of production: The case of nanotechnology. Paper prepared for the Development, Governance and Nature panel discussion, Cornell University, April 4, 2005. http://www.einaudi.cornell.edu/files/calendar/4951/GouldCornellDGNnanotech.pdf

Hard, M. and Jamison, A. (2005). *Hubris and Hybrids: A Cultural History of Technology and Science*, New York: Routledge.

Harris, J. (2007). Enhancing *Evolution? The Ethical Case for Making Better People*, Princeton: Princeton University Press.

Hottois, G. (1984). *Le signe et la technique. La philosophie à l'épreuve de la technique*, Paris: Aubier Montaigne.

Hughes, J. (2003). Rediscovering utopia. http://www.betterhumans.com (last accessed May 26, 2009)

Hughes, J. (2004). *Citizen Cyborg*. Cambridge, MA: Westview Press.

Hunt, G. (2006). Nanotechnoscience and complex systems: The case for nanology. In Hunt, G. and Mehta, M. (eds) *Nanotechnology. Risks, Eethics and Law,* UK and USA : Earthscan, pp. 43-58.

Joy B. (2000). Why the future doesn't need us, *Wired* 8.04, April 2000, http://www.wired.com/wired/archive/8.04/joy.html (last accessed May 26, 2009)

Jones, R. (2004). *Soft Machines*, Oxford: Oxford University Press.

Jones, R. (2008). The production of knowledge, *Nature Nanotechnology,* 3, pp. 448-449.

Kastenhofer, K. (2007). Converging epistemic cultures? A discussion drawing on empirical findings. *Innovation: The European Journal of Social Science Research* 20(4), pp. 359-373.

Kearnes, M. B. (2006). Chaos and control: Nanotechnology and the politics of emergence, *Paragraph* 29(2), pp. 57-80

Knorr-Cetina, K. (1999). *Epistemic Cultures. How the Sciences Make Knowledge*, Cambridge Massachussets: Harvard University Press.

Kufer, S. K., Puchner E. M., Gumpp H., Liedl T., and Gaub H. E. (2008). Single-molecule cut-and-paste surface assembly, *Science*, 3, pp. 594-596.

Kurzweil, R. (1999). *The Age of Spiritual Machines: When Computers Exceed Human Intelligence*, London: Viking Press.

Kurzweil, R. (2000). Promise and the peril, *Interactive Week*, October 23, 2000.

Kurzweil, R. (2005). Nanoscience, nanotechnology, and ethics: Promise and peril. Mitcham, C. (ed) *Encyclopedia of Science, Technology and Ethics,* MacMillan Reference Books, p. xli-xlviii.

Lacey, H. (1999). *Is Science Value Free? Values and Scientific Understanding*, London and New York: Routledge.

Latour, B. (1987). *Science in Action: How to Follow Scientists and Engineers Through Society*, Cambridge: Harvard University Press.

Maskin, E. (2007). Mechanism Design Theory: How to Implement Social Goals. http://nobelprize.org/nobel_prizes/economics/laureates/2007/maskin-slides.pdf (last accessed May 26, 2009).

Meridian Institute (2006). Overview and comparison of conventional water treatment technologies and water nano-based treatment technologies. Background paper for the International Workshop on Nanotechnology, Water and Development. www.merid.org/nano/watertechpaper (last accessed May 26, 2009)

Moore, G. (1965). Cramming more components onto integrated circuits, *Electronics Magazine,* 19 April.

Moriarty, P. (2008). Reclaiming academia from post-academia, *Nature Nanotechnology,* 3, pp. 60–62.

National Nanotechnology Initiative (2007). *Strategic Plan 2007.* www.nano.gov/NNI_Strategic_Plan_2007.pdf

National Science and Technology Council (1999). *Shaping the World Atom by Atom.* http://itri.loyola.edu/nano/IWGN.Public.Brochure (last accessed May 26, 2009)

National Science Foundation (NSF) (2001). *Societal Implications of Nanoscience and Nanotechnology,* National Science Foundation Report, March 2000. http://www.wtec.org/loyola/nano/NSET.Societal.Implications/ (last accessed May 26, 2009)

Nordmann, A. (2004). Collapse of Distance. Epistemic Strategies of Science and Technoscience. http://www.unibielefeld.de/ZIF/FG/2006Application/PDF/Nordmann_essay2.pdf

Nordmann, A. (2005). Noumenal Technology: Reflections on the Incredible Tininess of Nano, *Techné* , 8, 3, http://scholar.lib.vt.edu/ejournals/SPT/v8n3/pdf/nordmann.pdf (last accessed 26.05.09)

Nordmann, A. (2007). If and then: A critique of speculative nanoEthics, *Nanoethics*, 1(1), pp. 31-46.

Nordmann, A. (2008). Technology naturalized: A challenge to design for the human scale. In Vermaas, P. E., Kroes, P. A., Light, A., Moore, S. (eds) *Philosophy and Design*, New York, Berlin: Springer, pp. 91-104.

Nordmann, A. and Schwarz, A. E. (2008). The lure of the "yes": The seductive power of technoscience. In Kaiser, M., Kurath, M., Maasen, S. and Rehmannn-Sutter, C. (eds) *Assessment Regimes of Technology. Regulation, Deliberation & Identity Politics of Nanotechnology*, Dordrecht, NL: Springer.

Núñez-Mujica, G. (2006). The ethics of enhancing animals, specifically the great apes, *Journal of Personal Cyberconsciousness,* 1(1).

Parens, E. (2005). Autheticity and ambivalence: Toward understanding the enhancement debate, *Hastings Center Report,* 35(3), pp. 34-41.

Rip, A. (2001). Contributions from social studies of science and constructive technology assessment. In Stirling, A. (ed), *On Science and Precaution in the Management of Technological Risk. Volume II. Case Studies,* Sevilla: Institute for Prospective Technology Studies (European Commission Joint Research Centre), November 2001, pp. 94-122

Roco, M. C. and Bainbridge, W. (eds) (2002). *Converging Technologies for Improving Human Performance: Nanotechnology. Biotechnology, Information Technology and Cognitive Science,* Arlington, VA: National Science Foundation.

Royal Society (2004). *Nanoscience and Nanotechnologies: Opportunities and Uncertainties*, London: Royal Society. http://www.nanotec.org.uk/finalReport.htm (last accessed May 26, 2009)

Salamanca-Buentello, F., Persad, D. L., Court, E. B., Martin, D. K., Daar, A. S. and Singer, P. A. (2005). Nanotechnology and the Developing World, *PLOS.*http://medicine.plosjournals.org/perlserv/?request=get-document&doi=10.1371/journal.pmed.0020097&ct=1 (last accessed May 26, 2009)

Saraniti, M. (2008). Artificial cells: Designing biomimetic nanomachines, *Nature Nanotechnology,* 3, pp. 647 – 648.

Schafmeister, C. E. (2008). Lego mit Molekülen, *Spektrum der Wissenschaft*, Dossier 1/08 Moleküle im Wandel, pp. 12-19.

Schwarz, A. (2008). Grüne nanotechnologie? In Köchy, K., Norwig, M. and Hofmeister, (eds) *Nanobiotechnologien. Philosophische, anthropologische und ethische Fragen*, Freiburg/ München: Verlag Karl Alber, pp. 85-106.

Smalley, R. (2001). Of chemistry, love and nanobots, *Scientific American*, 285, pp. 76-77.

Songmei, W., Gonzalez M. T., Huber, R., Grunder, S., Mayor, M., Schönenberger, C., Calame, M. (2008). Molecular junctions based on aromatic coupling, *Nature Nanotechnology*, 3, pp. 569-574.

Stein, D. (2007). Nanopores: Molecular ping-pong, *Nature Nanotechnology*, 12, pp. 741-742.

Theis, T., Parr, D., Binks, P., Ying, J., Drexler, K. E., Schepers, E., Mullis, K., Bai, C., Boland, J. J., Langer, R., Dobson, P., Rao, C. N. and Ferrari M. (2006). Nan'o.tech.nol'o.gy n, *Nature Nanotechnology*, 1, pp. 8-10.

UNESCO (2006). *The Ethics and politics of nanotechnology*, UNESCO. http://unesdoc.unesco.org/images/0014/001459/145951e.pdf (last accessed 26.05.09).

Varga (2007). Nano Solar News: Global Warming & 2015 Cost Trends. In Nanotechnology Now, http://www.nanotech-now.com/columns/?article=038 (last accessed May 26, 2009).

Vogt T., Baird, D. and Robinson, C. (2007). Opportunities in the 'post-academic' world, *Nature Nanotechnology,* 2(6), pp. 329-332.

Whitesides, G. (2001). The once and future nanomachines, *Scientific American*, September, pp. 78-83.

Whitesides, G. (1995). Self-assembling materials, *Scientific American*, 273, pp. 146-149.

Wickson, F. (2008). Narratives of nature and nanotechnology, *Nature Nanotechnology*, 3, pp. 313-315.

Wiggins, P., van der Heijden, T., Moreno-Herrero, F., Spakowitz, A., Phillips, R., Widom, J., Dekker, C. and Nelson, P. C. (2006). High flexibility of DNA on short length scales probed by atomic force microscopy, *Nature Nanotechnology,* 1, pp. 137-141.

Woodrow Wilson International Center for Scholars (2007). Nanotechnology and Life Cycle Assessment. Syhtesis of results obtained in a workshop at Washington D.C., 2-3 October 2006. www.nanotechproject.org/file_download/168 (last accessed May 26, 2009).

Wood, S., Jones, R., Geldart A. (2003). *The Social and Economic Challenges of Nanotechnology,* London: Report to the Economic and Social Research Council (ESRC).

World Bank (1991). Agricultural biotechnology: The next green revolution? *Technical Papers of the World Bank*, N. 133,. http://ideas.repec.org/p/fth/wobate/133.html.

Xu, J. and Lavan D. A. (2008). Designing artificial cells to harness the biological ion concentration gradient, *Nature Nanotechnology*, 3, pp. 666 - 670

Yong-Beom, L., Yoon, Y.-R., Lee, M. S. and Lee, M. (2008). Filamentous artificial virus from a self-assembled discrete nanoribbon, *Angewandte Chemie International Edition*, 47(24), pp. 4525-4528.

Zhang, W. and Zhang, X. (2003). Single molecule mechanochemistry of macromolecules, *Prog. Polym. Sci.*, 28(8), pp. 1271-1275.

Chapter 14

It's Perfect and I Want to Leave

Lupin Willis

The air outside was dusty dry.

Eye grazingly, throat scorchingly, brain tighteningly, dusty dry. It circled around the old lady as she gently closed the door. Another reminder of the life outside, she thought. Outside these walls. Outside this body. Outside these beliefs. But the boy breezed by her without a care in the world.

"How can you bare that air?" she gasped as he passed.

"What? It's great out there!" the boy dismissed, "You should come this afternoon, the launch is going to be awesome!"

That damn launch, she thought. It was all anyone thought about. They were right though, it was indeed time to leave. Leave before it became time to flee.

"No thanks Sugar, I'm going another way."

The boy began busily searching the kitchen cupboards, flinging them open one at a time and scanning the contents for something he approved of.

"Got any Peshcrofts in here? Wisklocks? Atendaires? Anything at all that isn't green and wilting already?" She sat down, disheartened as usual. Drowning under a wave she felt like she was watching envelop her slowly. Slow enough that sometimes she even forgot to notice. Notice it seeping in through the pores of her skin, sliding into her blood and whirling through her brain. Again and again and again and again.

Nano Meets Macro - Social Perspectives on Nanoscale Sciences and Technologies
by K L Kjølberg & F Wickson

www.panstanford.com
978-981-4267-05-2

"It's this air" she sighed. "It wilts everything…it's heavy with too many molecules."

"Now you're talkin' crazy Gran!" the boy laughed. "Come on, I'll make you a cup of tea. I'm sure you've got some homemade cookies around here somewhere…"

He was a good boy, she thought for the thousandth time in her life, a good boy to her.

The tea came steaming and sweating with scent. Honey and ginger today, she thought, how sweet of him to remember.

"You know the air's perfect Gran" the boy said sitting down, enormous bunch of chocolate chip cookies in hand. "The sensors monitor everything – particulates, UV, humidity, you name it. I checked mine like an hour ago – they're all in acceptable range."

"That's not what I sense" she grumbled. "You know, with my eyes, ears, nose, you name it." He gave her a friendly scowl. "Doesn't my experience of the world count for anything? And who decided what an 'acceptable range' is anyway? Don't I get a say in that?" she lamented with a hint of aggression.

"I really don't know what you've got against the biosensors Gran. They just give us more info, you know?" The boy continued dunking his cookies, sucking them up just before they slipped.

"I'm just saying that I have senses too Sugar, and they are telling me a different story. A story that is dry and heavy and penetrating. Tell me, how does the air smell to you my boy?" He shook his head smiling at her. Of course he couldn't smell anything, she thought, he'd forgotten how. The sense of smell was losing its importance, losing the evolutionary struggle, being let go as unnecessary for survival. No need to know what something smelled like when you could get a full technical readout on its condition in an instant. Any instant, anywhere. It was perfect, of course.

"It smells like solar trees to me" he winked at her. She could tell by the glint in his eye that he was teasing her. They had covered this ground before, many times. His retort made her smile. A weak, worn and weary smile, but a smile nonetheless.

"Don't get me started on those things!" she huffed in a half laugh, "You know how I feel about them."

"Yeah, but I still don't really get it Gran" the boy pressed. "Solar cells, shaped like trees, giving us energy, what can possibly be wrong with that? Seriously. What is wrong with forests of solar cells?"

"We used to have forests of trees!" she spat indignantly.

"Now we have forests of solar trees!" the boy bounced back.

"No! Now we have parking lots full of Solrest corporation sun farms, not forests my boy. Forests have birds and bees and badgers, beavers and bears, not just boxes of battery power. Tell me, what do you know of beavers my boy?" He stifled a giggle.

"But Gran! What incredible advances we have made in energy storage! Do you know what I heard the other day? "

"Please don't tell me Sugar, I don't want to know. I want to know what you know about beavers." She whimpered a little and took the last long drink of warm tea, attentively sensing it streaming into her belly, relaxing her in waves. The boy waited quietly. He knew he had touched a nerve.

"Wood was a form of energy storage too you know" she began again softly. "Wood that we had to work hard for, that we had to thank the tree for, the soil for, the air for…"

"I thank the Solrest Corporation every night when I use the Activision Gran, I promise!" He winked at her again and she could tell that he wasn't going to drop this. "Besides, the energy storage of wood was the forests' problem, right? That's what got 'em cut down. That and all the space they took up. The dynanocells can be printed cheaply, put anywhere and work at higher efficiency than plant leaves. Everybody's using them! It's all good Gran, I tell you!"

"And what happened to the leaf shaped beetle Sugar? Do you remember him?"

"Oh no! Not again with the leaf shaped beetle!" the boy laughed good humouredly. He loved arguing with his grandmother, she always pushed his weaknesses, sweet stubbornly insistent old lady that she was.

"I told you, I'll grant you that we lost *something* when we lost that beetle, but you have to admit that we *gained* something with the solar tree!"

"My question, my boy, is what is it exactly that you think we have gained? Have you ever asked yourself that? Did we actually want it? Need it? Appreciate what we lost? What happens if I call it a solar machine instead of a solar tree? Do you feel differently about it then?"

"That's more than one question Gran. We might need more tea."

He was a good boy. Good to her.

She let her gaze drift out the window as the boy returned to the kitchen. In some way, he was right of course. She couldn't see anything wrong with the air. It looked perfect. Nothing but blue sky and light breeze blowing across scattered clouds. At least I can still see the sky she mused gratefully. See the sky and feel the soil. For those small gifts she would always be grateful.

"Why don't you at least start using smart packaging Gran" the boy yelled from the kitchen, head immersed in the frigerator. "There is no way to tell how old some of this stuff in here is! With smart packaging it's so simple." He began singing the jingle: "If it's yellow, let it mellow. Blue? Well it's waiting just for you! Red and I'm a little more than ready, while green means going going gone!"

Her head started pounding. She was tired again, really feeling her age today. The weight of all those years, the effects of all their wear and tear. She wiped a dirty tear from her leaking eyes and thought about lying down for a while. She thought the boy might not mind. She turned to look at him and her heart filled with love. He was turning over her vegetables one by one, inspecting them in his own special way.

"What is this anyway?" he asked, holding one up for her to see.

"Jonathon. James. Earthmore. Wilson" she chastised him loudly. "That, is a cucumber! And if you don't know that, you should have spent many, more years in these fields rather than running away to that so-called school of yours!"

"Cucumber…yeah right" he mumbled, "maybe I remember it now…Seedy little beast, right? Man, it totally freaks me out that we just used to plant any old seed and eat whatever the result was. How uncivilized we were! No quality control at all."

"Your grandfather and I knew exactly what a quality seed was, thank you my boy!" She was feeling particularly argumentative with him today. "Our years of breeding experience, and life just living on this land, taught us all we needed to know and you used to be quite happy eating the result, I can tell you!"

"But crop control makes sure we know exactly what we're growing. Cloning keeps all the plants the same, the hydroponic ponds work with the biosensors to deliver everything the plants need, and the genetic fix has all the insects and diseases covered. It's perfect!"

"And the taste, my boy, how does it taste?"

"Taste, smell, whatever Gran! It'll taste like whatever you want to add to it." She knew she was never going to convince him on the importance of food having great taste before additives. She had tried before, many times.

"Okay then, how do you know that the plant is getting everything it needs? What if it needs love, or companionship, is it getting those things?"

"You can't prove to me that a plant needs love Gran! Maybe you need to love them, but they don't need you."

"We all need each other!" she pounded on the table. Surprised at her sudden outburst, they sat in silence for a while.

"Your tea is getting cold" he eventually offered in subtle surrender. They continued to sit quietly in each others company. Thinking their own thoughts. Generations apart. Worlds apart. Bound by love and intertwined lives.

The old lady started rubbing her temples. Remixing her memories as her mother used to say. Let us bring forth the memory of the birth then, she whispered softly to herself. Closing her eyes she saw it all replay. The pain. The joy. The relief. Bundled in blankets, those tiny rosy red cheeks glowing in the snow. Such beauty. Such innocence. Her tears. Tears of exhausted, enamored wonder. The feeling of holding that little beating heart. That sweet smell of her new life just arrived. Her perfect baby girl. She remembered watching her breathe, stroking the tiny wisps of hair around her ears. She remembered swallowing her outrage. A Forbidden? How could this perfect little girl be a Forbidden? Just for being free-born - born from a woman untested, uncontrolled, unquarantined, and unguaranteed. The guilt she had endured hiding her 'irresponsible risk to reproduction'. The constant fear and guilt, just for bearing her perfect little free-born baby. She remembered thanking Gaia for her mother's love and support. So wise in the ways of the world she was. How she had wanted to be just as good a mother to her little bundle as her own mother had been to her. Her mother who had held her, hidden her, and helped her give birth. Her mother who had handed her this precious little baby girl. "And what happened to that perfect little baby girl?" the mean part of herself asked. "She left", she snapped back. "No", she thought more slowly, "she is here with you in the form of the boy".

He seemed to be waiting for permission to speak.

"Sugar" she asked softly. "What did your mother tell you about why she stopped speaking to me?"

"Oh come on Gran, you know what she told me. She told me you had gone crazy, that she couldn't reason with you anymore, and that she thought that made you dangerous. Happy?"

"She was perfectly right of course. But why did you come back to me after she died? Didn't you believe her?"

"I wanted to see for myself, I suppose. I remembered you Gran, from when I used to come by as a kid, and I kind of always thought you were a bit crazy, but that was okay, because I liked your kind of crazy." The boy was fidgeting, not looking at her.

"You didn't think I was dangerous?" He laughed out loud. "Look at yourself Gran, no ones ages like you anymore! Dangerous?! How funny…"

"Yes…" she laughed weakly. "You're right of course"

"You're not to blame for what happened Gran" he said in a suddenly serious tone, turning to face her. "Mum died because she lost control of that vehicle. What you said to each other before she left and what state she was in, none of that matters. It was her choice to leave and that's it. End of Story. Life over." She felt tears welling, felt like her heart would break. Would she ever have the courage to tell him? Tell him that it was her fault? Her fault that his mother had not been tested. Her fault that she, as it turned out, had not been perfect. Not perfect at all.

"Let's talk about something else Gran" he said gently taking her hand. "Why don't you come tonight, come and watch the launch with me? It's going to be a real moment in history you know - *One astounding moment in time!* He quoted the commercials and looked at her hopefully. "Don't you want to experience it?" She thought about this for a while, wiping at her eyes with the edge of her sleeve.

"I agree that tonight will hold a special moment in time my boy, and I will mark it in my own way, I promise. I am an old lady though, not some booted up threenager! I will celebrate it in a special way, right here in the garden, and I will miss you. That is for sure."

"I'll miss you too Gran" he said, giving her a sideways squeeze, "but I'll see you next week old woman, so let's not get all teary about it". He wiped another dirty tear inching its way down her cheek. "These eyes…" he muttered in frustration, trailing off to give her a firm embrace. The old lady sighed gratefully and softened in her grandson's arms; at least he had stopped pushing her to have her eyes done. Regeneration work was never for her but he usually needed convincing of that too. She sure wasn't much fun to be around anymore. No more running, no more climbing, no more skiing. No more of any of that for her anymore, ever. She sighed heavily.

"Don't worry Gran, I'll photosnatch it for you and show you on my capsule next week. No worries. What are you going to do anyway?"

"Sleep, my boy" she said quietly. "I am just going to sleep."

They said their goodbyes in the usual way, him turning to wave as he hit the bottom of the stairs.

"See you next week!" he called. She, as always, blew him a kiss in reply.

"Yes of course" she called after him. "Perfect."

He was a good boy. Good to her. She wondered once again if she should leave him a letter.

After the sun melted and stars coated the sky, she heard the loudspeaker begin the countdown. At this signal, she stood up from the table and made her way out to the coat rack as she had done so many nights before. She took out her long woolen blanket wrap, buttoned from top to toe, slipped on her boots and slid out the back door. Slowly, carefully and breathing deeply, she made her way out through the garden. When she reached the fig tree she greeted it like an old friend. Leaning against it for support, she slid off her boots and placed them carefully beside her in the hollow of the tree's tall roots. With some effort, she slowly lowered her weary body and lay down in the burrow-like bed she had made for herself. The ground embraced her as it had every night since her daughter's burial and she sighed in relief. As the countdown concluded and a stream of firing thunder roared towards the stars, she lay in the earth and looked at the sky.

"Perhaps now, I am ready to die" she whispered, gently closing her eyes.

Questions for Reflection:

1. Could you be unhappy in a perfect world?
2. To what extent do you think our sensory perception of the world can/should be replaced by technology?
3. Do we value nature and technology differently? Why/Why not?

The Metamorphosis of Forms

Giuliana Cunéaz

Images of the nano-world obtained with powerful microscopes allowed me to see wonderful forests, flowers and dust clouds. I saw the cosmos mirrored in the microcosm, similiar to the way the branched-out structure of minerals mirrors the peripheral area of a neuron, or thc atomic structure of certain polymers, or even the shape of a tree or a coral. Portuguese author Fernando Pessoa has described the idea of 'one single multitude' and I was inspired to show the coexistence of different creative forms that do not necessarily lead to unity in a work of art.

In 2006 I created a series of *screen paintings*, where I combined the two processes of video and painting, and wondered about the metamorphosis of forms, their chasing after one another in different and unpredictable environments. If, through self-replication (a powerful promise of nanotechnology), molecules can spawn copies of themselves and thus abolish diversity, with my painted insertions I wanted to follow the opposite course, and bring diversity back. I enhanced the ambiguity of the image through the addition of elements drawn from the animal or vegetable world and painted directly on the screen so as to multiply the image as if in a Zen ritual.

Guiliana Cunéaz began exploring nanoscale images in 2003, attracted to a secret, invisible and mysterious universe that transcended the most evident appearances of matter. Her whole exploration has always involved attempts at crossing the dividing line that artificially separates our way of perceiving reality between the conscious and the unconscious, the visible and the invisible. Complexity, in all its different aspects, has always fascinated her.

Question for Reflection:

Do you think that atoms dream?

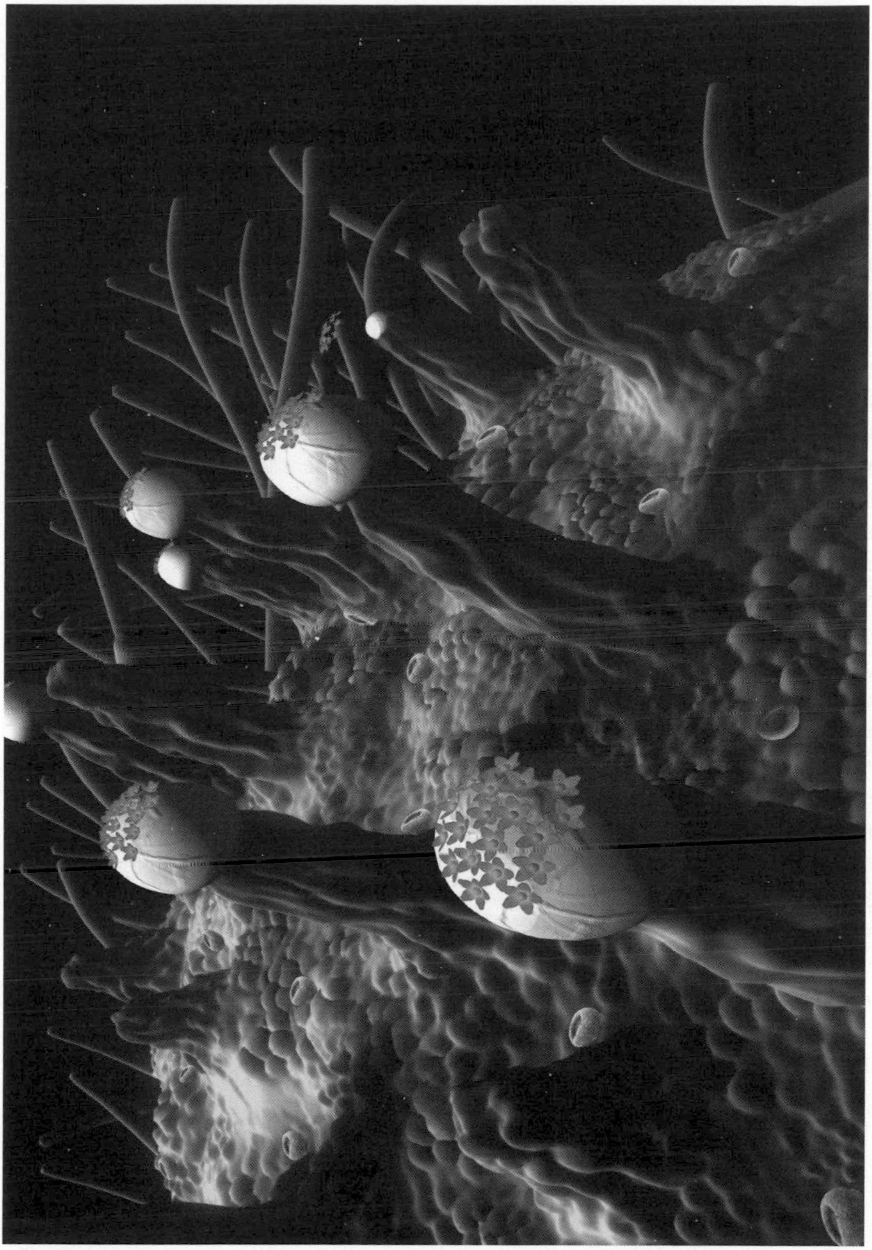

For colour reference turn to page 552.

The Crucifixion of Nemesis

Lukas E. Sivak

Nothing would exist without limit. All truth, morality, language and thought, the very concept of existence itself, is only possible because of limit, because we draw a line between one thing and another. From limit derives approximation and mortality, from these is born our consciousness, creativity and compassion - our humanity.

Nemesis is the goddess of moderation, of proportion. With a measuring rod in one hand and a sword in the other, she is the embodiment of the concept of limit. There was a time when Nemesis was feared and respected; a time when it was understood that she would punish, proportionally, those whose arrogance drove them to attempt to become gods. Perhaps this a myth, but imagining it holds up a mirror to our own time. Nemesis is no more to us now than a quaint archaism, a dismembered lump of marble that we may still marvel at, but no longer possess the faculty to feel the presence of. We rage against the limits that surround us in petulant fits of excess, demanding 'more', demanding 'beyond', demanding 'forever'. We no longer wonder what the automatic consequences of excess will be. We put such considerations down to cowardice, conservatism or lack of vision. We believe that the faculty of understanding is given to us so that we can manipulate bigger and bigger elements in the universe, smaller and smaller molecules, that we can take control, gain power. Yet Nemesis is still there, woven into the web of laws which govern all things. Every time our pride, ignorance, vanity, greed and desire for power push us too far, we are crucifying her again. Perhaps she understands our frustration and out of compassion endures. Yet she will only be stretched so far and when that limit is reached, her punishment is as inevitable as our transgression.

Question for Reflection:

How does this relate to nanoscale science and technology?

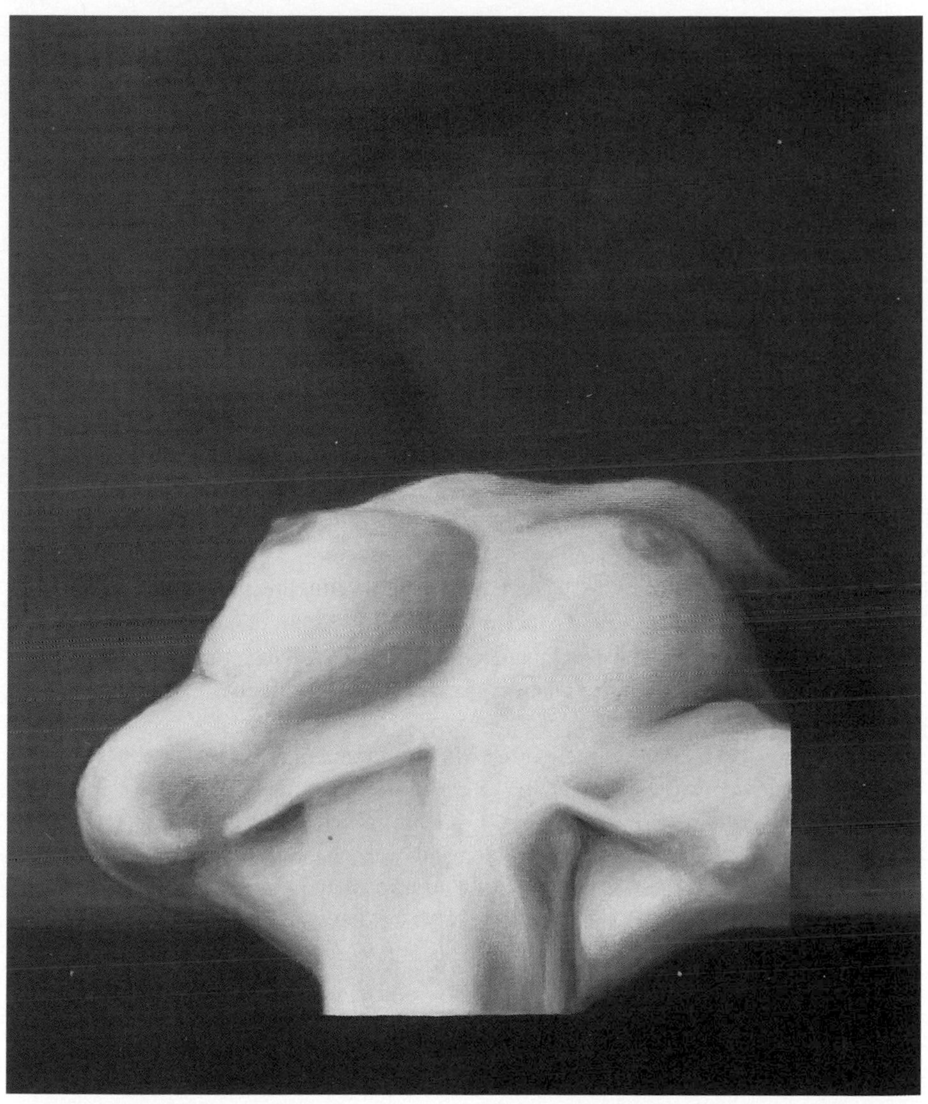

Entanglio

Jack Mason

Images of all things atomic are not photographic; they are inherently digital. The atomic scale is smaller than the wavelengths of light, therefore we are only able to perceive and represent molecular reality indirectly, i.e. by sensing the minute forces of nanoscale objects. This information must then be translated into visual expression. My visual art sometimes seems to be a collaboration with the very atoms, molecules and nanoscopic ghosts that I've encountered in my source material. These fundamental components of physical reality are, after all, effectively immortal: all atoms have existed since the dawn of time, and are continually being recycled and re-used. In some ineffable way, I feel they are seeking to express themselves, and I am merely their conduit. Each giclee represents my own interpretation of digital images of the nanoscale. My technique is similar to the composition of techno music, in that I combine multiple molecular images, transfiguring these until they reveal new characteristics, patterns or motifs. The process reflects experiments being conducted by nanotech researchers and is an attempt to build a visual vocabulary for a landscape of staggering smallness.

John Hastings (Jack) Mason grew up in Minnesota and New Jersey, and is a graduate of Brown University (B.A., Semiotics) and Columbia University's Graduate School of Journalism. Along his varied way, Jack has been a pole vaulter, an actor, a screenwriter, a journalist, a web producer, and a keyboard player. Since 2004, he has worked in strategic communications and social media at IBM in New York City.

Question for Reflection:

Can you imagine the journey of an atom since the dawn of time?

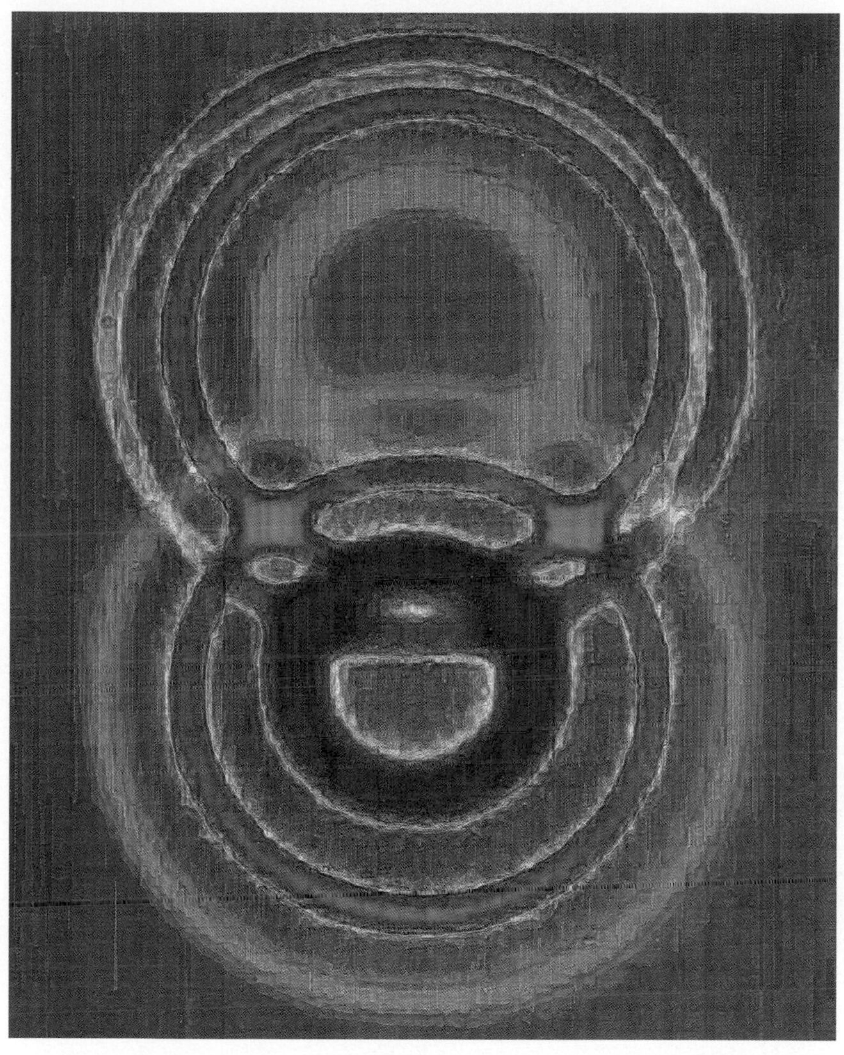

For colour reference turn to page 554.

Nano Meets Macro:
In the Tough Decisions

This section presents perspectives related to the management, regulation and governance of nanoscale sciences and technologies. The contributors relating to this node are specifically concerned with how nanoST are approached in the processes and institutions for decision-making. It is a central part of much of the work around this node to show why and how these kinds of decisions should extend traditional political institutions to include enhanced public and/or stakeholder involvement. Some of the central concepts in this section are: institutions, expertise, risk, uncertainty, public participation, public engagement, responsibility, power, values, decision-making, trust and law.

The notion of governance is frequently used in the academic contributions in this node and is a term that is commonly associated with demands of transparency and broader involvement in decision-making processes. With increased transparency, regulatory processes are being exposed to those outside the political institutions involved. The general idea with broader involvement in decision-making is that these processes should also be the object of input from outside traditional regulatory institutions. For nanoST, as this section shows, these ideas have become extremely important.

In detail, this section contains six academic contributions, one short fiction story and two artworks. In chapter 15, 'Smoke that Thunders: Risk, Confusion and Regulatory Frameworks', Diana Bowman and Geert van Calster outline the current situation regarding the regulation of nanoST, especially focusing on risks related to nanomaterials in cosmetics. The authors' main attention is European legislation, but they also draw some lines to what happens in other parts of the world. They speculate on whether what they identify as a current lack of regulatory adjustments for nanoST will soon be followed by a sudden cascade of legislation aimed at addressing regulatory gaps opened up by the emerging technology.

Gian Carlo Delgado, in chapter 16, focuses on the governance of nanoST in a Latin American context, and specifically in Mexico. Delgado is mainly concerned with economic aspects of nanoST, and the possibilities for Latin American countries to profit from the enormous investments made in nanoST industrialisation. One of the main issues in this respect is how to take part in the economic benefits, while at the same time avoiding the risks and ensuring that the development takes place in an accountable, transparent and inclusive manner.

Georgia Miller and Rye Senjen in chapter 17, 'Nanotechnology in Food and Agriculture', are interested in the consequences of a massive investment in nanoST solutions in agriculture and food production. They are especially interested in its role in steering development in the direction of an even more industrialised and globalised agricultural sector. The authors point to the fact that broader social consequences of this development are seldom discussed. They pose the question of whether nanoST in agriculture is the solution to major challenges related to nutrition and food-safety, and argue that in their opinion it rather increases these crises.

In chapter 18, Fern Wickson, Frøydis Gillund and Anne Ingeborg Myhr focus on scientific advice and the process of risk-assessment in decision-making for nanoST. They use the example of nanoparticles to describe different types of scientific uncertainty, and argue that in the current regime of risk-assessment as decision-support, there are areas of uncertainty not being taken into account. The authors continue to describe a number of alternative ways forward for decision support wherein both quantitative and qualitative uncertainties may be acknowledged.

Sarah Davies, Matthew Kearnes and Phil Macnaghten have authored chapter 19, entitled 'Nanotechnology and Public Engagement: A new kind of (social) science?' They give an account of the occurrence of public engagement and lay involvement in the development of nanoST. The authors draw attention to the way in which the increased tendency to want to govern nanoST in terms of their social relations, is one of its main characteristics. In analysing this trend, they discuss the role for social sciences in this picture, and the way that the opportunity to be involved with nanoST at an early stage of development puts new demands on this line of research. They suggest that social sciences working with nanoST should use this opportunity to, among other things, be more exploratory in its approaches, much like nanoST itself.

Hope Shand outlines, in chapter 20, the role that civil society organisations have had in initiating debate and critical engagement with nanoST development and regulation. Shand uses the Canadian civil society organisation ‘the ETC group’ as an illustration, and gives examples of how through their campaigns they have been pioneers in debates about nanoST governance. The ETC group is particularly famous for arguing for a moratorium on nanoST until more is known about their consequences. Shand also emphasises the importance of seeing nanoST in a broader context as one field within number of technologies that are drivers of societal change.

In the short fiction story called ‘Moving In’ Lachlan Atcliffe describes a future in which nanoscale technologies have become commonplace. He explores this world through the eyes of a young lawyer and points to some of the aspects of daily life that may change, as well as what is likely to remain the same, in a potential world dense with nanoST. Through the legal cases presented to us by the lawyer in the story, the reader is forced to reflect upon the many small and large tough decisions that new technologies inflict upon us.

In her artwork ‘Quoiazander’, Leda C. Sivak portrays the way humans and their technologies are embedded in the environment. She alludes to the way technologies are used in indigenous cultures and to the notion of sustainability. In this, she causes us to reflect upon the way that decisions sometimes imperceptibly and over long periods of time steer us in one direction or another – until it seems like it was unavoidable that we ended up exactly where we did.

“Discovery,” is an artwork created by David Hylton. Hylton places the viewer outside, but within reach of, the possibilities of the nanoscale. In doing so he challenges their curiosity and desire to know, despite their rational awareness of the possible risks involved in attaining that knowledge. The viewer is encouraged to face the tough decision of balancing these conflicting desires.

The topic of governance of nanoST is a large issue of attention that is growing as the innumerable aspects of the tough decisions we are facing become increasingly apparent. One of the additional topics we would have liked to present in this section but were unable to include was a perspective specifically focused on the issue of poverty. The notion of a nano-divide, between industrialised and industrialising countries, and between privileged and less privileged inhabitants of the same country, is an important challenge for nanoST governance. In many respects, the questions of distributive justice confront us with some of the toughest decisions of all for nanoST. We would also have liked to present a chapter dealing with the various approaches to technology assessment, and their take on the challenges of nanoST governance. A third issue we feel belongs in this node of attention but which we were unable to include is that taking the point of view of industry, and outlining the internal attempts of corporations for self governance, for instance through voluntary codes of conduct.

To sum up, this section is about the fact that decisions affecting the development of nanoST are made all the time. Both 'big' and 'small' decisions may be tough, and sometimes it is difficult to know for sure whether a decision is big or small! A recurring notion in this section is that of 'responsibility', mentioned by most, and in some way or another discussed by all of the authors in this section. In taking tough decisions for an uncertain future, in terms of nanoST development, as well as in other fields of life, one of the real challenges is to know what it means to behave responsibly.

Chapter 15

Smoke that Thunders: Risk, Confusion and Regulatory Frameworks

Diana Bowman & Geert van Calster

'....Mosi-oa-Tunya luring the distant day-tripper,
The smoke that thunders, so scary,
The smoke that reviews His greatness...'
Victoria Falls the Silent Magician, Blessing Charuka

Consumer demand for new and improved products has encouraged industry to experiment with new products and processes, including nanotechnologies. While current products incorporating engineered nanomaterials are relatively mundane, it is anticipated that the next generation of nano-enabled products will radically transform traditional products and industries. While the use of nanotechnologies offers a range of benefits, there has been increasing debate over the potential risks associated with the technology's use, in particular in 'up close and personal' care products, including cosmetics. The focus of this chapter is on the operation and adequacy of regulation for effectively managing the potential risks posed by the inclusion of nanomaterials in cosmetic products.

Nano Meets Macro - Social Perspectives on Nanoscale Sciences and Technologies
by K L Kjølberg & F Wickson

www.panstanford.com
978-981-4267-05-2

1. Introduction

Known to locals as "Mosi-oa-Tunya" (*the Smoke that Thunders*), the 1855 'discovery' of a formidable waterfall in the Zambezi River by Dr David Livingstone represented a significant milestone in history. The since re-labelled "Victoria Falls" - with its awesome force and thick 'smoke' - symbolised a new frontier for European explorers. Today, Victoria Falls remain important to the two nations that they span, with visitors to the region contributing significant economic resources to both jurisdictions.

It is not surprising that *the Smoke that Thunders* instills awe and mystique, fear and fascination, as visitors to the banks of the Zambezi attempt to reconcile the very real risks posed by the Falls. To our thinking, new technologies, particularly nanotechnologies, give rise to similar feelings of awe and fascination, as well as doubts and concern. Nanotechnologies are viewed by some as simply a re-labelling of the old, a convergence of different domains, or an area of unrivalled economic potential. To others, the technology may be viewed as being a powerful force with limited, or insufficient, controls. It is a technology which induces excitement and apprehension. Or a world of limitless possibilities to advance the fate of down-stream users, which due to the ubiquitous nature of nanotechnologies will, arguably, be all members of society. Moreover, just as blankets of mist cloud the underlying characteristics of Victoria Falls, questions of safety, risks, and the adequacy of regulatory regimes continue to cloud the nanotechnologies landscape.

Moreover, literally just above the steep drop of the Falls, rock formations have created a pond-like structure. Swimmers are regularly seen bathing in this oasis of calm, surrounded by frantic water masses. This is reminiscent of the current regulatory landscape for nanotechnologies, as we set out in detail below. Indeed whilst a variety of studies have reviewed the appropriateness of the existing legal regimes to deal with the human and environmental safety aspects of this emerging technology, none of these jurisdictions (Australia, United Kingdom (UK) United States (US), and the European Union (EU)) had by December 2008 considered it necessary to introduce 'nano-specific' amendments to their legislative frameworks.[1] Thus the current regulatory landscape feels like a calm pond surrounded by frantic review. Whether it will eventually lead to a "waterfall" of regulatory initiatives within the European Union (EU) or other jurisdictions remains to be seen.

The aim of this chapter is to demystify the current regulatory debates associated with the use of engineered nanomaterials in certain areas and applications. While the focus of the chapter is on the regulatory concerns being raised within the context of the EU, it is acknowledged that these feelings of awe, fascination, concerns and confusion sit within a broader landscape. For this reason, the chapter provides an overview of developments occurring in other jurisdictions, including the US and Australia. It also highlights transnational developments. Readers are then provided with a brief introduction to regulatory concepts and regulatory models in the second section of the chapter. Strengths and weaknesses of the different regulatory models are highlighted. The third section of the chapter commences with an overview of recent EU policy and governance activities for nanotechnologies. Key regulatory frameworks for managing nanotechnologies within the EU are outlined in order to illustrate the complex regulatory web already being employed to manage the regulatory risks posed by nanotechnologies. The adequacy of these frameworks for regulating cosmetic products containing nanomaterials is examined in light of recent European Commission regulatory review and the Scientific Committee on Consumer Safety's (SCCP) report. The final section synthesises the strengths and weaknesses of the regulatory regime examined, as well as the regulatory regimes more generally. Particular attention is paid to the common challenges and limitations. It concludes by making a number of recommendations designed to encourage the responsible development of nanotechnologies, including both hard and soft regulatory responses.

---

[1] This situation has since changed, with the European Parliament having proposed nano-specific text in the recast of the regulation on cosmetic products on 24 March 2009. However, an analysis of these provisions is beyond the scope of this chapter.

2. The Known Unknowns

Nanotechnologies are rapidly being incorporated into production processes and products across a range of industry sectors. Yet despite the economic potential and promised benefits of the technology, there is an increasing body of scientific literature that suggests that some nanomaterials may, by virtue of the novel physico-chemical properties that appear at the nanoscale, pose a risk to human and environmental health under certain conditions (see for example, Maynard 2006; Powell & Kanarek 2006; Oberdörster *et al.* 2007; Poland *et. al.* 2008). Scientific uncertainty exists, for example, in relation to potential hazards posed by some nanomaterials and the risk of human exposure to these particles (Dreher 2004; Scientific Committee on Emerging and Newly Identified Health Risks (SCENIHR) 2006). There is also limited knowledge over the adequacy of conventional risk assessment paradigms for calculating risks. This has challenged the research community's ability to adequately determine or predict risks associated with certain nanomaterials (Maynard 2006; Kandlikar *et al.* 2007).

It is therefore not surprising that as nanotechnology-based products continue to make their way into the market, commentators have begun to question the capacity of governments to adequately govern all facets of the technology. The concerns raised by these commentators acknowledge that while current nanotechnologies processes and products fall under an extensive web of regulatory instruments, these may not be appropriate for managing the potential risks. This is in part due to the fact that the frameworks may, for example, regulate specific substances which do not differentiate between substances of the same chemical composition but in different size or crystalline forms. Alternatively, the trigger for regulatory oversight may be a mass or volume threshold which may not be appropriate for nanoscale substances (Chaudhry *et al.* 2006; Ludlow *et al.* 2007).

Within the EU, non-governmental organisations such as the Soil Association and Friends of the Earth Europe (in partnership with Friends of the Earth Australia) have expressed their concern in relation to the use of nanotechnologies within the agri-food industry, as well as the cosmetics industry (see for example, the opinions of Soil Association

2008; Friends of the Earth 2006; Miller & Senjen 2008). The areas of concern identified by these commentators are areas in which products undergo limited (or no) authorisation processes prior to their entry onto the EU market. Such an approach is similar to that of many other jurisdictions.

There now appears to be a growing consensus among stakeholders that policy and regulatory action is required to ensure that nanotechnologies are adequately and effectively regulated. Such action, it has been argued, is required to ensure human and environmental safety as the technology become more pervasive. However, as illustrated by the range of activities that have been initiated to date, the precise nature and scale of any such changes, remains a subject of debate. These activities have included:

- formal regulatory reviews to determine the adequacy of current regulatory arrangements (see, for example, Health and Safety Executive 2006; Chaudhry *et al.* 2006; Ludlow *et al.* 2007; Food and Drug Administration 2007; Environmental Protection Agency (EPA) 2007; European Commission 2008a; Food Safety Authority of Ireland 2008; Food Standards Agency 2008)

- amendments to council ordinances expressly to include nanotechnologies (Berkeley City Council 2006)

- voluntary information gathering programs (National Industrial Chemical Notification and Assessment Scheme (NICNAS) 2007; 2008; Department of Environment Food and Rural Affairs (DEFRA) 2006; 2008; EPA 2008)

- mandatory reporting requirements and data call-ins (Californian Department of Toxic Substances Control 2009)

- multi-lateral co-operation and co-ordination to address a range of research priorities including standardisation, metrology, toxicology and ecotoxicology (International Organization for Standardization (ISO) 2008; OECD 2008)

- the development and implementation of risk governance and management frameworks (International Risk Governance Council 2006; Environmental Defense & DuPont 2007), and
- the development of voluntary codes of conduct (Coalition of Non-Governmental Organizations 2007; European Commission (EC) 2008b; Royal Society *et al.* 2007).

While the actors, scope and instruments vary, the overarching objective of each of these initiatives would appear to be the same - the responsible development of nanotechnologies. But these and the many other initiatives have not been enough to quell the concerns of some commentators. Calls continue for legislative instruments to be revised so as to include new, nano-specific provisions, including from the European Parliament.[2] Recommendations for more substantial legislative and regulatory responses have also been voiced (ETC Group 2003; Friends of the Earth Australia 2006; Standing Committee on State Development 2008). Such calls recognise the fundamental role that regulation can have in managing both potential scientific and societal risks.

There are however a range of instruments and mechanisms that can be employed to regulate - not all of which involve government as the central driving force. It is therefore fundamental, when considering Current and future approaches to regulating different facets of nanotechnologies, to understand not only the purpose of regulation, but also the inherent strengths and weaknesses of different regulatory options. It is to these concepts that this chapter now turns.

3. Regulation and Regulatory Frameworks: A Theoretical Overview

What is regulation? And what purpose does it serve? Black's definition of regulation provides a useful starting point:

---

[2] European Parliament legislative resolution of 24 March 2009 on the proposal for a regulation on cosmetics products, A6-0484/2008; European Parliament legislative resolution of 24 March 2009 on the proposal for a regulation on novel foods, A6-0512/2008.

> *Regulation is the sustained and focused attempt to alter the behaviour of others according to defined standards or purposes with the intention of producing a broadly identified outcome or outcomes, which may involve mechanisms of standard-setting, information-gathering and behaviour-modification* (Black 2002, p.19).

As such, while regulation can be undertaken by a range of different actors – though most commonly associated with governments – and through the employment of a range of instruments, the primary purpose of regulation is often to alter behaviour. In many instances, this will be done for the purpose of managing or minimising the scientific risks associated with certain actions or activities.

A practical although necessarily simplified[3] way to think about the different forms of regulation is to view them as existing along a continuum. The actors, instruments, and oversight will vary along the continuum, as too the level of compliance required and the potential sanctions for non-compliance. At one end sits the most well known regulatory form commonly referred to as state-based or command and control regulation. Command and control regulation is enforced through, for example, Acts of Parliaments, Government Regulations and Guidelines. It is therefore characterised by its compulsory nature, the appearance of strong accountability, the creation of a level playing field, and high degree of certainty. While this traditional form of regulation may be popular with the concerned public and regulators, a number of commentators have argued that it suffers from a range of weaknesses. These include, for example, its often slow and cumbersome nature, a lack of flexibility, and its perception as being resource intensive (see, for example, Moran 1995; Gunningham and Rees 1997; Sinclair 1997; Utting 2005). This regulatory form will not always therefore be the most appropriate or desirable option.

---

[3] One needs to bear in mind that such basic distinctions, instructive as they may be to get a grasp of the regulatory landscape, do suffer from over-simplification. For instance in the EU, 'command and control' has never been characterised by the kind of rigidity and flexibility typically associated with it, neither have alternative instruments necessarily displayed the advantages for which they are in theory applauded (see Van Calster 2009; Lee 2005).

At the other end of the continuum sit different forms of self-regulation, or civil regulation. As with state-based regulation, different forms of self-regulation exist and may include, for example, voluntary and third-party regulation. The state may or may not play a minor role under these approaches (Gunningham & Rees 1997; Van Calster & Deketelaere 2001; Vogel 2006a; 2006b; Levi-Faur & Comaneshter 2007; Van Calster 2009). By enabling non-state actors to play an active role in crafting the regulatory arrangements, it has been suggested that self-regulation promotes regulatory innovation and creativity. It has also been suggested that this form of regulation may not only be more responsive, but also less-resource intensive than command and control regulation (Sinclair 1997). However the limited or negligible role of the state in the regulatory arrangement, and the varying degrees of monitoring and enforcement underpinning the various approaches has resulted in some commentators raising questions over the perceived legitimacy and efficacy of this type of regulation within particular sectors (see for example, Webb & Morrison 1996; Gunningham & Rees 1997).

Co- or hybrid-regulation sits in the middle of the regulatory continuum, and it is therefore unsurprising that this form of regulation draws upon the strengths of the two regulatory forms outlined above. This broad and diverse category of regulation may include various forms. These vary from meta-regulation to enforced self-regulation and thereby allow governments to have varying levels of involvement and oversight in the regulatory approach. For Sinclair (1997, p.544), co-regulation enables industry to "shape regulatory outcomes, with government still retaining general design oversight".

As highlighted in Section 2, each of these three broad regulatory approaches have already been employed to regulate different stages and dimensions of nanotechnologies. As noted above, these have included the use of existing legislative instruments, codes of conduct administered by individual organisations or trade associations, and voluntary risk governance frameworks. It is important to recognise too that each regulatory form sits against a backdrop of other regulatory approaches, and that no one singular form exists in isolation. This, as Gunningham and Sinclair remind us in relation to environmental regulation, highlights the fact that the various mechanisms are complementary in nature and that:

> *'single instrument' or 'single strategy' approaches are misguided, because all instruments have strengths and weaknesses, and because none are sufficiently flexible and resilient to be able to address all environmental problems in all context[s]* (Gunningham & Sinclair 1999, p. 50).

Appropriate regulation of nanotechnologies will be dependent on finding a sensible and workable balance between the different regulatory approaches.

Effective regulation of nanotechnologies will also be dependent on regulators being able accurately to evaluate potential risks. In this sense, the term 'risk' most commonly refers to scientific risks, in which the risk assessment is undertaken through an analysis of exposure and hazard related data. However, as articulated by Bowman and Hodge (2008, p.196), "applying conventional risk assessment paradigm to engineered nanoparticles in order to determine risks is conceptually simple, but also unfortunately problematic at this early stage in the technology's development".

Leading scientists such as Maynard (2006), Oberdörster *et al.* (2007) and Weisner *et al.* (2006) have however reminded us that until the scientific community has a better understanding of the potential hazards, levels of exposure and the adequacy of conventional risk assessment paradigms, qualifying and prioritising risks posed by certain nanomaterials will be challenging.

While regulation may be primarily focused on minimising scientific risks, it is important to recognise that broader societal risks may also be considered within the regulatory paradigm. These are likely to vary between jurisdictions (Jasanoff 2005; Van Calster 2008). The current debate has already been framed by certain commentators so as to include ethical and social questions (see, for example, Anderson *et al.* 2005; Wood *et al.* 2007). This suggests that a broader view of regulation is already part of the regulatory debate within jurisdictions such as the EU.

Indeed the EU's "Code of Conduct" (EC 2008c), for instance, includes such values as human rights, human dignity, and inclusiveness, hereby also reflecting the broad scope of the EU's risk management approach. While recognising the importance of societal risks within the regulatory paradigm, the focus of this chapter is however on the narrow, scientific view of risks and the adequacy of current EU regulatory instruments and matrices to manage these risks.

4. EU Regulatory Frameworks and their Applicability to Nanotechnologies

In light of the anticipated economic benefits associated with nanotechnologies across the entire value chain (i.e. from manufacturing through to end products) (Lux Research 2005) it is unsurprising the EU has continued to show leadership in its investment in research and development (R&D) activities. This is arguably best illustrated by its commitment of approximately €570 million to the area in the first year (2007) of the Framework 7 Program alone (Hullmann 2008). While a significant proportion of this funding has been dedicated to fundamental scientific research activities, the EU has also invested in a number of research activities aimed at examining the broader ethical, social and legal dimensions. In this respect, the Commission has acknowledged that investment within these areas will "help to reach good governance in nanotechnology" (Tokamanis, as cited in Hullmann 2008, p.3).

This research has been complemented by a number of EC Communications on nanotechnologies, including "Towards a European Strategy for Nanotechnology" (EC 2004) and "Nanosciences and Nanotechnologies: An Action Plan for Europe 2005-2009" (EC 2005). For the purposes of this chapter, it is worth noting that while the former of these two Communications acknowledged the need for a regulatory framework for nanotechnologies, it did not provide insights into what the components of the framework should be. In the latter Communication, the Commission stated that the R&D of the technology must be carried out in a "responsible manner" (EC 2005). The Communication suggested that while the industry itself must take a lead role in the responsible development of the technology, it would not be adverse to the "examin[ation] and, where appropriate, propose adaptations of EU regulations in relevant sectors" (EC 2005, p.10).

The focus on the responsible development of nanotechnologies has been reiterated in the Commission's own "Code of Conduct for Responsible Nanosciences and Nanotechnologies Research" (EC 2008c). Launched in early 2008, the voluntary code aims to encourage and promote safe, responsible and sustainable nanotechnology-related research, while balancing competing pressures associated with innovation and commercialisation.

The recent publication of an in-house review by the Commission (2008a) arguably provides the greatest insight into the institution's current views and research priorities for the regulation of products and processes incorporating nanotechnologies. The wide-ranging review examined the suitability of a number of key EU regulatory frameworks for nanomaterials including industrial chemicals, products such as cosmetics and food, and the environment. A summary of the key legislative instruments identified by the Commission for each of the frameworks is set out in Tables 1 and 2 below.

While none of the key legislative instruments identified by the Commission within the context of these frameworks specifically referred to the term "nano", the review highlighted the fact that nanomaterials are already regulated at the Community level. This is achieved through a broad and complex matrix of legislative instruments, which are further supported by an extensive body of implementing documentation. In their view, these instruments were considered to be adequate and appropriate for regulating nanomaterials. In relation to, for example, worker protection the Commission stated that,

> *The Framework Directive and the above-mentioned daughter Directives present a comprehensive package of legal requirements aiming at ensuring a high level of protection of workers health and safety. These requirements, whilst they do not make explicit mention of nanomaterials and nanotechnologies, define a legislative*

Table 1: Key Legislative Instruments for Regulating Nanomaterials in the EU

Products							
Plant Protection Products	*Biocides*	*Cosmetics*	*Aerosol Dispensers*	*Medical Products*	*Cars*	*Food*	*Consumer Products not Covered by Specific Legislation*
• Directive 91/414/EEC	• Directive 98/8/EC	• Directive 76/768/EEC	Directive 75/324	• Regulation (EC) No 726/2004 • Directive 2001/83/EC	• Directive 2007/46/EC	• Regulation 178/2002 • Regulation 258/97 • Regulation (EC) 1935/2004 • Directive 89/107/EEC; 2002/46/EC; 1925/2006; 89/398/EEC; 2001/15/EC; 96/25/EC; 79/373/EC; 82/471/EEC • Regulation 1831/2001	• Directive 2001/95/EC

Adapted from: European Commission (2008a)

Table 2: Key Legislative Instruments for Regulating Nanomaterials in the EU

Chemicals	*Worker Protection*	*Environment*						
		Integrated Pollution Prevention and Control	*Major Accidents, Seveso II Directive*	*Water*	*Waste*	*Air Quality*	*Soil*	*Environmental Liability*
• Regulation (EC) No 1907/2006 (REACH)	• Directive 89/39/EEC and a number of Daughter Directives	• Directive 2008/1/EC	• Directive 96/82 • Directive 67/548/EEC • Directive 1999/45/EC	• Directive 2000/60	• Directive 2000/12/EC • Directive 91/689/EEC	• Directive 96/62/EC and a number of Daughter Directives		• Directive 2004/35/EC

Adapted from: European Commission (2008a)

> *Framework that applies to most occupational risks including those arising from the presence of nanomaterials* (EC 2008a, p. 12).

In relation to consumer products not covered by specific regulation, the EC stated that:

> *The mechanisms foreseen in the GPSD [Directive on general product safety] allow risks in relation to products containing or consisting of nanomaterials and nanotechnologies to be taken into account* (EC 2008a, p. 28).

The review stressed that the fundamental issue associated with the regulation of nanomaterials was not the adequacy of the Directives and Regulations *per se*, but rather their implementation in light of the current lack of understanding on the potential risks posed by the nanoscale substances. The current gaps in knowledge relating to measurement, characterisation, potential hazards and exposure, and the adequacy of risk assessment paradigms, was seen to have a negative impact on, for example, risk assessment processes, authorisation procedures, technical documentation, and other forms of implementing documentation (EC 2008a). In light of these substantial limitations and knowledge gaps, the Commission (2008a, p. 37) outlined an extensive program of research designed to "support legislative work in the field of environment, health and safety". Recognition of the need for an extensive program of fundamental research across multiple fields highlights the complexities associated with adequately regulating not only nanomaterials as individual substances but also when used within different classes of regulated products.

But how does the Commission's review reconcile with the findings of other commentators who have examined aspects of the EU's regulatory frameworks? Or that of the Commissions' own Scientific Committee on

Consumer Safety (SCCP) which has provided their own opinion on the safety of nanomaterials in cosmetic products?

As noted above, the inclusion of nanomaterials within a range of cosmetic products including anti-ageing creams, make up, and sunscreens[4] has arguably been one of the most contentious and hotly debated areas so far (see, for example, Grobe *et al.* 2008). This is not surprising given the 'up close and personal' nature of these products. To date, the specific scientific concerns raised by commentators in relation to the use of nanomaterials within these topically applied products have focused on the lack of scientifically sound data on:

1. the ability of nanoscale substances, in particular insoluble or biopersistent nanomaterials to penetrate the stratum corneum (the outer most layer of the skin), pass into the dermis (the middle layer of skin) and enter the vascular (or circulatory) system, and

2. the potential consequence, in terms of hazards, should absorption and translocation occur (Hoet *et al.* 2004; Oberdörster *et al.* 2005; Bowman & Fitzharris 2007).

As illustrated in Table 1, these products are governed primarily by Council Directive 76/768 EEC (the Cosmetics Directive) and its subsequent amendments. As noted by the EC (2007, p. 2) itself, the current regulatory regime consists of a "patchwork of more than 45 amendments with no set of definitions and no coherent terminology." In accordance with its main objective—"the safeguarding of public health..."—the Directive and its amendments set out the legal requirements and principles pertaining to cosmetic products within EU Member States, including those relating to "composition, labelling and packaging of cosmetic products..." (Council Directive 76/768/EEC,

---

[4] It is important to note that a sunscreening product within the EU is considered to be a cosmetic product, and regulated as such. This can be contrasted to jurisdictions such as Australia, where sunscreens are defined as therapeutic goods, and regulated under the therapeutic goods regulatory framework.

Preamble, at 3[5]). As established by the Cosmetics Directive, any cosmetic product put onto the market within the European Community must not pose a risk to human health when used "under normal or reasonably foreseeable conditions" (Article 2). This is achieved through the use of pre-marketing requirements including risk assessments, and the use of 'positive', 'negative' or 'restricted' lists of chemical substances in the Directive's Annexes. The legal instrument does not establish any pre-market registration, approval or review requirements for regulatory agencies in determining the safety of a cosmetic product prior to its entry on the Community market. This is consistent with the regulatory approach adopted in other jurisdictions such as Australia and the US.

Bowman and Van Calster (2008) have argued that, *prime facie*, it is possible to assume based on the structure and function of the regulatory regime, that all cosmetics products incorporating engineered nanomaterials available in the Community market are safe and should not pose a risk to the health of the consumer when used for the purpose for which they were intended. However, the authors went on to suggest that perhaps the reality is somewhat more complicated due to the current scientific uncertainties associated with hazard and exposure.

The European Commission is more optimistic in the operation of its regulatory regime. In its review, the Commission noted in relation to the overarching EU legislative framework that:

> *On the basis of the obligation to carry out a risk assessment and the possibility to lay down through implementing legislation detailed conditions of use for certain ingredients, risks in relation to nanomaterials and nanotechnologies can, therefore, in principle be dealt with in an appropriate way* (EC 2008a, p. 17).

However, it would appear that current concerns are not directly related to the question of whether or not cosmetics incorporating

[5] Council Directive of 27 July 1976 on the approximation of the laws of the Member States relating to cosmetic products.

nanomaterials are regulated, but rather the adequacy of conventional safety assessment methods for active ingredients. These methods are set out in the SCCP's "Notes on Guidance for the Testing of Cosmetic Ingredients and their Safety Evaluation" (Guidance Notes). It is this 'enabling instrument' that therefore sets out the evaluation requirements for industry for establishing the safety of their cosmetic products.

The Guidance Notes apply to all active ingredients contained within a cosmetic product, including nanoscale particles such as titanium dioxide (TiO_2) and zinc oxide (ZnO). However basic information requirements for the toxicological dossiers to be evaluated by the SCCP for the purposes of the safety evaluation do not expressly include information pertaining to the size of the chemical ingredient being assessed. Rather the conventional safety evaluation process, which includes a number of toxicological tests, primarily relies on mass metrics. As such the Guidelines in their current form do not distinguish an ingredient on the basis of its size, but rather chemical name. The implication of this is that the toxicity testing regime used to determine the toxic potential of a TiO_2 particle with an average diameter of 50 nm will be the same as that applied to a TiO_2 particle with an average diameter of 800 nm. This is despite differences in, for example, their surface to volume ratios and other physico-chemical characteristics of the TiO_2, each of which may be important in determining the substance's toxicity. This suggests that a potential exists, albeit small, for potentially unsafe cosmetic products which incorporate specific types of nanomaterials, such as insoluble or biopersistent nanoparticles, to be placed onto the Community market.

However, for harm to occur, it is important to remember that several steps would have to occur. First, the nanomaterials would have to first penetrate the stratum corneum, pass into the epidermis and enter the vascular system. Second, the particle itself would have to be hazardous, and be present at a concentration high enough to cause harm. As noted above, a lack of sound scientific data exists in relation to these requirements.

In light of the current knowledge gaps, it is not surprising that in their final opinion, the SCCP (2007, p.38) concluded that "there are large data gaps in risk assessment methodologies with respect to nanoparticles in cosmetic products." The SCCP also noted that safety assessment methods developed for traditional cosmetic ingredients may not be appropriate for all nanomaterials being used. The Committee identified insoluble nanoparticles as the family of particles of greatest concern at that time. The SCCP went on to state that existing assessment methods need to be validated for nanomaterials and new methods which take into account the specific physio-chemical characteristics of the substance should be developed where necessary.

While one initial step to address this gap could be, for example, to modify the current safety requirements set out in the SCCP's Guidance notes, so as to require additional information and risk assessments to be undertaken on insoluble nanomaterials. Such action would not require a wholesale change of the regulatory framework, but rather an amendment to the Guidance Notes.

While this may satisfy some people's concerns, any such tweaking of the current framework would not in itself address the more fundamental issue – that the current risk assessment methodologies may be in themselves inadequate for assessing the risks posed by these particles due to their primary reliance on mass metrics. These concerns have been raised by the European Union's Scientific Committee on Emerging and Newly Identified Health Risks (2006) and the SCCP (2007). It would therefore appear that until validated *in vivo* and *in vitro* risk assessment methodologies are developed for engineered nanomaterials, including insoluble and bio-persistent nanoscale substances, and incorporated into the regulatory framework, there is a *potential* for certain commercially available cosmetic products which contain nanomaterials to pose a risk to human safety. While any such risk may be small, it is difficult to contend that it does not exist.

Unquestionably, consumers and citizens may be prepared to accept a degree of risk in relation to, for example, therapeutic applications as the benefits have the potential to outweigh the associated risks. However it is argued that consumers should not have to be prepared or required to accept any level of risk of harm, including minimal risk, in relation to their use of cosmetic products. It is not surprising therefore that organisations such as Which? have proposed that nanomaterials "should be independently assessed before they can be used in cosmetic products" and that:

> *The new EU Regulation should include an annex of permitted nano materials that can be used in cosmetics based on a positive safety assessment by the relevant independent expert committee (currently the SCCP). If they aren't on the list, they shouldn't be used* (Which? 2008, p. 8).

5. Tranquil or Turbulent Waters Ahead?

In contrast to the overall findings of the EC's in-house regulatory review, a number of independent examinations of current EU regulatory frameworks and instruments have suggested that the current frameworks may not be appropriate for managing all of the potentials risks posed by certain engineered nanomaterials *if* they are proven to be hazardous (see for example, Royal Society & Royal Academy of Engineering 2004; Chaudhry *et al.* 2006; Chaudhry *et al.* 2008; Bowman & Van Calster 2007; Miller & Senjen 2008; Van Calster *et al.* 2008). These reviews have gone on to suggest a number of inadequacies in the current regulatory framework, many of which the authors of this chapter would argue need to be addressed.

One area of commonality noted in each of the reports - including the European Commission's (2008a) review - was the acknowledgement that at present, there is simply insufficient scientific data to evaluate the potential hazards and risks posed by some engineered nanomaterials to human and environmental health through the particle's life-cycle. That is of course despite the fact that these particles are already being manufactured and incorporated into a range of products. The lack of scientific data and how this feeds into the regulatory cycle could be relatively easily remedied by taking the necessary risk management steps if and when more scientifically robust data on hazards and exposure becomes available.

To date, the EU has opted for an incremental approach to addressing the regulatory concerns being expressed by stakeholders (although the European Parliament has voiced its doubts and may yet succeed in imposing a moratorium in particular on insoluble nanoparticles present in food[6]). Under this approach, the EU has proposed the addition of what we would call "nano-hooks" into several regulatory instruments. These proposed nano-hooks are designed to gather information on, and differentiate between, for example, particles of different sizes and would appear to be a first step in specifically differentiating and therefore regulating nanomaterials. The addition of these proposed hooks to legislation has occurred within the context of broader re-casts of existing regulatory instruments, such as the Cosmetics Directive and the Novel Foods Directive. Importantly, the proposed recast of these instruments has occurred upon review of the regulations which had been scheduled for other reasons and not specifically to address the challenges presented by nanotechnologies. One notably exception is the amendment to the Registration, Evaluation, Authorisation and Restriction of Chemicals (REACH) Regulation 2006/1907/EC, to reflect the concern that carbon and graphite in their nanoscale form were piggy-backing on the exemption for naturally occurring substances.[7] The advantage of having these hooks or regulatory triggers ready for use is evident, especially in the EU, where the preparation and adoption of new legislative instruments may be a slow, as well as hotly contested, process.

The proposed inclusion of the term 'particle size' as a parameter or regulatory trigger in respect to the risk assessment process for cosmetics within the EU would seem to focus the regulatory response to nanotechnologies firmly in the liability sphere, along the lines of the recently adopted REACH Regulation. A primary objective of REACH is to reverse the burden of proof from the regulators to the regulatee. Under REACH therefore, the manufacturer/importer/supplier of the substance is required to prove that the substance is safe prior to its entry onto the Community market. This is not in itself a poor regulatory strategy,

---

[6] Note 2 above.

[7] Regulation 987/2008 (OJ [2008] L268/14] amends Annex IV to the REACH Regulation (this is an Annex which includes substances considered to carry minimum risk), removing carbon and graphite form the Annex '*due to the fact that the concerned Einecs and/or CAS numbers are used to identify forms of carbon or graphite at the nano-scale, which do not meet the criteria for inclusion in this Annex*'. The lifting of the exemption simply means that notification—and risk analysis—will certainly be required. It does not address the issues linked to such analysis.

provided the authorities at the same time efficiently pursue the soundness of accompanying technical regulations, such as testing methods.

While governments and commentators have to date largely focused on evaluating existing legislative frameworks and their adequacy to regulate nanotechnologies, fewer commentators have focused on the potential role or roles that other 'softer' regulatory mechanisms outline in Section 4 may play in regulating nanotechnologies. This is not surprising given the embryonic nature of the technology, and the uncertainties that exist in relation to its developmental trajectories. However, as illustrated above, considerable doubt exists as to the extent to which state-based regulatory regimes alone are suitable for regulating all aspects of the technology. It therefore appears prudent for stakeholders to examine a range of alternative and arguably more flexible regulatory mechanisms, including those falling under the co-regulation and self-regulation banners (Grobe *et al.* 2008). These regulatory forms could be utilized in the first instance, and built upon and/or modified as required in response to the evolving scientific knowledge. In this way, they could be utilized as methods for promoting the responsible development of nanotechnologies.

6. Conclusion

Even at this early stage it is reasonably foreseeable that at least in some jurisdictions the challenges posed by nanotechnologies will promote intensive regulatory innovation, under which stakeholders will devise new forms of regulatory approaches to supplement traditional 'command and control' regulation.

In the EU, the picture is less clear. For a variety of reasons (Van Calster 2008), the public's trust in national regulators remains low. In response, it would appear that the European institutions, including the EC, have stepped into this void and sought to strengthen the intervention of the EU within the regulatory process. This appears to be especially the case in relation to regulation which specifically relates to human health and safety, and the environment. The publics', member states', and EU regulators' appetite for transfer of regulatory oversight to softer forms of regulation than that of command and control approaches would appear to be in all likelihood very low.

Whilst a flurry of studies currently review and question the appropriateness of the existing legal regimes to deal with the environment, health and safety aspects of nanotechnologies, as noted above, no one of these jurisdiction has so far has considered it necessary to introduce or modify legislation in order to specifically address the identified challenges or regulatory gaps.

The regulatory landscape at the moment feels like *Smoke That Thunder's* calm pond surrounded by frantic review. Whether it will eventually lead to a waterfall of regulatory initiatives remains to be seen.[8]

Questions for Reflection:

1. What role, if any, do you believe that self-regulation (soft law, voluntary codes of conduct etc.) can play in regulating nanotechnologies?
2. How could public concerns feed into regulatory processes?
3. What aspects of nanotechnology development should be regulated on a local, national, international or global level?

[8] The authors would like gratefully to acknowledge the support of the Research Fund K.U. Leuven and the Australian Research Council Nanotechnology Network.

Bibliography

Anderson, A., Allan, S., Petersen, A. and Wilkinson, C. (2005). The framing of nanotechnologies in the British newspaper press, *Science Communication*, 27(2), pp. 200-220.

Berkeley City Council (2006). *Agenda - Berkeley City Council Meeting*, 5 December, Berkeley: Berkeley City Council.

Black, J. (2002). Critical reflections on regulation, *Australian Journal of Legal Philosophy*, 27, pp. 1-36.

Bowman, D. M. and Fitzharris M. (2007). Too small for concern? Public health and nanotechnology, *Australian and New Zealand Journal of Public Health*, 31(4), pp. 382-384.

Bowman, D. M. and Hodge, G. A. (2008). A big regulatory tool-box for a small technology, *NanoEthics*, 2(2), pp. 193-207.

Bowman, D. M. and Van Calster, G. (2007). Does REACH go too far? *Nature Nanotechnology*, 1, pp. 525-526.

Bowman, D. M. and Van Calster, G. (2008). Flawless or fallible? A review of the applicability of the European Union's cosmetics directive in relation to nano-cosmetics, *Studies in Ethics, Law, and Technology*, 2(3), pp. 1-35.

Californian Department of Toxic Substances Control (2009). *Chemical Information Call, in, Carbon Nanotubes*, 22 January, Sacramento: DTSC.

Chaudhry, Q., Blackburn, J., Floyd, P., George, C., Nwaogu, T., Boxall, A. and Aitken, R. (2006). *Final Report: A Scoping Study to Identify Gaps in Environmental Regulation for the Products and Applications of Nanotechnologies*, London: Defra.

Chaudhry, Q., Scotter, M., Blackburn, J., Ross, B., Boxall, A., Castle, L., Aitken, R. J. and Watkins, R. (2008). Applications and implications of nanotechnologies for the food sector. *Food Additives & Contaminants*, 25(3), pp. 241-258.

Coalition of Non-Governmental Organizations (2007). *Principles for the Oversight of Nanotechnologies and Nanomaterials.* http://nanoaction.org/nanoaction/doc/nano-02-18-08.pdf (last accessed Dec 4, 2008).

Department of Environment Food and Rural Affairs (2006). *UK Voluntary Reporting Scheme for Engineered Nanoscale Materials*, London: Defra.

Department of Environment Food and Rural Affairs (2008). *The UK Voluntary Reporting Scheme for Engineered Nanoscale Materials: Seventh Quarterly Report,* August, London: Defra.

Dreher, K. L. (2004). Toxicological Highlight: health and environmental impact of nanotechnology: Toxicological assessment of manufactured nanoparticles, *Toxicological Sciences*, 77(1), pp. 3-5.

Environmental Defense and DuPont (2007). *Nano Risk Framework,* New York: EDF.

Environmental Protection Agency (2007). *EPA Nanotechnology White Paper,* Washington, D.C.: Prepared for the US Environmental Protection Agency by members of the Nanotechnology Workgroup, a group of EPA's Science Policy Council.

Environmental Protection Agency (2008). Notice: Nanoscale Materials Stewardship Program, *Federal Register*, 73(18), pp. 4861-4866.

ETC Group (2003). *No Small Matter II: The Case for a Global Moratorium Size Matters!* Ottawa: ETC Group.

European Commission (2004). *Towards a European Strategy for Nanotechnology - Communication from the Commission COM (2004) 338*. Brussels: E.C.

European Commission (2005). *Nanosciences and Nanotechnologies: An Action Plan for Europe 2005-2009,* Brussels: E.C.

European Commission (2007). *Public Consultation Paper on the Simplification of Cosmetics Directive 76/768/EEC,* 12 January. Brussels: E.C.

European Commission (2008a). *Regulatory Aspects of Nanomaterials: Summary of legislation in relation to health, safety and environment aspects of nanomaterials, regulatory research needs and related measures {COM(2008) 366 final},* Brussels: E.C.

European Commission (2008b). *European Commission Adopts Code of Conduct for Responsible Nanosciences and Nanotechnologies Research, IP/08/193,* 8 February, Brussels: E.C.

European Commission (2008c). *European Commission Adopts Code of Conduct for Responsible Nanosciences and Nanotechnologies Research, IP/08/193,* 8 February, Brussels, E.C.

Food and Drug Administration (2007). *Nanotechnology - A Report of the U.S. Food and Drug Administration Nanotechnology Task Force,* Washington, D.C.: FDA.

Food Safety Authority of Ireland (2008). *The Relevance for Food Safety of Applications of Nanotechnology in the Food and Feed Industries*, Dublin: FSAI.

Food Standards Agency (2008). *A Review of the Potential Implications of Nanotechnologies for Regulations and Risk Assessment in Relation to Food,* London: FSA.

Friends of the Earth Australia (2006). *Nanomaterials, Sunscreens and Cosmetics: Small Ingredients, Big Risks,* Sydney: FoE Australia and FoE US.

Grobe, A., Renn, O. and Jaeger, A. (2008). *Risk Governance of Nanotechnology Applications in Food and Cosmetics,* Geneva: International Risk Governance Council

Gunningham, N. and Rees, J. (1997). Industry self-regulation: An institutional perspective, *Law & Policy*, 19(4), pp. 363-414.

Gunningham, N. and Sinclair, D. (1999). Regulatory pluralism: Designing policy mixes for environmental protection, *Law & Policy*, 21(1), pp. 49-76.

Health and Safety Executive (2006). *Review of the Adequacy of Current Regulatory Regimes to Secure Effective Regulation of Nanoparticles Created by Nanotechnology: The Regulations Covered by HSE,* London: HSE.

Hoet, P., Bruske-Hohlfeld, H. I. and Salata, O. (2004). Review: Nanoparticles - known and unknown health effects, *Journal of Nanobiotechnology*, 2(12), pp. 1-15.

Hullmann, A. (2008). *European activities in the Field of Ethical, Legal and Social Aspects (ELSA) and Governance of Nanotechnology,* Brussels: DG Research, E.C.

International Risk Governance Council (2006). *White Paper on Nanotechnology Risk Governance - Towards an Integrative Approach,* Geneva: IRGC.

International Organization for Standardization (2008). *TC 229 - Nanotechnologies.* http://www.iso.org/iso/iso_catalogue/catalogue_tc/catalogue_tc_browse.htm?commid=381983&published=on&development=on (last accessed Nov 12, 2008).

Jasanoff, S. (2005). *Designs on Nature: Science and Democracy in Europe and the United States*, Princeton: Princeton University Press.

Kandlikar, M., Ramachandran, G., Maynard, A. D., Murdock, B. and Toscano, W. A. (2007). Health risk assessment for nanoparticles: A case for using expert judgment, *Journal of Nanoparticle Research*, 9, pp. 137-156.

Lee, M. (2005). *EU Environmental Law,* Oxford: Hart.

Levi-Faur, D. and Comaneshter, H. (2007). The risks of regulation and the regulation of risks: The governance of nanotechnology. In Hodge, G. A., Bowman, D. M. and Ludlow, K. (eds) *New Global Frontiers in Regulation: The Age of Nanotechnology,* Cheltenham: Edward Elgar, pp. 149-165.

Ludlow, K. A., Bowman, D. M. and Hodge, G. A. (2007). *A Review of Possible Impacts of Nanotechnology on Australia's Regulatory Framework,* Melbourne: Monash Centre for Regulatory Studies.

Lux Research Inc (2005). *Nanotechnology: Where Does the US Stand? Testimony before the Research Subcommittee of the House Committee on Science,* New York: Lux Research Inc.

Maynard, A. D. (2006). *Nanotechnology: A Research Strategy for Addressing Risk.* Washington, D.C.: Project on Emerging Nanotechnologies, Woodrow Wilson International Centre for Scholars.

Miller, G. and Senjen, R. (2008). *Out of the Laboratory and on to Our Plates: Nanotechnology in Food & Agriculture,* Melbourne: FoE Australia, FoE Europe and FoE US.

Moran, A. (1995). Tools of environmental policy: Market instruments versus command-and-control. In Eckersley, R. (ed), *Markets, the State and the Environment: Towards Integration,* South Melbourne: Macmillan Education.

National Industrial Chemical Notification and Assessment Scheme (2007). *Summary of Call for Information and the Use of Nanomaterials,* Canberra: NICNAS.

National Industrial Chemical Notification and Assessment Scheme (2008). Industrial nanomaterials: Voluntary call for information 2008, *Australian Government Gazette*, 7 October, pp. 8-24.

Oberdörster, G., Oberdörster, E. and Oberdörster, J. (2005). Nanotoxicology: An emerging discipline evolving from studies of ultrafine particles, *Environmental Health Perspectives*, 113(7), pp. 823-839.

Oberdörster, G., Stone, V. and Donaldson, K. (2007). Toxicology of nanoparticles: A historical perspective, *Nanotoxicology*, 1(1), pp. 2-25.

OECD (2008). *Nanotechnologies at the OECD,* Paris: OECD.

Poland, C. A., Duffin, R., Kinlock, I., Maynard, A. D., Wallace, W., Seaton, A., Stone, V., Brown, S., MacNee, W. and Donaldson, K. (2008). Carbon nanotubes introduced into the abdominal cavity of mice show asbestos like pathogenicity in a pilot study, *Nature Nanotechnology*, 3, pp. 423-428.

Powell, M. C. and Kanarek, M. S. (2006). Nanomaterial health effects - Part 1: Background and current knowledge, *Wisconsin Medical Journal*, 105(2), pp. 16-20.

Royal Commission on Environmental Pollution (2008). *Novel Materials in the Environment: The Case of Nanotechnology*, London: HM Government.

Royal Society, Insight Investment, Nanotechnology Industries Association, Nanotechnology Knowledge Transfer Network (2007). *Responsible Nanotechnologies Code: Consultation Draft - 17 September 2007 (Version 5),* London: Responsible NanoCode Working Group.

Royal Society and Royal Academy of Engineering (2004). *Nanoscience and Nanotechnologies: Opportunities and Uncertainties,* London: RS-RAE.

Scientific Committee on Consumer Products (2005). *Request for a Scientific Opinion: Safety of Nanomaterials in Cosmetic Products,* Brussels: Health & Consumer Protection Directorate-General, E.C.

Scientific Committee on Consumer Products (2007). *Opinion on Safety of Nanomaterials in Cosmetic Products*, Brussels, Health & Consumer Protection Directorate-General, E.C.

Scientific Committee on Emerging and Newly Identified Health Risks (2006). *Modified opinion (after public consultation) on the Appropriateness of Existing Methodologies to Assess the Potential Risks Associated with Engineered and Adventitious Products of Nanotechnologies,* Brussels: E.C.

Sinclair, D. (1997). Self-regulation versus command and control? Beyond false dichotomies, *Law & Policy*, 19(4), pp. 529-559.

Standing Committee on State Development (2008). *Nanotechnology in New South Wales,* Sydney: Legislative Council.

Soil Association (2008). *Press Release: Soil Association First Organisation in the World to Ban Nanoparticles — Potentially Toxic Beauty Products that get Right under your Skin*, 17 January, Bristol: SA.

Utting, P. (2005). *Rethinking Business Regulation - From Self-Regulation to Social Control. Technology, Business and Society Programme,* Geneva: United Nations Research Institute for Social Development.

Van Calster, G. and Deketelaere, K. (2001), The use of voluntary agreements in the European Community's environmental policy. In Orts, E. and Deketelaere, K. (eds) *Environmental Contracts – Comparative Approaches to Regulatory Innovation in the United States and Europe*, London: Kluwer, pp. 199-246.

Van Calster, G. (2008), Risk regulation, EU law and emerging technologies: Smother or smooth? *NanoEthics*, 2(1), pp. 61-71.

Van Calster, G., Bowman, D. M., and D'Silva, J. (2008). Sufficient or deficient? A review of the adequacy of current European legislative instruments for regulating nanotechnologies across three industry sectors, Paper presented at the *TILTing Perspectives on Regulating Technologies* Conference, 10-11 December, Tilburg.

Van Calster, G. (2009). 'An overview of regulatory innovation in the European Union', *Cambridge Yearbook of European Legal Studies,* Oxford, Hart Publishing, forchoming.

Vogel, D. (2006a). *The Private Regulation of Global Corporate Conduct,* Centre for Responsible Business Working Paper Series, Berkeley: Department of Political Science, University of California.

Vogel, D. (2006b). *The Role of Civil Regulation in Global Economic Governance,* Oxford: Global Economic Governance Programme, Oxford University.

Webb, K. and Morrison, A. (1996). The Legal Aspects of Voluntary Codes, Paper presented at the *Exploring Voluntary Codes in the Marketplace Symposium*, 12-13 September, Ottawa.

Weisner, M. R., Lowry, G. V., Alvarez, P., Dionysiou, D. and Biswas, P. (2006). Assessing the Risks of Manufactured Nanomaterials, *Environmental Science and Technology*, 15, pp. 4337-4345.

Which? (2008). *Small Wonder? Nanotechnology and Cosmetics,* London: Which?

Wood, S., Jones, R. and Geldart, A. (2007). *Nanotechnology: From the Science to the Social - The social, Ethical and Economic Aspects of the Debate,* London: Economic and Social Research Council.

Chapter 16

Economic and Political Aspects of Nanotechnology Governance in Latin America: The Case of Mexico

Gian Carlo Delgado

This chapter aims to evaluate the current situation of nanotechnology governance in Mexico and Latin America regarding the role, interests, actions and contradictions of governmental, business and scientific spheres. Considered as key elements for any endogenous process of industrialisation, these 'spheres' will help to outline the organization of the entire chapter. By identifying their main characteristics, limitations and potentialities, the chapter offers a panoramic but polished assessment of Mexico's nanotechnology governance, particularly from an economical and political perspective. In this context it is worth clarifying that given the limited flow of information related to nanoscience and nanotechnology (N&N) in the country, most of what is presented here is a product of extensive field research and an analytical and interpretative exercise that seeks to construct an overview; not only by describing facts but mainly by giving them a logical interpretation in terms of their 'best' explanatory power in the Mexican context. To conclude, a general prospective of Mexico's normative 'best' scenario is included, especially with respect to a regulatory framework for: nanotechnology R&D; nano-toxicology evaluation and precautionary measures; introduction of nanomaterials into industrial processes; handling of imports nanotechnology products; and on actively informing and dialoguing with the general public.

Nano Meets Macro - Social Perspectives on Nanoscale Sciences and Technologies
by K L Kjølberg & F Wickson

www.panstanford.com
978-981-4267-05-2

1. Introduction: Science and Technology in Rich and Poor Countries

As recognized in contemporary history, there are three key elements for a successful development of science and technology (S&T) aimed to industrialize any given country, to promote its economic growth and increase its competitiveness: the Nation state (civil and military sections); the national economic units or business community; and the S&T research entities (public and private).

The synergies generated between such 'nodes' or corners of what could be called the *triangle of science and technology industrialization* (Delgado 2008a), are key aspects of any innovation agenda of a competitive economy nowadays. There are of course other relevant aspects, strategic features as well as further actors that facilitate the process of synergy; for example, industrial lobbies, expert consultancy groups, think tanks, etc. See Figure 1.

The more complex the knitted network of the *triangle*, the more advanced the innovation-industrialization system and, consequently, the economy (however not necessarily the social equity). Accordingly, the *triangle* is a typical structure of developed countries, of course, each one with its own particularities and geographical, societal and cultural realities (Chalmers 1982; Delgado 2002; 2006a; 2007a; McGrath 2002). However, the general characteristics shared by all successful experiences are the consolidation, regulation and protection (i.e. through policies, subsidies, espionage and technology acquisition agreements, etc.) of a set of *strategic industries* - civilian and military - as well as the core of the scientific & technological apparatus that feeds them (Melman 1970; Libicki 1989).

Chalmers Johnson, expert on Asian capitalism and advisor of several multinational corporations, accurately realizes that:

> *...leaving aside the former Soviet Union, the main developed countries –Britain, the United States, Germany, France, Sweden, Belgium, the Netherlands, Switzerland, Japan, and the East Asian NICs (South Korea, Taiwan, and Singapore) – all got rich in more or less the same way. Regardless of how they justified their policies, in actual*

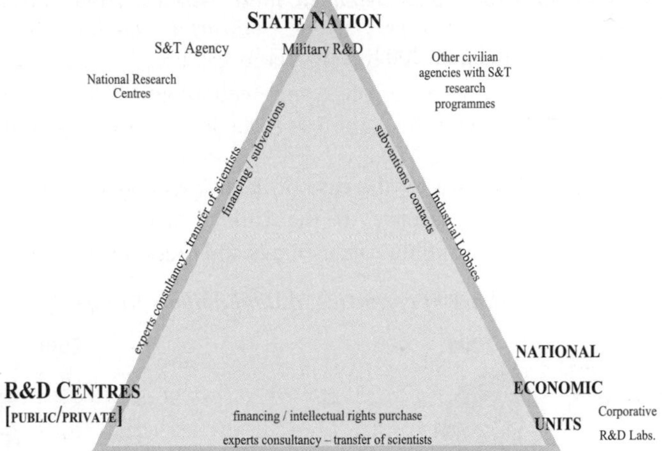

Figure 1: *Triangle of Science and Technology Industrialization*

Source: Delgado, Gian Carlo (2006). Aspectos éticos, sociales y ambientales de la nanotecnología. PhD Dissertation Thesis. Autonomous University of Barcelona. Spain.

practice they protected their domestic markets using high tariff walls and myriad 'nontariff barriers' to trade[1] *[...] All these "developing" nations begged, bought, or stole advanced technology from countries that first pioneered it and then, through reverse engineering and targeted investment, improved on it. They used the state power to support and protect efficient capitalists within their own national boundaries who had the potential to become exporters. They poured subsidies into uncompetitive industries in order to substitute domestically produced goods for imports, often at almost any price. Some of them captured overseas markets through imperial conquest and*

[1] Tariff barriers are a tax that a government collects on goods coming into a country. Non-tariff barriers are those measures aimed to restrict imports under the argument of ensuring conformity of imported products with technical standards, measures for phytosanitary and environmental protection, measures for ensuring food safety, etc.

colonialism and then defended these markets from other would-be conquerors, using powerful navies and armies...(Chalmers 2004, p. 263).

1.1. The 'triangle' in action

The result is well known for the case of developed countries. Data from 2006 shows the preeminence of the United States (US), European countries and Japan regarding patent filings and grants (see Figure 2).

Share of countries in total patents filings

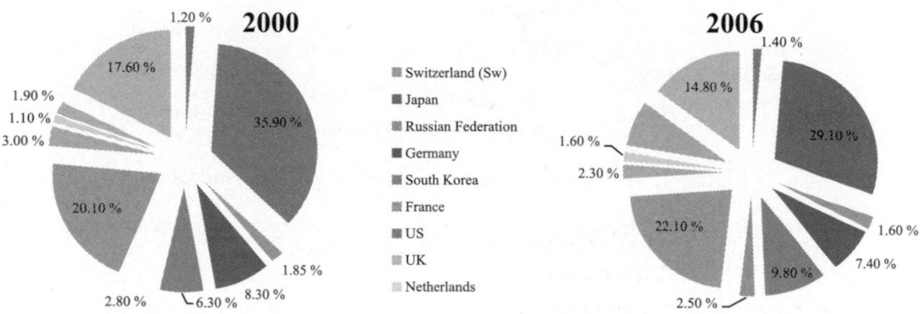

Share of countries in total patents grants

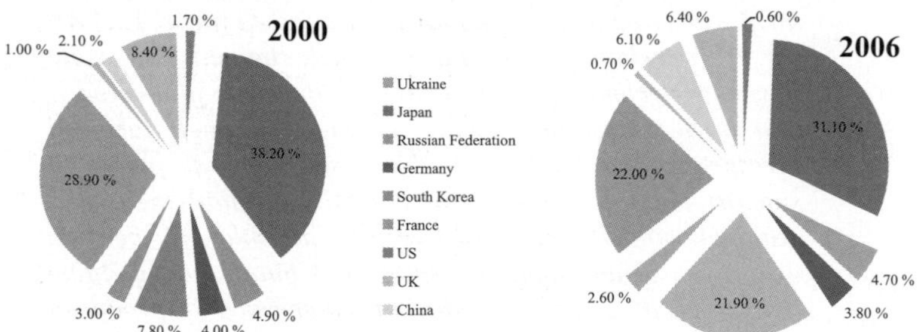

Figure 2 For colour reference turn to page 564.

* WIPO's data breaks down the amounts for each European county, a procedure that dilutes Europe's role on patenting tendencies worldwide.

Source: WIPO (2008). *World Patent Report: A Statistical Review 2008*. Switzerland.

It is important to clarify that the data corresponds only to one year, an aspect to take into account since new active countries aren't necessarily, in general terms, leaders. They might be strongly catching up, but not as a result of a consolidated and potent scientific and technological R&D platform but, instead, as an outcome of the formation of an efficient 'triangle-like' structure focused only on some specific technological niches. This seems to be the case of China and South Korea, for example, given that they are still far away from being global technological powers (WIPO 2008; Delgado 2006a; 2008b). Statistics from the Organization for Economic Co-operation and Development OECD) confirm the latter given that it estimated that in 2001, France, Germany, UK, US and Japan owned 83.6% of total patents worldwide (OECD 2005).

Hitherto, data concerning specific areas seems to be helpful for a more refined picture of the current state of the situation. For example, the US tends to control patents on electricity, electronics, instruments and chemical; the European countries (mainly European Union - EC) those of processes, mechanical goods, and consumer goods; and, Japan is relatively strong in the electric and electronic sectors (after the US) (EC 2003, p. 337).

Such innovation strength reflected in patent domination can be, for instance, corroborated in terms of market control. Data from 2003 demonstrates Europe's domination over the pharmaceutical domain with 30-34% of the world market, while the US holds just nearly 25%. In contrast, the US leads the communications sector with 24% being closely followed by Japan. The aerospace sector is controlled by the US, with practically half of the market. Europe holds just around 30%. In the scientific instruments area, the market is in US hands with 49%. Europe is second with around 28-31% of the worldwide instruments market (NSF 2004). To the point, for the National Science Foundation – NSF(US) the high-tech market is controlled by the US with 32-33% of the worldwide production, pursued by the European Union with 22.8% and Japan with 12.9% (NSF 2004). The just mentioned is of major relevance if it is considered that the high technology industry is strategic. This is not only because it fosters the systematic application of knowledge altering or transforming the goods and production arrangements throughout the economy, but mainly because it allows a strong global position (Libicki 1989).

1.2 The scientific-technological matrix

When one of the "corners" of the triangle mentioned earlier doesn't exist, is weak or when it is partially replaced by foreign actors, one is in the presence of a *truncated* S&T industrialization scheme (Fajnzylberg 1983). I define this as a *scientific-technological matrix* of constrained strength that, in the best case, might only generate a relative presence in the international market with specific products or applications. This type of 'matrix' or structure functionality is typical of underdeveloped countries. The usual weaknesses are: reduced and uncoordinated public spending on S&T and education, and consequently, poor numbers on patent filings and patents granted; the lack of supportive and stimulating policies or, in a wider sense, of a long-term national project; the absence of strong and delineated bonds between industry and research centres; and the tendency of the few national industries with the potentiality of commercializing innovations of relinquishing their R&D needs to foreign entities or establishing their R&D centres abroad (Delgado 2007a).

Latin America is a case in point. This includes all major actors, Brazil, Mexico and Argentina, which together represent more than three quarters of the S&T of the region. Data for these countries reveals that, on one side, these were the main actors capturing direct foreign investment in the region, which increased more than 3 times from 1991 to 2006. On the other side, it is recognized that 40 to 50% of the 500 major companies in those countries are foreign companies or subsidiaries (Delgado 2007a). What's more, Latin American industries are mainly dedicated to extractive or to low technological input activities going from mining and cement fabrication, all the way to food products, retail stores and telecommunication services (Delgado 2007a).

Hence, in general terms and as shown by all indicators, Latin America lags a long way behind with respect to science and technology R&D. For example, in developed countries the number of researchers per thousand labour force (full time equivalent) for the year 2003, was between 10 and 15 times higher than in Latin America (IADB 2006). In the number of PhDs per hundred thousand inhabitants, the average for the region was 1.6, while in the US it was 10 and in European countries like Spain, 14 (IADB 2006). The same tendency is registered in relation to S&T journal articles published per hundred thousand inhabitants where the average for Latin America in 2003 was 3.1, while in US it was around 72 (IADB 2006).

Moreover, the brain drain phenomenon is very high in Latin America. It is estimated that 80% of the graduate faculty in Haiti, Guyana and Jamaica live abroad, mainly in the US. Mexico, Argentina, Nicaragua and Honduras have between 30 to 35% (Özden 2006). In the case of Mexico, it is considered that around 15% of its university graduates have migrated to rich countries (Goldin & Reinert 2007); 475 000 of them to the US. At least 30% of the total PhDs that Mexico generates emigrate. It should be added that 79% of Mexican sciences and engineering postgraduate students living abroad never come back. The direct economic cost for the country of losing those scientists and engineers has been estimated at several hundred millions of dollars, while the direct and indirect costs (i.e. loss on tax receipts, on long term productivity, etc.) at US$ 32.5 billion or 5.23% of the gross domestic product of 2001 (Hernandez 2007).[2] In this context, it is not surprising that medium-low technological input activities in Mexico correspond to at least 90% of industrial gross domestic production (Clemente & Durán 2008).

Taking the above into account, it can be said that the lack of a *triangle*-like structure in the region is obvious. This is true even when considering relative successful paradigms such as the Brazilian aerospace industry (small airplanes and medium resolution satellites) or the biotechnological innovations of Cuba (Delgado 2007a; 2007b).

2. Nano-Revolution, Potentialities and Promises

Nanotechnology has recently arrived with a burgeoning future prospect. Its potential covers fields from new materials useful for textiles, packaging, food or transportation industries; nanodevices

---

[2] The total wealth transfer from poor to rich countries represented by the brain drain stands somewhere between 45 to 60 billion dollars according to World Bank researchers (Goldin & Reinert 2007).

and nanomaterials for sophisticated medical procedures and treatments, all the way to more effective security and military innovations (Roco & Bainbridge 2001; 2003; National Research Council 2002; European Commission 2005; Berube 2006; Delgado 2008a).

The growing rhythm of public and private spending on research and development of nanoscience and nanotechnology N&N) isn't surprising. Public spending worldwide has been estimated from US$430 million in 1997, to US$ 4.6 billion in 2004; and US$6.4 billion in 2006 (Delgado 2008c). Private spending is calculated at US$4 billion for 2004 and at US$6 billion in 2006. The latest data available is for 2008 when private and public spending combined was estimated at around US$18.1 billion (personal communication with Mihail Roco, September 11, 2009).

The US National Nanotechnology Initiative (NNI) expenditure for 2010 is estimated to be nearly US$1.6 billion, an amount to which is to be added the funding coming from State Initiatives and other indirect contributions. The European Union's 7th Framework Programme (2007-2013) approved €3.45 billion for "nanosciences, nanotechnologies, materials and new production technologies". This corresponds to only a third of the total European spending on nanotechnology (Delgado 2008a). In addition, the German annual spending is set at around €330 million; the French at € 150 million and the British at €50 - 75 million annually (Delgado 2008a). The 3rd Science and Technology Basic Plan of Japan (2006-2010) has allocated US$ 5 billion for nanotechnology research and development; that is around a billion per year (Delgado 2008c).

A growing impetus is also registered on nanotechnology patents. A 15% average annual growth rate is estimated since the mid 1990s at both the European Patent Office (EPO) and the US Patent and Trademark Office (USPTO) (Igami & Okazaki 2007). However, countries' shares in nanotechnology patents from 1992 to 2001 at the USPTO and the EPO give to US the first position, followed by Europe and Japan (see Figure 3).

Nanotechnology spending worldwide concentrates on application areas such as the health sector and the life sciences (18%) and chemicals (12%). Other areas include energy, communication and information

Nanotechnology Patents (1992-2001)

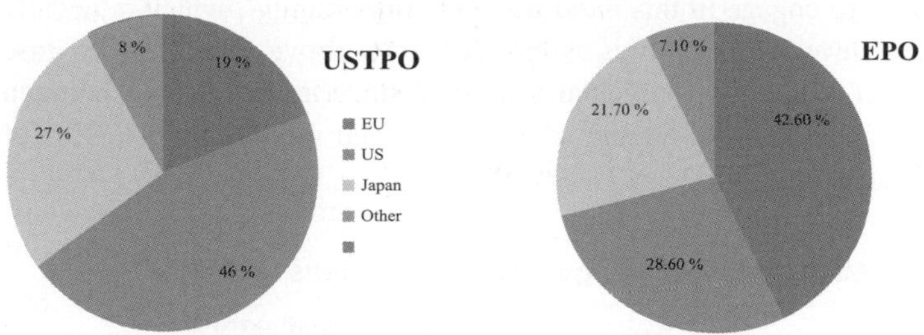

Figure 3 For colour reference turn to page 564.

Source: Glánzel et al.(2003). *Nanotechnology Analysis of an Emerging Domain of Scientific and Technological Endeavour*. Steunpunt O&O Statistieken Belgium July:46.

technologies, transportation, and environmental applications (see Figure 4). As a result, more than 400 products from companies from more than a dozen countries and around 600 nanotechnology raw materials, intermediate components and industrial equipment items used by manufacturers have been identified (Maynard 2006).

This data seems to be consistent with market share and estimations on product sales that at some level use nanotechnology. Diverse sources calculate that the nanotechnology market by 2004 was around US$ 12.9 billion and by 2006 nearly on US$ 50 billion (Delgado 2008a). Estimations for 2010 are at half a trillion dollars and between 1 and 2.5 trillion dollars by 2015 (Delgado 2008c).

This context of enthusiasm, or what has been called *nano hype* (Berube 2006), clearly entails a social and technological construction of a defined *imaginary*[3] that can, or can not, accomplish its own forecast and that may, or may not, produce unexpected outcomes, either positive or negative. In any case, what seems to be clear now is that investment in nanotechnology is real and that it may produce profound effects on the economy, society and the environment. Latin America, with all its

[3] Understood as the coming to be of certain forms of organization, production and reproduction of societies. It usually can be identified as a set of values, interests, symbols, institutional organization entities, regulations and prospectives common to a particular *social group* and the corresponding society and production system.

limitations, constraints and potentialities (its S&T 'truncated' platform), is trying to engage in this *nano wave* of 'opportunity', which is notably lead by developed countries as briefly shown above. The Mexican case can be considered of particular relevance since it is at a point between the Brazilian enthusiasm and advancement and the meagre actuation of Venezuela or Chile (Delgado 2008a).

Main Nanotechnology Application Areas (2007)

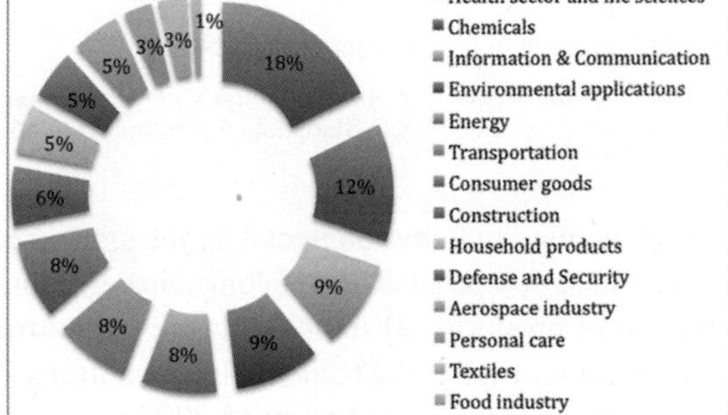

Figure 4 For colour reference turn to page 565.

Source: Nanoposts (2007). *Government Policy and Initiatives in Nanotechnology Worldwide 2007*. Canada: 39.

3. Nanoscience and Nanotechnology in Mexico

By the end of 2008 in Mexico there wasn't a national nanotechnology initiative or program as there was in Brazil or Argentina, which established in 2004-2005 the Argentinean Foundation for Nanotechnology (FAN) followed by the approval of the national strategic plan (*Plan Estratégico de Desarrollo de las Microy Nanotecnologías*) (Delgado 2008a). This lack of a nanotechnology program in Mexico is astonishing, particularly because of the proximity to the US, the assumed awareness of its national nanotechnology initiative (NSTC 2000; NRC 2002), and the global and certainly Latin American tendency of "getting on the *nano-train*" under the rationalization of the so called *knowledge economy* and the necessity of increasing national competitiveness as one of the main tools for development (for the case of Mexico, see Presidencia de la República 2007).[4] Notice that this is the typical conceptualization of a *linear model* that presumes the following automatic scheme:

> more research on (novel) S&T and improvement of infrastructure → more innovation → more competitiveness → more growth and, thus, better social and environmental indicators.

Yet, public and private spending on S&T, and thus on technological research and development in Mexico, are very low and, in the long run, decreasing in real terms (when considering inflation).[5] See Table 1 for comparative data on selected countries.

Leaving aside such contradictions between discourse and practice, it seems also unreasonable to have a lack of effective planning, coordination and implementation of a national N&N agenda in Mexico when, on one side, there has been since 2002 a financing formula for S&T in which 'sectorial' resources are allocated under the criteria of accomplished objectives and priorities established between the government, scientists, engineers, and companies - national and foreign - and, on the other side, while the country's *Special Program on Science and Technology 2001 – 2006*, considers N&N as a strategic area.[6] This was also explicitly acknowledged in the S&T cooperation agreement formally signed between Mexico, Canada and the US under the umbrella of the *Security and Prosperity Partnership of North America* (SPP 2005).[7]

---

[4] The *National Development Plan 2007-2012* states that: "...not taking advantage of new technologies and not contributing to their development will imply both, leaving aside a significant resource of structural development and losing economical competitiveness" (Presidencia de la República 2007, p. 35).

[5] In 2000 Mexico's investment in technological research and development was 0.37% of the GDP; by 2006, it was 0.47%, a progress that mainly corresponded to the private sector, including foreign actors (DIARIO OFICIAL DE LA FEDERACIÓN 2008). In comparative terms, Mexico is one of the OECD's members with the lowest investment rate in S&T since the Organization average is 2.25% (for the year 2005) (DIARIO OFICIAL DE LA FEDERACIÓN 2008).

[6] The program, however, did not present an implementation plan choosing to simply leave the statement as such.

[7] The SPP 2005 report states: "Explore opportunities for collaboration in other key areas,

However, it is to be documented that the Mexican National Council of Science and Technology (CONACYT) funded up to 152 R&D projects related to N&N during the period of 1999 to 2004 (53% in the area of materials) (CIMAV-SE-FUNTEC 2008). They involved more than fifty research institutions and totalized a poor erogati on of US$14.4 million (CIMAV-SE-FUNTEC 2008). This was just a fraction of Brazil's N&N

Table 1

Investment in technological research and development (IDE)⚜

Country	Cumulative IDE Millions of dollars (current)⊚		
	1991-2000	*2001-2005*	*1991-2005*
Brazil	95,500	67,346	162,846
Canada	118,738	100,314	219,052
Mexico	23,491	22,719	46,210
US	1,998,911	1,483,537	3,482,448

Source: DIARIO OFICIAL DE LA FEDERACIÓN (2008). *Programa Especial de Ciencia y Tecnología 2008-2012*. Mexico. December 16th.

expenditure during 2001 – 2006 that totaled almost 140 million reales or approximately US$55 million. Both cases, though, are trivial when compared with the US, which just between 2000 and 2004 spent up to US$3.28 billion on N&N.

In 2006 CONACYT created two national N&N laboratories with a two million dollar budget each. It also established megaprojects on strategic areas of research through the funding of five institutions with US$ 10,000 each for that purpose. At the same time, the Ministry of Economy (*Secretaría de Economía* – SE) was designated to be in charge of "promoting the use and commercialization of nanotechnology" (this happened exactly in the same way as it did in Brazil and Argentina where the respective ministries of economy, and not of S&T, took the main role). For that purpose, the SE formed a group of officials

including biotechnology, nanotechnology, supply chains and logistics management...small and medium-sized enterprises (SMEs)... and on an approach to emerging markets" (SPP 2005, p. 11).

responsible for scrutinizing the state of development and potential of N&N research and commercialization in Mexico.[8]

As what could be seen as a product of the actions mentioned above, Mexico's *Special Program on Science and Technology 2008-2012* (PECYT) determined nanotechnology, mechatronics and material sciences as some of the priority and strategic scientific and technological areas to be supported. In particular it established as key areas those of catalysis, polymers, nano(bio)materials, thin films, nanoceramics, simulation and characterization of nanomaterials (DIARIO OFICIAL DE LA FEDERACIÓN 2008). It is still to be seen if such a 'support' will be limited and disarticulated as it has been in the past or, in other words, if such a special program will or will not be a mere discourse of intentions.

3.1 Nanotechnology R&D and the Mexican innovation geography

By now, Mexico has several cooperation and technology transfer agreements, as well as local, regional and international R&D networks as the country is considered to have significant potential for developing some N&N niches (mainly those of polymers, metallurgy, textiles, electric appliances, oil, food processing and environmental remediation). Data from 2004 indicates that there were 20 groups performing N&N research in Mexico, totaling around two hundreds scientists (Volker Lieffering / Malsch TechnoValuation 2004). More precise figures from 2008 identified 449 researchers in 56 R&D centres carrying out N&N activities. Fourteen CONACYT institutions comprise 29% of N&N researchers; eight entities of the National Autonomous University (UNAM) 18%; the National Institute of Oil (IMP) 15%; six entities of the National Polytechnic Institute (IPN) 8%; while the rest of the researchers are spread out over 31 other entities (CIMAV-SE-FUNTEC 2008).

There are also documented 157 labs and 17 pilot plants that work on more than 340 lines of N&N research. Some pilot plants or projects are private ventures or associations between academic institutions and

[8] The group worked with scientists and businessmen until the summer of 2008 when the head of the Ministry was removed by the President; an action that caused the redistribution of some officials, including those in charge of N&N promotion. Since then and until March, 2009, the destiny of this group was uncertain.

companies. As stated in July 2008 by an official of the SE: "...there are 13 pilot projects being developed at this time and 11 more on their way for its industrial escalation" (Ortiz 2008). However, it is worth pointing out that those institutions with the most advanced equipment in the country are UNAM, IMP, the National Metrology Institute (CENAM) and some CONACYT R&D facilities (Institute of Scientific and Technological Research of San Luis Potosí—IPICYT, Research Centre on Applied Chemistry—CIQA, and the Centre of Research and Advanced Studies - CIMAV).

The relative concentration, geographically and institutionally, of the major R&D centres in the country corresponds to historical centralization of economic activities in Mexico. Even though some efforts of decentralization of R&D activities have been made (through for example the CONACYT's centres that are mostly spread throughout the country), the top institutions are located in Mexico City and its metropolitan area, the State of Mexico, and the cities of Guadalajara and Monterrey. More recently also in Zacatecas, Querétaro and along the border with US, the later very much related to *maquila*[9] activities. For instance, records from 2004 indicate that the highest degree of potential innovation corresponded precisely to the following Mexican states (note that 1 means the highest capacity in national terms): Mexico City (0.794); State of Mexico (0.735); Nuevo León (0.655); Jalisco (0.538); Guanajuato (0.423); Puebla (0.402); Chihuahua (0.388); and Coahuila (0.293) (Clemente 2008). These figures can also be corroborated in terms of patents granted. From 1994 to 2004, Mexico City concentrated, in average, more than a third of the patents granted nationally. Nuevo león, Jalisco, and the State of Mexico followed, concentrating around 10% of the patents granted by 2004. In the same period, Chihuahua and Coahuila increased their participation from little more than 1% to more than 4% (Mendoza *et al.* 2008).

---

[9] Maquila refers to the specialization of assembling imported components (and some national ones) for the subsequent export of the final products. The intention: to reduce production costs and avoid strong environmental and/or labor regulations that usually operate in the home country of most multinational corporations. It is the other side of the so-called out-sourcing process of production.

Such *geography of innovation* in Mexico comprises nano-R&D activities. Therefore, the *geography of N&N innovation* is pretty much the same since the main regions indicated above for S&T in general are the same for N&N (see Figure 5). Yet, Mexico's northern states, like Chihuahua and Baja California, seem to have an additional relevance within national geography of N&N innovation, a phenomena that will be explained below.

3.2 N&N collaboration schemes, networks and technological corridors

There are, as indicated before, some networks and international cooperation programs in Mexico. Since 2003, the UNAM has run the Network of Research Groups on Nanoscience (REGINA), which works on synthesis, characterization, theory and simulation, and potential applications of some nanomaterials. It consists of more than 60 researchers and 15 labs. The University's Project on Environmental Nanotechnology (PUNTA), another UNAM network, was formally set up in 2005 and links diverse nanoscientists and engineers with the purpose of developing cheap and efficient nano-catalysis materials and processes for air and water applications. Other technological outputs include the development of endogen research equipment. PUNTA consists of around 100 researchers and postgraduate students and in the first three years of operation has two pending patents. Finally, in 2008, the intention of creating a network that could link all UNAM's N&N activities was announced (Gaceta UNAM 2008). If it takes place, the initiative is expected to work on targeted applications in areas such as agriculture, biomedicine, energy and environment by coordinating efforts and resources, transferring knowledge and patent rights and thus incubating *start ups* (Gaceta UNAM 2008). Its relevance is due to the fact that UNAM generates almost half of the R&D activities in the country and, as already indicated, roughly a fifth of the national N&N research. Still, some operative, financial and political issues figure as major obstacles.

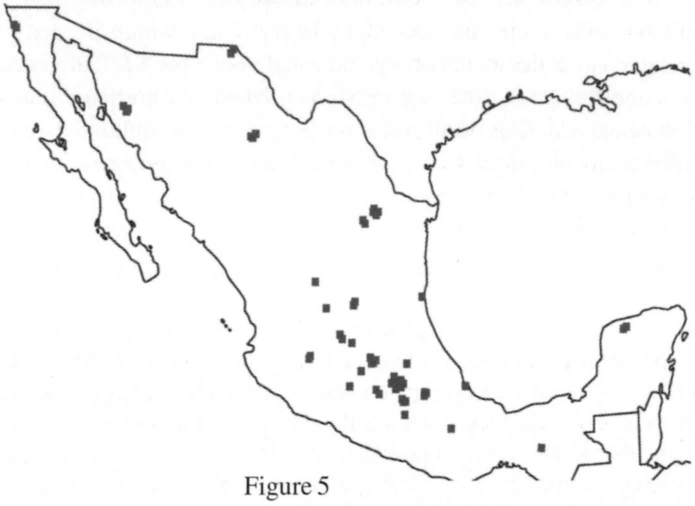

Figure 5

There is also the *National Thematic Network on Nanoscience and Nanotechnology*, being significantly financed by CONACYT since 2008/2009. It mainly concentrates and seeks to coordinate efforts of CONACYT, the National Institute of Nuclear Research (ININ), UNAM, IPN and the Metropolitan Autonomous University CUAM R&D centres; nonetheless, it also links with other national institutions and seeks to celebrate cooperation agreements with international ones. Another network is the *International Network of Nanoscience and Nanotechnology* (INN), consisting of individual researchers of diverse national and foreign institutions.[10] Its articulation seems to respond to

[10] UAM, IMP, BUAP, Universidad Iberoamericana, ININ, Superior School of Mechanical Engineering of the IPN, National Institute of Neurology and Neurosurgery (INNN), University of Texas, University of California, University of Chile, University of Concepción (Chile), University of Parma (Italy), Superior Technical Institute (Portugal), Institute of New Materials (Germany), Technical University of Istanbul (Turkey), and the University of Pablo de Olavide (Spain)

personal initiatives and again, to open up possibilities for international collaboration in specific research projects.

For the national support scheme of N&N the latter is of great importance due to weak government financial support. This has been clearly seen for several years with major international cooperation agreements on N&N with R&D institutions and government agencies from the US (including the US-Mexico Foundation for Science); Canada (carried out mainly by the Trade Commissioner of the Embassy of Canada in Mexico); Europe (mainly through bi-national agreements and European Commission programs or projects); the Russian Federation, among others (for further details see: Delgado 2008a).

Above and beyond national and international networks, CONACYT has been promoting the formation of technology parks, some with N&N incubator projects. Probably the most sound is the Nanotechnology Incubator at the Research and Innovation Technology Park (PIIT) in Monterrey, Nuevo León—the so-called *city of knowledge* (www.mtycic.com.mx). The park is a 70 ha complex dedicated to technology research and transfer in five priority areas: biotechnology, nanotechnology, mechatronics, information and communications technologies and health (*www.piit.com.mx*).

The nano-incubator, formally inauqurated on September 2009, is a collaborative scheme between Mexican research centres, Arizona State University and the IC2 of the University of Texas in Austin; all located at the park. It has, five incubator activities or pilot plants for the production of nanoparticles, nanotubes, polymer composites and other nanostructurated nanomaterials. Most pilot plants will have, an average capacity of production between 0.5 to 1kg per hour. The technology for such production has been partially transferred and Mexico will pay for the rights to use it.[11]

Along with a parallel biotech-incubator, the amount being allocated for 'phase one', is up to US$10 million, including the construction of two research facilities of 2,000 - 2,500 m^2 for each incubator (Gobierno del

[11] For example it will transfer a physical chemical method technology from Xetacomp (US) and a technology platform for the production of nanocomposites with polymeric matrix from the Centre for Applied of Nanotechnology (Hamburg, Germany).

Estado de Nuevo León 2008; Haurwitz 2008).PIIT owners and tenants comprise several national R&D entities[12] the National laboratory of Advanced Informatics (LANIA); an institution specialized on business incubation and commercialization of new technologies of the University of Texas in Austin (IC2 INSTITUTE); a N&N research facility of the Arizona State University; and THE TEXAS A&M UNIVERSITY. It also links national and foreign companies such as: Cemex (Mexico), Sigma (Mexico), Motorola (US), AMD (US), Xignux (Mexico), DeAcero (Mexico), Bosch (Germany), Cydsa (Mexico), Lamosa (Mexico), Vitro (Mexico), Lania (Mexico), Owens Corning (US), Pepsico-Gamesa (US)[13] etc. (Gobierno del Estado de Nuevo León 2008).

Considering other local R&D initiatives, what seems to be taking place is the formation of a technology bi-national corridor that stretches from Houston (even Louisiana), towards Austin and San Antonio in the US, then continuing into Mexico towards Monterrey, Saltillo and San Luis Potosí. Such a phenomenon suggests that Mexico's S&T development is being highly coupled to bi-national and other foreign business interests. If this scheme indeed is in some degree fine for big business (the vision since 2004 of the Mexican Ministry of Economy and the FUMEC),[14] it is not desirable if it is visualized as 'the' main blueprint for Mexico's high-tech development agenda given that, in the medium-long run,

---

[12] Five CONACYT's centres (Centre for Research on Advanced Materials – CIMAV; Centre for Engineering and Industrial Development – CIDESI; Centre for Scientific Research and Higher Education – CICESE; Centre for Research and Assistance on Technology and Design – CIATEJ; and the Centre for Research in Mathematics - CIMAT); the Centre for Innovation and Strategic Product Design – CIDEP of the Technological Institute of Superior Studies of Monterrey (ITESM); the Centre for Innovation, Research and Development of Technology and Engineering – CIIDIT of the Autonomous University of Nuevo León; and a biotech and medical physics research hub of the Centre for Research and Advanced Studies – CINVESTAV.

[13] It is worth noting that Pepsico-Gamesa has explicitly expressed its interest in developing nanotechnological food processes, certainly one of the few projects in the country in such an area.

[14] For the SE and the Foundation, the main innovation regions are Sonora, Baja California, Chihuahua, Nuevo León and Tamaulipas. The technology niches are mainly those related to the automotive industry, MEMS and NEMS, and IT technologies.

it seriously deters any endogenous industrialization national project (for a detailed argument see Delgado 2007a).

This tendency is of great significance, especially if it is kept in mind that other major technology parks are being deployed in Mexico, some also inspired as continuous bi-national technological corridors. Those are the cases of the Innovation and Technology Transfer Park (PIT2) of ITESM and the Paso del Norte MEMS/NEMS Packaging Cluster, both in Chihuahua[15]; and the Silicon Border Development Science Park located in Mexicali and the so-called 'Condominio Tecnológico Tijuana', both in the state of Baja California[16] (see map in Figure 6).

3.3 Nanotechnology and the Mexican S&T-matrix reality

From the information that has been presented, it is evident that Mexico (as is common in all Latin American countries) doesn't have a triangle S&T industrialization structure. Even more, it can be argued that the country's S&T matrix is increasingly weak. Its R&D activities are considerably disarticulated, geographically condensed and quite

---

[15] *PIT2* is an initiative of the government of the Mexican state of Chihuahua, the SE and the ITESM-Chihuahua. The construction of the 7,500 m^2 facility started on January 2008. It will offer technological services to industries in the areas of software, automation and electronics. Some companies already involved are ITNova, Visteon, TGC and Dinámica. *Paso del Norte* was established in February 2004. It's a collaboration and joint effort between academia, government, and industry/business of the Paso del Norte region with the goal of developing MEMS/NEMS packaging applications, but also, design, development and commercialization. The role of the Bi-National Sustainability Laboratory (BNSL) is considered a key aspect.

[16] *Silicon Border Development Science Park*, a 30 km^2 facility, is dedicated to semiconductors, solar/photovoltaic, medical devices, nanotechnology, flat panel display and aerospace technologies. Strongly involved from the Mexican side are: the Autonomous University of Baja California, the Technological University of the State, and the Centre of Nanoscience and Nanotechnology of UNAM. The park has been selected by Q-Cells (Germany), a huge solar cell manufacturer, for its next major expansion with an initial investment of US$ 3.5 billion. *Condominio Tecnológico Tijuana* is a US$20 million initiative formally arranged on February 2008 between CONACYT, the Mexican state of Baja California and the local and international business community. The technology-park-type facility will integrate specialists from the Autonomous University of Baja California, Technology Institute of Baja California, Technological University of the State, among other R&D entities. It is expected to be operative by 2011 when it will offer technology services on the areas of polymers, mechatronics, metrology, energy efficiency and advanced materials.

dependent on international cooperation agreements and technology transfer schemes. Its S&T policy fails, on one hand, to connect R&D priorities with national necessities, and on the other hand, to stimulate a long-term endogenous high-tech industrialization process because it is assumed that the best way for increasing 'national' industrial competences is through foreign direct investment. Consequently, Mexico's technology balance of payments (TBP) is negative as the country imports more technology than it exports (this is even considering *maquila* processes). Data indicates that in 1985 Mexico had a negative TBP of US$146.9 million. In 1990 it grew to US$307 million and by 2004 to US$511 million. In 2007, the country exported US$43 million, less than a tenth of what was imported (OECD 1999 and 2005). Hence, the small number of patents registered by national residents (131 in 2005), not to say those related to N&N that are expected to total a few dozen at most, is not surprising.

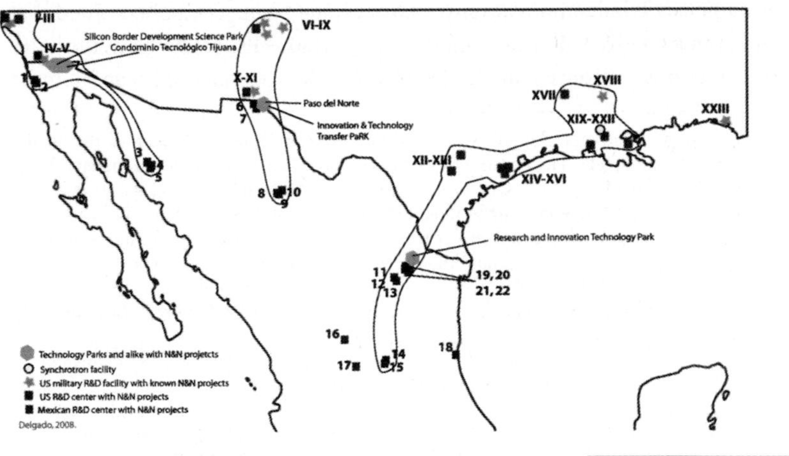

Figure 6 For colour reference turn to page 565.

Feasible bi-national N&N corridors

I. University of California – Los Angeles / Integrated Nanosystems Research Facility at Irvine / NASA's Institute for Cell Mimetic Space Exploration; II. Caltech; III. Naval Air Warfare Centre Weapons – Point Mugu / Naval Surface Warfare Centre – Seal Beach / US Air Force Space Missile Centre – El Segundo; IV. University of California – San Diego; V. Space and Naval Warfare Systems Centre – San Diego; VI. Centre for Integrated Nanotechnologies at Alamos National Laboratory; VII. Centre for Integrated Nanotechnologies at Sandia National Laboratory); VIII. Kirtland Airforce Research Laboratory; IX. University of New Mexico; X. New Mexico State University –Las Cruces; XI. White Sands Missile Range – Army Research Laboratory; XII. University of Texas –Austin; XIII. Texas A&M University; XIV. Rice University; XV. Baylor College of Medicine; XVI. University of Houston; XVII. Louisiana Tech University; XVIII. Engineer Research and Development Centre – US Army Corps of Engineers; XIX. Louisiana State University – Baton Rouge; XX. University of New Orleans Research Park; XXI. University of Louisiana – Lafayette; XXII; Centre for Advanced Microstructures & Devices – Baton Rouge; XXIII. Naval Surface Warfare Centre – Panama City, FL / Air Armament Centre.

1. Centro de Nanociencias y Nanotecnología UNAM; 2. Centro de Investigación Científica y de Educación Superior de Ensenada; 3. Universidad de Sonora; 4. Centro de In vestigación en Alimentación y Desarrollo, A.C.; 5. Instituto Tecnológico de Hermosillo; 6. Universidad Autónoma de Ciudad Juárez; 7. Instituto Tecnológico de Ciudad Juárez; 8. Instituto Tecnológico de Chihuahua; 9. Centro de Investigación en Materiales Avanzados, S. C; 10. Universidad Autónoma de Chihuahua; 11. Corporación Mexicanade Investigación en Materiales, S.A. de C.V: 12. Centro de Investigación en Química Aplicada; 13. Instituto Tecnológico de Saltillo; 14. Universidad Autónoma de San Luis Potosí; 15. Instituto Potosino de Investigación Científica y Tecnológica A.C; 16. Universidad Autónoma de Zacatecas; 17. Universidad Politécnica de Aguascalientes; 18. Instituto Tecnológico de Ciudad Madero; 19. Instituto de Innovación y Transferencia de Tecnología de Nuevo León; 20. Universidad Autónoma de Nuevo León; 21. Instituto Tecnológico de Estudios Superiores de Monterrey; 22. Universidad de Monterrey.

In addition, it should be mentioned that the equipment and instruments for N&N are limited and, in some cases, need to be transferred as is the case of a super computer donated in 2004 by Motorola (US) to the INAOE with the intent of allowing the fabrication of integrated circuits (Delgado 2008a). Furthermore, in spite of the implementation of incentives for S&T development and commercialization, mainly by setting up new technology parks, capturing foreign direct investment and transferring foreign novel technology, endogenous entrepreneurial understanding of nanotechnology and its potentialities is very poor in Mexico. For instance, in 2008, little more than a hundred companies were identified as involved or interested in N&N; 25% of them located in the Mexican state of Nuevo León; 14% in Jalisco; 11% in San Luis Potosí and 8% in Zacatecas. Of those, just a few are actually carrying out R&D activities.[17] However, in order to have a correct image of the scope of the presence of nano-products in Mexico, one must also add all the products produced abroad that are imported and sold in the country, such as L'Oreal products and those offered by Nano-Soluciones, the Mexican subsidiary of the German nanotechnology company Percenta AG.

Consequently, private spending on S&T in Mexico is also limited and in numerous cases it is associated with industrial application processes, which are for the most part foreign. In addition, some national companies are more interested in offering technological services to those big fabricants, than on developing their own products. Of course there are some exceptions, as for instance some of the biggest Mexican companies (such as CEMEX). Nevertheless, such Mexican companies do not always trust the national innovation system capacity, a main reason for establishing their R&D centres abroad. This is again the case of CEMEX

---

[17] For example: 1) Kaltex (Mexico) produces a nanofibre that is used in socks in order to eliminate bad odors; 2) Comex (Mexico) anti-graffiti paint –nanostructured- can be already found in the market; 3) Viakable (Mexico) uses a nanostructured isolator with improved mechanical properties; 4) Burlington Industries (US) is producing anti-wrinkle and anti-stain textiles; 5) Nemak/Alfa (US) is improving mechanical properties of automotive components while reducing their weight through aluminum-nanomaterials alloys; 6) A. Schulman (US) produces nanostructured plastic for the so called "plasticulture" (greenhouse agricultural production) that needs plastics with improved light transmission, UV light blocking, proper diffusion of light, water avoidance condensation, etc.

that has one of its main R&D centres in Switzerland. Such a lack of confidence, however, is not completely unfounded since for the case of N&N, national long-term capacity for preparing sufficient new generations of nanoscientists, nanotechnologists and specialized engineers still seems to be lacking.[18]

4. Nanotechnology Implications: Debate and Regulation in Mexico

Environmental, ethical and societal aspects of nanotechnology have been deliberated in the international context by governments, scientists and businessmen, mainly from the US and Europe (Roco & Bainbridge 2001 and 2003; EPA 2002; Roco & Tomellini 2003; Royal Society and the Royal Academy of Engineering 2004; Hett 2004; Lauterwasser 2005; Berube 2006; Maynard 2006; NNI 2007; ISO 2008). Independent of perspectives and rationales, such aspects have been seen, and even officially recognized, as key issues in N&N policy-making.

Yet, in Mexico and Latin America, the lack of an organized and compelling appraisal of such nanotechnology aspects seems to be a constant. This is also true for the governmental and business arenas even when considering some preliminary steps that are being taken. In Mexico for instance, there are few but significant efforts, most of them coming out of or being articulated by the scientific-academic arena.

In 2006 a first book on such issues was published (Foladori & Invernizzi 2006). In November of the same year, the *Latin American Network of Nanotechnology and Society* (RELANS) was founded by two academic institutions; one based at the UAZ and other at the UNAM. Since January 2007 UAZ has remained as the only coordinator of the network choosing to study social, political and ethical aspects of N&N from the point of view of the social sciences, while UNAM`s Interdisciplinary Research Centre on Sciences and Humanities (CEIICH) resolved instead to work separately.

[18] To illustrate, the IPICYT until 2007 had the only national PhD on N&N, however, it is to be recognized that a few N&N bachelor programs are already active or on their way, for example at the Universidad de las Américas, Universidad de la Ciénaga – Michoacán, the CNYN-UNAM, etc.

At the beginning of 2008, the CEIICH created an interdisciplinary 'consortium' with the purpose of stimulating dialog ue while generating inter-, and multi-disciplinary knowledge. "NanoMe x" consortium was formalized as an association of experts on societal and humanistic issues, the CEIICH and two R&D centres carrying out N&N projects: the Centre of Nanosciences and Nanotechnology (CNYN) and the Center of Applied Sciences and Technological Development (CCADET).

During its first working year, NanoMex organized a major international and interdisciplinary event on nanoscience and nanotechnology—*NanoMex08* (www.ceiich.unam.mx/nanomex08). Scientists, engineers, social scientists, businessmen and policy makers from all over the country and from abroad attended, with about one hundred speakers and more than two hundred and fifty participants. This event was also the platform for the presentation of both "Mundo Nano", probably the first interdisciplinary magazine on N&N in Mexico, edited by the 'consortium', and the video-documentary "En la frontera de lo pequeño: nanociencia y nanotecnología en la UNAM" which divulgates a panoramic vision of what is being done at the UNAM regarding N&N, including societal, ethical and environmental aspects.

By the beginning of 2009, NanoMex consortium became "nanoUNAM" with the purpose of expanding and formalising joint efforts of nearly a dozen of UNAM entities working on scientific, social and humanistic issues of N&N. "nanoUNAM" determined to work for the next few years on increasing an interdisciplinary N&N public awareness through diverse activities, such as working on a proposal for the introduction of N&N issues into undergraduate programs; the organization of international events (following NanoMex events); the promotion and implementation of *ad hoc* workshops for businessmen, academics, government officials and the general public; the publication not only of the magazine *Mundo Nano*, but also of books and booklets on diverse aspects of N&N; and the execution of research needed for the construction of local 'barometers' on N&N awareness (starting with the UNAM community).

From the government there are two main issues that are being addressed. First, and the most advanced effort, is the participation of the CENAM at the IPO group for the standardization of nanomaterials. Thus, there is an eventual commitment for developing national standards, practices and specifications of nanomaterials, a responsibility that is closely linked to nanomaterials safety regulation. Second, a preliminary evaluation on environmental and health aspects of N&N in Mexico is already being carried out by the IPICYT under the request of the National Institute of Ecology (INE).The Federal Commission for the Protection Against Sanitary Risks (COFEPRIS,) a branch of the Ministry of Health, also expressed its interest in the topic at the end of 2008, however, no official project has been launched by them as yet.

What should follow is the governmental endorsement of an instructive agenda on nanotechnology implications and risks aimed to engage, above all, key policy makers and other stakeholders needed for the promotion of a responsible nanotechnology development, including recycling and disposal. This means that the Mexican government should develop as soon as possible nanotechnology regulations.

Given the commercialization of dozens of nano-products in Mexico, it is also necessary to follow up and catalogue nanotechnology products being produced and sold. This should include production processes and proper safety measures for workers. In this sense, characterization, standardization as well as nanotoxicological studies are required for nanomaterials being produced and/or used in the country. For financial limitations it may be wise to specifically focus on those materials with a major propensity to be industrialized in the country, as for example silver, given Mexico's large reserves. In the process, the nation can contribute to the effort of developing not only appropriate national, but also, international protocols for toxicity and environmental impact evaluations. Subsequently, a first and major step for nanotechnology governance in Mexico should be, among others, the creation of an "Interagency Working Group on Nanotechnology Environmental and Health Implications". This is a figure that already exists in the US providing and exchanging information and establishing working programs on relevant areas such as nano-toxicology.

With or without a national nanotechnology initiative, and with or without a long-term endogenous industrialization project, more studies on nanotechnology risks should be carried out and sponsored nationally, not only regarding technical and scientific aspects as mentioned before, but also those concerning social, political and ethical aspects; an issue that until the end of 2008 was not formally recognized and supported by the government. The effort should, however, be of a long-term nature and as a result, should avoid the Brazilian experience that supported a national network on societal and environmental aspects of nanotechnology (RENANOSOMA) only to see it disappear a few years later.

5. Conclusion

Economical, political and security problems in Latin America, such as unemployment, low wages, lack of social security, criminality, etc., are the main issues discussed in the public domain. As a result, others tend to be silenced and when they are not, they are usually placed in a second level priority. In this context, it is remarkable that even though S&T are considered one of the main motors of the economy, there is a generalized lack of public debate on how to implement an endogenous and long term program on S&T. Even more fragile is the consciousness and dialogue regarding its ethical, economical, societal and political implications in the short, medium and long term.

The prospective for N&N public perception and the dialogue on its implications is consequently very poor. This is true for S&T in general as it was reported in a national survey on 2005 when it was estimated that 39% of the population was interested on S&T after sports and culture but only 8.9% were well informed; 26% had moderate information and the rest were low or non-informed (CONACYT 2005).

Considering that this situation is very similar for the rest of Latin America, it is urgent to promote, not only at the national but also at least at the regional level, a growing public interest and public participation on S&T issues and implications; particularly those related to novel technologies such as nanotechnology. This exercise must include all societal spheres, including policy makers, politicians, educators, NGOs and the general public since it seems necessary, and in some sense mandatory, that a N&N regulation is socially dialogued and agreed. In other words, if N&N is supposed to be the major technological revolution in the 21st century despite the many unknowns that surround its development, what should be of major interest for a respectable nanotechnology governance in Mexico is not only the maximization of benefits considering the national reality and needs; but also the minimization of unnecessary or avoidable costs, as well as the development of best risk management practices that facilitate a responsible and dialogued development and socialization of N&N when it is considered useful for solving social needs and environmental problems.

Questions for Reflection:

1. Is there a nano-divide, in the sense of an asymmetrical distribution of the benefits and risks of nanotechnology innovation? If so, who is this divide between?
2. Do you think all countries should pursue endogenous and independent science and technology development? Why/Why not?
3. How might (developing) countries stimulate nanotechnology R&D, and at the same time, ensure that negative consequences are minimised?

Bibliography

Berube, D. (2006). *Nano-hype: The Truth Behind the Nanotechnology Buzz,* US: Prometheus Books.

Chalmers, J. (1982). *MITI and the Japanese Miracle. The Growth of Industrial Policy 1925–1975*, US: Stanford University Press.

Chalmers, J. (2004). *The Sorrows of Empire. Militarism, Secrecy and the End of the Republic,* Canada: Metropolitan Books.

Cimav-se-funtec (2008). Diagnóstico y Prospectiva de la Nanotecnología en México, *Centro de Investigación en Materiales Avanzados-Secretaría de Economía-Fundación Mexicana para la Innovación y Transferencia de Tecnología en la Pequeña y Mediana Empresa*, February, Mexico.

Clemente R. and Durán (2008). México: geografía económica de la nnovación, *Comercio Exterior*, Vol. 58, No. 11, November, pp. 756-768.

conacyt (2005). Percepción Pública de la Ciencia y la Tecnología en México, Mexico: conacyt.

Delgado, G. C. (2002). *La Amenaza Biológica,* Mexico: Plaza y Janés.

Delgado, G. C. (2006a). *Aspectos éticos, sociales y ambientales de la nanotecnología,* PhD Dissertation Thesis, Autonomous University of Barcelona, Spain.

Delgado, G. C. (2006b). China in der Hochtechnologie-Konkurrenz, *Das Argument*, No. 268, pp. 52 – 60.

Delgado, G. C. (2007a). Colonialidad Tecnológica y Desindustrialización en América Latina, *CONtexto Latinoamericano. Revista de Análisis Político*, Ocean Sur, No. 4. July-September, pp. 143-168,

Delgado, G. C. (2007b). Competencia intercapitalista en tecnología estratégica y su militarización: el caso del sistema satelital Galileo, *Revista de Sociologia e Politica*, No. 29, Universidad Federal de Paraná, Curitiba, Brazil, November, pp. 105-130.

Delgado, G. C. (2008a). *Guerra por lo Invisible. Negocio, implicaciones y riesgos de la nanotecnología,* Mexico: Ceiich, UNAM.

Delgado, G. C. (2008b). Tecnología China 2.0. Nuevo balance del complejo científico-tecnológico chino, *Realidad Económica,* No. 236, May 16th – June 30th, pp. 75-98.

Delgado, G. C. (2008c). Economía Política de la Nanotecnología, *Mundo Nano. Revista Interdisciplinaria en Nanociencia y Nanotecnología,* No. 1, Vol. 1, UNAM, November, pp. 87-94.

Diario Oficial de la Federación (2008). *Programa Especial de Ciencia y Tecnología 2008-2012*, December 16th, Mexico: DOF.

EPA - Environmental Protection Agency (2002). *Nanotechnology and the Environment: Applications and Implications*, August, Washington, D.C.: EPA.

European Commission (2003). *Third European Report on Science & Technology Indicators*, Brussels: E.C.

European Commission (2005). *NanoMedicine. Nanotechnology for Health,* Office for Official Publications of the European Communities, September, Brussels: E.C.

Fajnzylberg, F. (1983). *La industrialización trunca de América Latina*, Mexico: Nueva Imagén.

Foladori, G. and Invernizzi, N. (coord) (2006). *Nanotecnologías Disruptivas. Impactos Sociales de las Nanotecnologías*, Mexico: Miguel Angel Porrúa.

Gaceta UNAM (2008). Trabaja la UNAM para hacer de las nanociencias motor del desarrollo nacional, *Gaceta UNAM,* 2 de junio.

Gobierno del Estado de Nuevo León (2008). *PIIT Master Plan*, Mexico: GENL.

Goldin, I. and Reinert, K. (2007). *Globalization for Development. Trade, Finance, Aid, Migration and Policy*, Washington, D.C.: World Bank.

Haurwitz, R. (2008). UT planning to build a business innovation centre in Mexico, *Statesman.com* October 8th. www.statesman.com/news/ content/news/stories/ local/10/08/1008utmexico.html

Hett, A. (2004). *Nanotechnology: Small Matter, Many Unknowns*, Switzerland: Swiss Reinsurance Company.

Hernández Suárez, J. L. (2007). La migración de trabajadores calificados, *Revista Electrónica Zacatecana sobre Población y Sociedad,* Year: 7. No. 30, January-August.

IADB – Inter-American Development Bank (2006). *Education Science and Technology in Latin America and the Caribbean. A Statistical Compendium of Indicators,* Washington, D.C.: IADB.

Igami, M. and Okazaki, T. (2007). *Capturing Nanotechnology's Current State of Development via Analysis of Patents,* STI Working Paper 2007/4, OECD, DSTI/DOC(2007)4. Paris, May 23rd.

ISO – International Organization for Standardization (2008). *Nanotechnologies – Health and Safety Practices in Occupational Settings Relevant to Nanotechnologies,* October 1st, Switzerland: ISO.

Lauterwasser, C. (2005). *Opportunities and Risks of Nanotechnologies*, London: Allianz AG, Centre for Technology / OECD.

Libicki, M. C. (1989). *What Makes Industries Strategic*, Washington, D.C.: The Institute for National Strategic Studies, National Defense University.

Maynard, A. (2006). *Nanotechnology: A Research Strategy for Addressing Risk*, US: Woodrow Wilson International Centre for Scholars.

McGrath, P. (2002). *Scientists, Business and the State*, Chapel Hill: The University of North Carolina Press.

Melman, S. (1970). *Pentagon's Capitalism: The Political Economy of War, New York:* McGraw-Hill.

Mendoza, J. E., Torres P., Hugo, V., Polanco G., and Mayrén (2008). Desigualdad del crecimiento económico regional e innovación tecnológica en México, *Comercio Exterior*, Vol. 58, No. 7, July, pp. 507 – 521.

NNI – National Nanotechnology Initiative (2007). *Strategy for Nanotechnology Related Environmental, Health and Safety Research*, Washington, D.C.: NNI.

NRC – National Research Council (2002). *Small Wonders, Endless Frontiers. A Review of the National Nanotechnology Initiative*, Washington, D.C.: National Academy Press.

NSF – National Science Foundation (2004). *Science and Engineering Indicators 2004,* National Science Board, Vol. 1.

NSTC – National Science and Technology Council (2000). *National Nanotechnology Initiative. The Initiative and Its Implementation Plan,* Washington, D.C.: Committee on Technology. Subcommittee on Nanoscale Science, Engineering and Technology.

OECD – Organization for Economic Co-operation and Development (1999). *Science, Technology and Industry Scoreboard 1999*, Paris: OECD.

OECD – Organization for Economic Co-operation and Development (2005). *Science, Technology and Industry Scoreboard 2005*, Paris: OECD.

Ortiz R., Silvia (2008). Del laboratorio al mercado, *Expansión,* July 30th. www.cnnexpansion.com/manufactura/tendencias/del-laboratorio-al-mercado (last accessed 17.03.09).

Presidencia de la República (2007). *Plan Nacional de Desarrollo 2007-2012*, Poder Ejecutivo Federal, Mexico: Presidencia de la República.

Roco, M. and Bainbridge, W. (2001). *Societal Implications of Nanoscience and Nanotechnology,* Washington, D.C.: National Science Foundation.

Roco, M. and Bainbridge, W. (2003). *Nanotechnology: Societal Implications – Maximizing Benefits for Humanity,* US: National Nanotechnology Initiative.

Roco, M. and Tomellini, R. (2002). *Nanotechnology Revolutionary Opportunities and Societal Implications,* Brussels: European Commission.

Royal Society and the Royal Academy of Engineering (2004). *Nanoscience and Nanotechnologies: Opportunities and Uncertainties,* London: Royal Society and the Royal Academy of Engineering.

SPP Security and Prosperity Partnership of North America (2005). *SPP Report to Leaders*, Prosperity Annex, June. www.spp.gov/report_to_leaders/prosperity_annex.pdf?dName=report_to_leaders.

Volker Lieffering / Malsch TechnoValutaion (2004). *Study on the Nanotechnology and Microsystems Technology in Mexico*, Netherlands: Volker Lieffering / Malsch TechnoValutaion.

WIPO – World Intellectual Property Organization (2008). *World Patent Report: A Statistical Review 2008*, Switzerland: WIPO.

Özden, Ç. (2006). *Brain Drain in Latin America*, Department of Economic and Social Affairs.UN/POP/EGM-MIG/2005/10, Washington, D.C.: United Nations Secretariat.

Chapter 17

Nanotechnology in Food and Agriculture

Georgia Miller & Rye Senjen

Nanotechnology in food and agriculture is emerging at a time when global food systems are under unprecedented stress. Close to one billion people now experience extreme food insecurity, while dangerous climate change and the loss of arable land further threaten the viability of existing farmland. At a time when global society should be relocalising agriculture and supporting small scale farming to ensure the social and environmental sustainability of food systems, in this chapter we suggest that nanotechnology is likely to underpin the next wave of high-tech industrialisation and globalisation of agriculture and food systems. We further suggest that highly processed nanofoods will not help to redress widespread diet-related health problems in countries where there is over-consumption of nutrient poor food. Instead, greater public health benefits could be achieved by increasing consumption of minimally processed fruit and vegetables. Finally, we provide an overview of the early scientific studies indicating that many nanomaterials now in use by the food and food packaging industries pose serious new health and environment risks for workers, the public and ecological systems.

Nano Meets Macro - Social Perspectives on Nanoscale Sciences and Technologies
by K L Kjølberg & F Wickson

www.panstanford.com
978-981-4267-05-2

1. Nanotechnology has Entered the Food Chain, but Uncertainty Surrounds How to Define a 'Nano' Ingredient

Nanotechnology is increasingly being used in agriculture, food processing, food packaging and in the surveillance of the transport and sale of food and food ingredients. Manufactured nanoparticles, nano-emulsions and nano-capsules can be found in agricultural chemicals, processed foods, food packaging and food contact materials including food storage containers, cutlery and chopping boards.

In this chapter the term 'nanotechnology' encompasses a range of technologies that operate at the scale of the building blocks of biological and manufactured materials – the 'nanoscale'. Nanotechnologies have been provisionally defined as relating to materials, systems and processes which operate at a scale of 100 nanometres (nm) or less. Nanomaterials have been defined by the International Standards Organisation (ISO/TS 27687 2008) as having one or more dimensions measuring 100nm or less.

However this definition of nanomaterials is problematic and likely to be far too narrow for the purposes of health and environmental safety assessment (Miller & Senjen 2008a). Some particles greater than 100nm show similar anatomical and physiological behaviour to nanomaterials, including very high reactivity, bioactivity and bioavailability, increased influence of particle surface effects, strong particle surface adhesion and strong ability to bind proteins (Cedervall *et al.* 2007; Linse *et al.* 2007; Garnett & Kallinteri 2006). Particles <300nm in size have the capacity to be taken up into individual cells (Garnett & Kallinteri 2006). For these reasons we believe that particles up to 300nm in size should be treated as nanomaterials for the purposes of health and safety assessment, especially in high exposure applications such as agriculture and food.

There is considerable discussion and disagreement regarding which size-based definition is most likely to capture the range in which novel nano-specific properties are seen. Definitions of 200nm or even 1,000nm have been used by government regulators and food and pharmaceutical researchers (Medina *et al.* 2007; Sanguansri & Augustin 2006; U.K. DEFRA 2006; U.S. FDA 2006). In other instances food regulators have chosen not to offer a size-based definition at all (FSANZ 2008; U.S.

FDA 2007). While the size at which food and other ingredients are defined as 'nanomaterials' may appear a minor issue, we are concerned that if regulators do not use the <300nm definition, many food ingredients which pose novel nanotoxicity risks will escape effective regulation.

2. Current and Near-Term Applications of Nanotechnology in Food and Agriculture

Secrecy surrounds the commercial availability of nanofoods. Estimates of commercially available nanofoods vary widely; nanotechnology analysts estimate that between 150-600 nanofoods and 400-500 nano food packaging applications were on the market in 2006/07 (Daniells 2007; Helmut Kaiser Consultancy Group 2007a; 2007b; Reynolds 2007; Cientifica 2006). Many of the world's largest food and agriculture companies have active nanotechnology research and development programs (Innovest 2006). By 2010 it is estimated that sales of nanofoods will be worth almost US$6 billion (Cientifica 2006).

The term 'nanofood' describes food which has been cultivated, produced, processed or packaged using nanotechnology techniques or tools, or to which manufactured nanomaterials have been added (Joseph & Morrison 2006). Examples of nano-ingredients and manufactured nanomaterial additives include nanoparticles of iron or zinc, and nanocapsules containing ingredients like co-enzyme Q10 or Omega 3. Nano additives are now marketed for use in margarine, soft drink, dairy products, sausages and other processed food. Nano packaging is used extensively, especially in plastic soft drink and beer bottles, but also to package confectionery, cheese, meats and other goods.

An investigation by Friends of the Earth, Australia, Europe and United States (Miller & Senjen 2008b) found on sale world wide 104 foods, food packaging, kitchen and agricultural products in which manufacturers acknowledged nanomaterial ingredients.

Table 1 provides examples of current applications of nanotechnology in food. The purpose of many of these applications is to increase the efficiency of modern production methods, although they are often marketed – at least in food industry publications if not on product labels -

as responding to 'consumer demand'. More broadly, applications include:

- Stronger nano-formulated or nano-encapsulated flavourings, colourings and nutritional additives (eg zinc, iron, lycopene, Omega 3).
- More potent nano-formulated processing aids to increase the pace of manufacturing and to lower costs of processing.
- Nanoscale emulsions to improve flow properties or food texture.
- Packaging and edible coatings to increase food shelf life by increasing barriers to oxygen and carbon dioxide exchange, or by giving UV protection or antibacterial properties.
- 'Track and trace' labelling and packaging.
- Antibacterial kitchen ware (refrigerators, chopping boards, cutlery, cleaning products) and food storage containers.
- Nano-formulation of on-farm inputs to produce more potent fertilisers, plant growth treatments and pesticides that respond to specific conditions or targets.

Future applications may include:

- Personalised 'interactive' foods to release nutrients or withhold allergenic substances.
- Reduction of processed foods' fat, carbohydrate or calorie content or increase of protein, fibre or vitamin content to enable foods such as soft drinks, ice cream, donuts or chips to be marketed as 'health' foods.
- Packaging to extend shelf life by detecting spoilage, bacteria growth, or nutrient loss, and releasing antimicrobials, flavours, colours or nutritional supplements in response.
- Nanobiotechnology as a tool in genetic engineering of seeds.
- 'Synthetic biology' to engineer artificial new organisms to produce colourings, flavourings and food additives, and ethanol from agrofuels.
- Nano surveillance systems to enable far-reaching automation of farm management.

Table 1: Examples of foods, food packaging and agricultural products that now contain nanomaterials (for detailed references see Miller & Senjen 2008b).

Product type	**Nano content**	**Purpose**
Mineral additive: Nutritional Drink Mixes	300nm particles of iron	Nano-sized iron particles have increased reactivity and bioavailability.
Food preservative: Sausage manufacture	30nm micelle (capsule) of lipophilic or water insoluble substances	Nano-encapsulation increases absorption of nutritional additives, increases effectiveness of preservatives and food processing aids. Used in wide range of foods and beverages.
Food additive: Omega -3	50nm 'nano-cochleates'	Effective means for the addition of highly bioavailable Omega-3 fatty acids to cakes, muffins, pasta, soups, cookies, cereals, chips and confectionery.
Food colourant: Synthetic lycopene	LycoVit 10% (<200nm synthetic lycopene)	Bright red colour and potent antioxidant. Sold for use in health supplements, soft drinks, juices, margarine, breakfast cereals, instant soups, salad dressings, yoghurt, crackers etc.
Food contact material: Antibacterial kitchenware	Nanoparticles of silver	Cutting boards, ladles, egg flips, serving spoons with increased antibacterial properties.
Food packaging: Plastic wrapping	Nanoparticles of silica in a polymer-based nanocomposite	Nanoparticles of silica in the plastic prevent the penetration of oxygen and gas of the wrapping, extending the product's shelf life. To wrap meat, cheese, long-life juice etc.
Food packaging: Plastic wrapping	Nanoparticles of zinc oxide	Antibacterial, UV-protecting food wrap.
Agricultural chemical: Plant growth treatment	100nm particle size emulsion	Very small particle size means mixes completely with water and does not settle out in a spray tank.

3. New Toxicity Risks for Human Health and the Environment

The use of manufactured nanomaterials in foods and beverages, nutritional supplements, food packaging and edible food coatings, fertilisers, pesticides and comprehensive seed treatments presents a whole new array of risks for the public, workers and ecological systems. These risks remain very poorly understood. However early scientific studies indicate that nanomaterials in commercial use by the food industry could be toxic for health and the environment, or cause longer-term pathologies.

3.1 Nanoparticles show greater reactivity, bioactivity and bioavailability than larger particles of the same substance

Nanoparticles have much greater access than larger particles to our bodies' cells, tissues and organs. Particles less than 300nm in size can be taken up by individual cells (Garnett & Kallinteri 2006), while those which measure less than 70 nm can be taken up by our cells' nuclei (Chen & Mikecz 2005; Geiser *et al.* 2005; Li *et al.* 2003), where they can cause major damage. Nanoparticles also have a large reactive surface area per unit of mass. In addition to greater bioavailability, nano-sized or nano-encapsulated vitamins, minerals and other active ingredients have greater bioactivity (Mozafari *et al.* 2006). If nano-nutritional additives and supplements provide an excessive dose of some vitamins or nutrients these may have a toxic effect or interfere with the absorption of other nutrients. Even substances which are not toxic in themselves can have a toxic effect if consumed in excessive quantities (Downs 2003; U.S. IOM 1998).

The high surface reactivity of nanoparticles can result in greater production of 'reactive oxygen species' (ROS) including free radicals. Production of ROS and free radicals is one of the key mechanisms of nanoparticle toxicity (Nel *et al.* 2006). Other properties of nanomaterials that influence toxicity include: chemical composition, shape, surface structure, surface charge, catalytic behaviour, extent of particle aggregation (clumping) or disaggregation, and the presence or absence of other groups of chemicals attached to the nanomaterial (Brunner *et al.* 2006; Magrez *et al.* 2006; Sayes *et al.* 2006). Many (although not all) manufactured nanoparticles are more toxic per unit of mass than larger particles of the same chemical composition (Oberdörster *et al.* 2005). The Project on Emerging Nanotechnologies at the Woodrow Wilson International Center for Scholars has suggested that the toxicological impact of 58,000 tonnes of manufactured nanomaterials might be the equivalent of 5 million or even 50 billion tonnes of conventional materials (Maynard 2006).

3.2 Preliminary studies show that nanomaterials could pose acute toxicity risks and long term pathological problems

Nanoparticles of titanium dioxide and zinc oxide are used as antimicrobial and UV protectors in food packaging and even as food additives. Yet *in vitro* studies have found that nanoparticle titanium dioxide destroyed plasmid DNA (Donaldson *et al.* 1996), produced damaging free radicals in brain immune cells (Long *et al.* 2006) and in the presence of UV radiation damaged cellular and sub-cellular skin cell structures (Vileno *et al.* 2007; Dunford et al. 1997). High concentrations interfered with the function of skin and lung cells (Sayes *et al.* 2006). An *in vitro* study found that for some cultured cells, zinc oxide nanoparticles were more cytotoxic than asbestos (Brunner *et al.* 2006). In preliminary feeding studies high oral doses of nanoparticle zinc (Wang *et al.* 2006), zinc oxide (Wang *et al.* 2007) and titanium dioxide (Wang *et al.* 2007b) caused severe toxicity or organ damage in test animals.

Nanoparticle silver is becoming widely used as an antimicrobial agent in food packaging and in food contact materials such as chopping boards, storage containers, cutlery and refrigerators. Colloidal silver that contains nanoparticles is also sold as a 'health supplement'. *In vitro* studies demonstrate that silver nanoparticles are highly toxic to rat brain cells (Hussain *et al.* 2006), mouse stem cells (Braydich-Stolle *et al.* 2005) and rat liver cells (Hussain *et al.* 2005).

Silica nanoparticles are used as food additives, for example in RBC Life Sciences' "SlimShake", and in food packaging, for example in plastic film sold by Bayer. Chen & von Mickecz (2005) found that 50nm and 70nm silica particles were taken up into the nucleus of cultured cells where they caused the onset of pathology similar to neurodegenerative disorders. Di Pasqua *et al.* (2008) found that both mesoporous silica nanomaterials and 250nm spherical particles of silica dioxide were also cytotoxic (toxic for cells) *in vitro*.

In addition to acute toxicity risks, the potential for long term pathological effects is also indicated by early studies. For instance, some clinical studies suggest that nanoparticles and small microparticles that are not metabolised can over time result in granulomas, lesions (areas of damaged cells or tissue), cancer or blood clots (Gatti 2004; Gatti *et al.* 2004; Gatti & Rivassi 2002; Ballestri *et al.* 2001). Scientists have suggested that nanoparticles and particles a few hundred nanometres in size in foods may already be associated with rising levels of irritable bowel and Crohn's disease (Ashwood *et al.* 2007; Schneider 2007; Gatti 2004; Lucarelli *et al.* 2004; Lomer *et al.* 2001).

Much discussion of toxicity risks associated with nanofood and agriculture focuses on those faced by consumers. However workers who handle, manufacture, package or transport foods and agricultural products that contain manufactured nanomaterials are likely to face higher levels of nano-exposure than the public and on a more routine basis. Scientists still do not know what levels of nano-exposure may harm workers' health, and whether or not any level of occupational exposure to nanomaterials is safe. Furthermore, reliable systems and equipment to prevent occupational exposure do not yet exist, and methods for measuring and characterising nanomaterial exposure have not yet been identified (Maynard & Kuempel 2005; U.K. HSE 2004).

3.3 Nanomaterials also pose ecological risks

The production, use and disposal of foods, food packaging and agricultural products containing manufactured nanomaterials will inevitably result in the release of nanomaterials into the environment.

Nanomaterials will also be released into the environment intentionally, for example as agricultural pesticides or plant growth treatments. Very few studies have examined the ecological effects of nanomaterials and their behaviour in the environment remains poorly understood. For example it remains unknown whether or not nanomaterials will accumulate along the food chain (Boxhall *et al.* 2007; Tran *et al.* 2005). However the preliminary evidence suggests that nanomaterials in commercial use by the agriculture and food industry may cause environmental harm.

Nanoparticle titanium dioxide has caused organ pathologies, biochemical disturbances and respiratory distress in rainbow trout (Federici *et al.* 2007) and caused mortality (Lovern & Klaper 2006) and behavioural changes (Lovern *et al.* 2007) in water fleas. Water fleas are used by regulators as an ecological indicator species. High levels of nanoparticle titanium dioxide have also been toxic to algae (Hund-Rinke & Simon 2006).

Byproducts associated with the manufacture of single-walled carbon nanotubes, mooted for future use in food packaging, caused increased mortality and delayed development of a small estuarine crustacean *Amphiascus tenuiremis* (Templeton *et al.* 2006). Earthworms exposed to double-walled carbon nanotubes produced significantly fewer cocoons in a dose-dependent response (Scott-Fordsmand *et al.* 2008). Exposure to high levels of nanoscale aluminium has been found to stunt root growth in five commercial crop species (Yang & Watts 2005).

Nanomaterials such as silver, zinc oxide and titanium dioxide are increasingly being added to food packaging and food contact materials for their antibacterial qualities. Products in which they are used include cling wrap, chopping boards, cutlery and food storage containers as well as a range of household appliances including dish washers, refrigerators and washing machines. If used on a large scale, nano-antibacterial agents could be harmful to microbes in the environment (Handy *et al.* 2008), or even disrupt the functioning of nitrogen fixing bacteria associated with plants (Throback *et al.* 2007). Any significant disruption of nitrification, denitrification or nitrogen fixing processes could have negative impacts for the functioning of entire ecosystems. There is also a risk that widespread use of antimicrobials will result in greater resistance among harmful bacteria (Melhus 2007).

The intentional release of nano-agrochemicals into the environment is a serious concern. Proponents of nanotechnology argue that because nano-agrochemicals are formulated for increased potency, they will be used in smaller quantities, delivering environmental savings (e.g. Parry 2008; Joseph & Morrison 2006). However in light of their far greater potency, we believe that nano-agrochemicals may be capable of causing even greater ecological problems than the chemicals they replace.

3.4 Production of nanomaterials has a huge environmental footprint

The manufacture of nanoparticles has a very high environmental footprint (Şengül *et al.* 2008). This is related to the highly specialised production environments, high energy and water demands of processing, low yields, high waste generation, the production and use of greenhouse gases such as methane and the use of toxic chemicals and solvents such as benzene. In one fullerene manufacturing process – which itself can be highly energy intensive and polluting - only 10% of the finished product was usable, with the rest sent to landfill (U.K. RCEP 2008). In a life-cycle assessment of carbon nanofibres, Khanna *et al.* (2008) found that producing carbon nano-fibres may have the potential to contribute to global warming and ozone layer depletion, and cause environmental or human toxicity that is as much as 100 times greater per unit of weight than those of conventional materials like aluminium, steel and polypropylene.

Nanoparticles are likely to be used in far smaller quantities than conventional substances, so a life-cycle assessment of the products they are used in would give a more accurate estimate of total energy and environmental impacts. Currently very few consumer products have undergone such an analysis. Exceptions include clay polypropylene nano-composites, nano-varnish with sol-gel technology, quantum dots and semiconductor crystals in light-emitting diodes (see Project on Emerging Nanotechnologies 2007). Nonetheless, these early nanomaterial life cycle assessments led the scientists to conclude that any environmental gains of nanoparticles (for example through their greater potency enabling smaller quantities of ingredients or materials to be used) may be outweighed by the environmental costs of their production.

4. Broader Implications of Nanotechnology in Food and Agriculture

Nanotechnology in food and agriculture is emerging at a time when global food systems are under unprecedented stress. Recent decades have revealed the high environmental costs associated with industrial scale chemical-intensive agriculture, including biodiversity loss, toxic pollution of soils and waterways, salinity, erosion, desertification and declining soil fertility (FAO 2007). The escalation of the global food crisis has also underscored the fundamental failure of global food and agriculture systems to meet the food needs of nearly a billion people.

Nanotechnology proponents (IFRI 2008) and even academics keen to promote the Millennium Development Goals (Salamanca-Buentello *et al.* 2005) have suggested that nanotechnology will deliver environmental sustainability and even eradicate extreme poverty and hunger. However Friends of the Earth Australia suggests that by entrenching our dependence on the industrialised, export-oriented agricultural system and the chemical and technology 'treadmills' that underpin it, nanotechnology is more likely to exacerbate the problems that caused the current global food crisis. We also suggest that the marketing of highly processed nanofoods as health promoting is likely to have negative public health outcomes for consumers world-wide. We recognise that greater public health benefits could be achieved through promoting diets that contain more minimally processed fruit and vegetables.

4.1 The context of the global food crisis

Prices for staple foods reached record highs during 2008. In the first quarter of 2008 alone, wheat prices increased by 130% and corn prices increased by 30% over 2007 prices (IFAD 2008). In some regions the price hikes were even more extreme. For example, in Timor Leste, the price of rice increased 300% over a six-month period (IRIN 2008); in Indonesia, the price of soy and tofu doubled in the space of a few weeks

(La Via Campesina 2008). Price rises have had the worst impact on poor people reliant on buying food. Food riots occurred in over 30 countries where the world's poorest people could no longer afford basic food.

There are a number of factors contributing to the price hikes that have exacerbated the food crisis. A leaked World Bank report admitted that 75% of the price increases in food crops may be caused by competition from agrofuels grown to fuel cars (Mitchell 2008). La Via Campesina acknowledges that competition from agrofuels and climate change-related droughts and floods has reduced food production in many areas, while speculation has driven up prices. However it observes that the global food crisis is primarily the result of many years of destructive trade policies, deregulation and 'structural adjustment' demanded by the International Monetary Fund and the World Trade Organization that have undermined domestic food production (La Via Campesina 2008). The removal of tariffs on imported foods, the growing retail domination by supermarkets that pay low prices to primary producers and the artificially low prices of subsidised imports or food dumped as 'aid' have undermined the viability of small scale farming in many regions. Food production for local communities has also faced competition from export-oriented luxury crops, loss of cropland to mining and urbanisation, contamination of water and soil, salinity and desertification. Many formerly self-sufficient countries are now dependent on importing food from the world market; the recent price hikes in wheat, corn, rice and soy beans have left the poor struggling to afford to eat.

4.2 The food crisis demands that we support small scale farmers to meet their own food needs, using sustainable agriculture

Recognition by governments, industry and inter-governmental forums of the right of small scale farmers to control food production to meet local food needs, - 'food sovereignty' - has been a key demand from farming and peasant communities (La Via Campesina 2008; Nyéléni - Forum for Food Sovereignty 2007). Around 75% of the world's hungry people live in rural areas in poor countries (U.N. FAO 2006). If rural communities can meet more of their own food needs via local production, they will clearly be less vulnerable to global price and supply fluctuations. La Via Campesina has argued that: "Small-scale family farming is a protection against hunger!" (La Via Campesina 2008). This view was supported by the four year International Assessment of Agricultural Science and Technology for Development which emphasised that to redress rural poverty and hunger, a key focus of agricultural policy must be empowering small scale farmers to meet their own food needs (IAASTD 2008).

Defending and reinvigorating sustainable small-scale farming requires action by governments to support agriculture that prioritises food production for local populations. This requires land reform, including control over and access to water, seed, credits and appropriate technology. It also requires the removal of trade policies and financial subsidies that preference industrial-scale farming for export or that promote the adoption of technologies or farming practices that will undermine the viability of small-scale farming.

Greater government support for agro-ecological and organic agriculture could help reinvigorate small scale farming and rural farming communities, maintain or augment yields, and also meet environmental objectives. Organic agriculture could help reduce small farmers' capital costs and reliance on agri-business companies. Because organic agriculture is labour-intensive, it could also result in job creation on farms. For example whereas the number of farm workers in conventional agriculture is in decline, organic farms have created an additional 150,000 jobs in Germany (Bizzari 2007). Agro-ecological initiatives in Brazil have delivered yield increases of up to 50%, improved incomes for farmers, restored local agricultural biodiversity and reinvigorated local rural economies (Hisano & Altoé 2002).

A key argument against supporting agro-ecological or organic agriculture has been that it will depress agricultural productivity. However on a global scale organic agriculture supports similar or increased yields compared to chemical-intensive industrial agriculture. In a study comparing yields between organic and conventional agriculture in 293 cases world wide, organic yields were comparable to conventional agriculture in the Global North and greater than those of conventional agriculture in the Global South (Badgley *et al.* 2007). A 22 year trial in the United States found that organic farms produced comparable yields, but required 30% less fossil fuel energy and water inputs than conventional farms, resulted in higher soil organic matter and nitrogen levels, higher biodiversity, greater drought resilience and reduced soil erosion (Pimental *et al.* 2005).

4.3 Nanotechnology will increase pressures faced by small operators and undermine efforts to relocalise food production

The potential role of new technologies in responding to the food crisis is controversial. As with genetically engineered (GE) crops, proponents have argued that nanotechnology will redress food shortages by promoting greater agricultural productivity (IFRI 2008). However the IAASTD (2008) report notes that whereas GE crops have had highly variable yields, they have also had negative broader economic consequences for farmers by concentrating ownership in agricultural resources and introducing new liabilities for farmers (IAASTD 2008). Similarly, Friends of the Earth Australia suggests that nano-agriculture is not required to achieve strong yields, but will add to the capital costs faced by small farmers and increase their reliance on technology, seed and chemicals sold by a small number of global agri-business companies.

By underpinning the next wave of technological transformation of the global agriculture and food industry, nanotechnology appears likely to further expand the market share of major agrochemical and seed companies, food processors and food retailers to the detriment of small operators (Scrinis & Lyons 2007).

Nano-encapsulated pesticides, fertilisers and plant growth treatments designed to release their active ingredients in response to environmental triggers, used in conjunction with nano-enabled remote farm surveillance systems, could enable even larger areas of cropland to be farmed by even fewer people (ETC Group 2004). By dramatically increasing efficiency and uniformity of farming, it appears likely that nano-farming technologies could accelerate expansion of industrial-scale, export oriented agricultural production which employs even fewer workers but relies on increasingly sophisticated technological support systems that have increasing capital costs (Scrinis & Lyons 2007; ETC Group 2004). Such systems could commodify the knowledge and skills associated with food production gained over thousands of years and embed it into proprietary nanotechnologies (Scrinis & Lyons 2007). It could also result in the further loss of small scale farmers and further disconnection of rural communities from food production, undermining efforts to achieve sustainable, relocalised food production.

Nanotechnology also appears likely to advantage global food processors and retailers over small operators. Nano track and trace technologies will enable global processors, retailers and suppliers to operate even more efficiently over larger geographic areas and through more complex supply chains, giving them a strong competitive advantage over smaller operators. Nano food packaging and edible coatings will extend food shelf life, enabling it to be transported over even further distances while reducing the incidence of food spoilage. This will significantly reduce the costs of global suppliers and retailers, giving them another competitive advantage over retailers selling fresh, local produce.

4.4 Nanotechnology could further erode our connection to local, seasonal food and promote food transport over ever longer distances

Nanotechnology is also likely to influence the eating habits of urban consumers, with associated public health and cultural implications. By enabling manufacturers to promote nano-reconstituted, nano-fortified or nano-packaged foods as delivering superior health benefits, hygiene or convenience, it is likely that nanotechnology will encourage even greater consumption of highly processed foods at the expense of minimally processed fruits and vegetables. By extending the shelf life of 'fresh' and processed foods, it is also likely that nanotechnology will further promote the eating of foods out of season and far from the place of their production. In this way, nanotechnology may further erode the relationship that exists (or once existed) between consumers and producers of foods, as well as peoples' cultural connection to traditional and minimally processed whole foods. The development of a cola drink that could be marketed as having the nutritional properties of milk is a case in point (Moss 2007). With the increasing use of nanotechnology to alter the nutritional properties of processed foods, we could soon be left with no capacity to understand the health values of foods, other than their marketing claims.

5. Nano-specific Regulation is Required to Ensure Both Food Safety and the Social Value of Nanofood Applications

Regulatory systems in the United States, Europe, Australia, Japan and other countries treat all particles the same; that is, they do not recognise that nanoparticles of familiar substances may have novel properties and novel risks (Bowman & Hodge 2007). Although we know that many nanomaterials now in commercial use pose greater toxicity risks than the same materials in larger particle form, if a food ingredient has been approved in bulk form, it remains legal to sell it in nano form. There is no requirement for new safety testing, food labelling to inform consumers, new occupational exposure standards or mitigation measures to protect workers or to ensure environmental safety. Incredibly in most countries, there is not yet even a requirement that the manufacturer notify the relevant regulator that they are using nanomaterials in the manufacture of their products. In an overdue yet positive first step, Food Standards Australia New Zealand has announced it will require companies to alert it to their use of nano ingredients (FSANZ 2008).

The need to amend regulatory systems to ensure appropriate management of new nanotoxicity risks has been recognised at the highest scientific levels. In 2004, the United Kingdom's Royal Society warned that nanomaterials could fall through regulatory 'gaps'. It called for nanomaterials to be treated as new chemicals, and subject to new safety testing and mandatory labelling before being allowed in products (U.K. RS/RAE 2004). In its 2006 report, the European Union's Scientific Committee on Emerging and Newly Identified Health Risks recognised the many systemic failures of existing regulatory systems to manage the risks associated with nanotoxicity (E.U. SCENIHR 2006). In 2008 the European Food Safety Authority also elaborated the shortcomings in existing risk assessment methodologies and the significant knowledge gaps regarding the biological behaviour of nanomaterials and their potential to cause harm (EFSA 2008). But even after four years of high level recognition of the new health and environmental risks associated with nanomaterials, their use in food and agriculture remains effectively unregulated. This is despite nanofood scientists calling for new regulations to ensure that all nanofood, nano food packaging and nano food contact materials are subject to nanotechnology-specific safety testing prior to being included in commercial food products (Lagaron *et al.* 2005; Sorrentino *et al.* 2007; IFST 2006).

Beyond the need for new regulation to manage the serious new toxicity risks associated with nanofood and nano agricultural products, Friends of the Earth Australia is calling for 'fourth hurdle regulation' to require manufacturers to demonstrate the social benefit of products they wish to sell. There is very rarely a requirement for product manufacturers to 'justify' risk exposures in terms of social benefits (Wynne & Felt 2007). Too often, it is an entrenched and unchallenged assumption that the market release of a new functional food or antibacterial product will necessarily deliver public health benefits. In many instances, putative benefits are argued by product proponents to justify or counterbalance the potential for new risks, despite potential benefits rarely being subject to the same kind of scrutiny and scepticism to which claims of potential risks are subject. Friends of the Earth Australia therefore supports the recommendations of Wynne and Felt (2007) for the inclusion of a social benefit test, supplementing the more usual investigations into efficacy, safety and environmental risk, as part of the regulation of nanotechnology in food and agriculture.

6. Civil Society Groups have Called for a Moratorium on Nanotechnology's Use in Food and Agriculture

Growing numbers of civil society groups have called for a moratorium on the commercial release of food, food packaging, food contact materials and agrochemicals that contain manufactured nanomaterials until nanotechnology-specific regulation is introduced to protect the public, workers and the environment from their risks. Some of these groups are also insisting that the public be involved in decision making. Groups calling for a moratorium include: Friends of the Earth Australia, Europe and the United States; Corporate Watch (UK); The ETC Group;

GeneEthics (Australia); Greenpeace International; International Centre for Technology Assessment (US); International Federation of Journalists; Practical Action; and The Soil Association UK. The International Union of Food, Agricultural, Hotel, Restaurant, Catering, Tobacco and Allied Workers' Associations, representing 12 million workers from 120 countries, has also called for a moratorium.

The Nyéléni Forum for Food Sovereignty was a civil society meeting of peasants, family farmers, fisher people, nomads, indigenous and forest peoples, rural and migrant workers, consumers and environmentalists from across the world. Delegates were concerned that the expansion of nanotechnology into agriculture will present new threats to the health and environment of peasant and fishing communities and further erode food sovereignty. The forum also resolved to work towards an immediate moratorium on nanotechnology (Nyéléni 2007 – Forum for Food Sovereignty 2007).

The organic sector is also beginning to move to exclude nanomaterials from organic food and agriculture. The United Kingdom's largest organic certification body announced in late 2007 that it will ban nanomaterials from all products which it certifies. All organic foods, health products, sunscreens and cosmetics that the Soil Association certifies will now be guaranteed to be free from manufactured nanomaterial additives (British Soil Association 2008). The Biological Farmers of Australia, Australia's largest organic representative body, have also moved to ban nanomaterials from products it certifies.

7. Conclusion

Nanotechnology poses broad threats to the development of more sustainable food and farming systems. At a time when organic farming and sales of organic food are experiencing sustained growth, nanotechnology appears likely to entrench our reliance on chemical and energy-intensive agricultural technologies and on highly processed foods that pose new health risks. Against the backdrop of dangerous climate change, there is growing public interest in reducing the distances that food travels between producers and consumers, yet long-life nano packaging and surveillance systems appear likely to promote the transport of foods over even greater distances. The potential for nanotechnology to further concentrate corporate control of global agriculture and food systems, to add to pressures faced by small operators and to further erode local farmers' control of food production is a source of concern. The serious nature of the global food crisis demands that we make every effort to support small scale farmers to meet their own food needs. Rather than investing significant sums of money in nanotechnology research, development, regulation, risk assessment and monitoring, we suggest that governments would be better off supporting ecologically sustainable small scale farming and promoting greater consumption of fresh foods based on a varied diet.

Questions for Reflection:

1. Do you think that nanotechnology-based agricultural systems would promote or discourage small-scale, localised agriculture? What do you think is desirable?
2. Can you envisage the coexistence of nanotechnology and organic agricultural production? If yes, how?
3. What is the relationship between food and culture? How might the use of nanotechnology impact this relationship?

Bibliography

Ashwood, P., Thompson, R. and Powell, J. (2007). Fine particles that absorb lipopolysaccharide via bridging calcium cations may mimic bacterial pathogenicity towards cells, *Exp. Biol. Med.*, 232(1), pp. 107-117.

Badgley, C., Moghtader, J., Quintero, E., Zakem, E., Chappell, M., Aviles-Vazquez, K., Salon, A. and Perfecto, I. (2007). Organic agriculture and the global food supply, *Renew Ag Food Systems,* 22 (2), pp. 86-108.

Ballestri, M., Baraldi, A., Gatti, A., Furci, L., Bagni, A., Loria, P., Rapana, R., Carulli, N. and Albertazzi A. (2001). Liver and kidney foreign bodies granulomatosis in a patient with malocclusion, bruxism, and worn dental prostheses, *Gastroenterol.,* 121(5), pp. 1234-8.

Beane, A., Freeman, L., Bonner, M., Blair, A., Hoppin, J., Sandler, D., Lubin, J., Dosemeci, M., Lynch, C., Knott, C. and Alavanja M. (2005). Cancer Incidence among Male Pesticide Applicators in the Agricultural Health Study Cohort Exposed to Diazinon, *Am. J. Epidemiol.,* 162(11), pp. 1070-1079.

Bizzari, K. (2007). The EU's biotechnology strategy: mid-term review or mid-life crisis? In Holder, H. and Oxborrow, C. (eds), *A scoping study on how European agricultural biotechnology will fail the Lisbon objectives and on the socio-economic benefits of ecologically compatible farming*, Brussels: Friends of the Earth Europe. Available at http://www.foeeurope.org/publications/2007/FoEE_biotech_MTR_midlifecrisis_March07.pdf (last accessed 26.05.09)

Bowman, D. and Hodge, G. (2006). Nanotechnology: Mapping the wild regulatory frontier, *Futures,* 38, pp. 1060-1073.

Bowman, D. and Hodge, G. (2007). A small matter of regulation: An international review of nanotechnology regulation, *Columbia Sci. Techno.l Law Rev., Volume 8,* pp. 1-32.

Boxhall, A., Tiede, K. and Chaudhry, Q. (2007). Engineered nanomaterials in soils and water: how do they behave and could they pose a risk to human health? *Nanomed.,* 2(6), pp. 919-927.

Braydich-Stolle, L., Hussain, S., Schlager, J. and Hofmann, M. (2005). In Vitro Cytotoxicity of Nanoparticles in Mammalian Germline Stem Cells, *Toxicol. Sci.,* 88(2), pp. 412-419.

British Soil Association (2008). *Press Release: Soil Association First Organisation in the World to Ban Nanoparticles - Potentially Toxic Beauty Products that get Right Under your skin.* Available at http://www.soilassociation.org/web/sa/saweb.nsf/848d689047cb466780256a6b00298980/42308d944a3088a6802573d100351790!OpenDocument (last accessed Nov 24, 2008).

Brunner, T., Piusmanser, P., Spohn, P., Grass, R., Limbach, L., Bruinink, A. and Stark, W. (2006). *In Vitro* cytotoxicity of oxide nanoparticles: Comparison to asbestos, silica, and the effect of particle solubility, *Environ. Sci. Technol.,* 40, pp. 4374-4381.

Cedervall, T., Lynch, I., Lindman, S., Berggård, T., Thulin, E., Nilsson, H., Dawson, K. and Linse, S. (2007). Understanding the nanoparticle-protein corona using methods to quantify exchange rates and affinities of proteins for nanoparticles, *Proceedings of the National Academy of Sciences,* 104(7), pp. 2050-2055.

Chen, Z., Meng, H., Xing, G., Chen, C., Zhao, Y., Jia, G., Wang, T., Yuan, H., Ye, C., Zhao, F., Chai, Z., Zhu, C., Fang, X., Ma, B. and Wan, L. (2006). Acute toxicological effects of copper nanoparticles *in vivo*, *Toxicol Lett,* 163, pp. 109-120.

Chen, M. and von Mikecz, A. (2005). Formation of nucleoplasmic protein aggregates impairs nuclear function in response to SiO2 nanoparticles, *Experiment Cell Res.*, 305, pp. 51-62.

Cientifica (2006). Homepage. Available at http://www.cientifica.eu/index.phppage=shop.browse&category_id=2&option=com_virtuemart&Itemid=80 (last accessed Dec 15, 2007).

Daniells, S. (2007). Thing big, think nano, *Food Navigator.com*, Europe, 19 December. Available at http://www.foodnavigator.com/news/ng.asp?n=82109 (accessed Dec 21, 2007).

Di Pasqua, A., Sharma, K., Shi, Y.-L., Toms, B., Ouellette, W., Dabrowiak, J. and Asefa, T. (2008). Cytotoxicity of mesoporous silica nanomaterials, *J. Inorgan. Biochem., 102,* pp. 1416–1423.

Donaldson, K., Beswick, P. and Gilmour, P. (1996). Free radical activity associated with the surface of particles: A unifying factor in determining biological activity? *Toxicol. Lett.,* 88, pp. 293-298.

Downs C. (2003). Excessive Vitamin A Consumption and Fractures: How Much is Too Much? Nutrition Bytes: 9(1). Available at http://repositories.cdlib.org/uclabiolchem/nutritionbytes/vol9/iss1/art1 (last accessed Jan 16, 2008).

Dunford, R., Salinaro, A., Cai, L., Serponc, N., Horikoshi, S., Hidaka, H. and Knowland, J. (1997). Chemical oxidation and DNA damage catalysed by inorganic sunscreen ingredients, *FEBS Lett.,* 418 pp. 87-90.

Ervin, D. and Welsh, R. (2003). Environmental effects of genetically modified crops: Differentiated risk assessment and management, Chapter 2a. In Wesseler, J. (ed) *Environmental Costs and Benefits of Transgenic Crops in Europe: Implications for Research, Production, and Consumption*, Dordrecht: Kluwer Academic Publishers.

ETC Group (2004). *Down on the Farm.* Available at http://www.etcgroup.org (last accessed Jan 17, 2008).

ETC Group (2005). *Oligopoly, Inc. 2005. Concentration in Corporate Power.* Available at http://www.etcgroup.org (last accessed Jan 17, 2008).

ETC Group (2007). *Extreme genetic engineering: An introduction to synthetic biology.* Available at http://www.etcgroup.org/upload/publication/602/01/synbioreportweb.pdf (last accessed Jan 17, 2008).

EFSA (2008). *Draft Opinion of the Scientific Committee on the Potential Risks Arising from Nanoscience and Nanotechnologies on Food and Feed Safety.* Available at http://www.efsa.europa.eu/cs/BlobServer/DocumentSet/sc_opinion_nano_public_consultation.pdf?ssbinary=true (accessed: Nov 24, 2008).

FAO (2006). *The State of Food Insecurity in the World 2006*, FAO, Italy. Available at http://www.ftp.fao.org/docrep/fao/009/a0750e/a0750e00.pdf (last accessed Nov 24, 2008).

FAO (2007). International conference on organic agriculture and food security 3-5 May 2007, FAO Italy. Available at http://www.ftp://ftp.fao.org/paia/organicag/ofs/OFS-2007-5.pdf (last accessed Dec 24, 2007).

Federici, G., Shaw, B. and Handy, R. (2007). Toxicity of titanium dioxide nanoparticles to rainbow trout (*Oncorhynchus mykiss*): Gill injury, oxidative stress, and other physiological effects, *Aquatic Toxicol.,* 84(4), pp. 415-430.

Foladori, G. and Invernizzi, N. (2007). Agriculture and food workers question nanotechnologies. The IUF resolution. Available at http://www.estudiosdeldesarrollo.net/relans/documentos/UITA-English-1.pdf (last accessed Jan 17, 2008).

FSANZ (2008). *Proposed Amendments to Part 3 of the FSANZ Application Handbook* October 2008. Available at http://www.foodstandards.gov.au/_srcfiles/ Proposed%20Amendments%20to%20Application%20Handbookoct%202008 1.pdf (last accessed Nov 24, 2008).

Garnett, M. and Kallinteri, P. (2006). Nanomedicines and nanotoxicology: Some physiological principles, *Occup. Med.* 56, pp. 307-311.

Gatti A. (2004). Biocompatibility of micro- and nano-particles in the colon. Part II, *Biomaterials* 25, pp. 385-392.

Gatti, A. and Rivasi, F. (2002). Biocompatibility of micro- and nanoparticles. Part I: In liver and kidney, *Biomaterials,* 23, pp. 2381–2387

Gatti, A., Tossini, D. and Gambarelli, A. (2004). Investigation of trace elements in bread through environmental scanning electron microscope and energy dispersive system, *2nd International IUPAC Symposium,* Brussels, October 2004.

Geiser, M., Rothen-Rutlshauser, B., Knapp, N., Schurch, S., Kreyling, W., Schulz, H., Semmler, M., Im, H., Heyder, J. and Gehr, P. (2005). Ultrafine particles cross cellular membranes by non-phagocytic mechanisms in lungs and in cultured cells, *Environ Health Perspectives,* 113(11), pp. 1555-1560.

Handy, R., Owen, R. and Valsami-Jones, E. (2008). The ecotoxicology of nanoparticles and nanomaterials: Current status, knowledge gaps, challenges, and future needs, *Ecotoxicol.,* 17, pp. 315–325.

Helmut Kaiser Consultancy Group (2007a). *Nanopackaging is Intelligent, Smart and Safe Life,* New World Study By Hkc22.com Beijing Office, Press Release 14.05.07. Available at http://www.prlog.org/10016688nanopackaging-isintelligent-smart-and-safe-life-newworld-study-by-hkc22-com-beijing-office.pdf (last accessed Jan 17, 2008).

Helmut Kaiser Consultancy Group (2007b). *Strong Increase in Nanofood and Molecular Food Markets in 2007 Worldwide.* Available at http://www.hkc22.com/Nanofoodconference.html (last accessed Jan 17, 2008).

Hisano, S. and Altoé, S. (2002). Brazilian farmers at a crossroad: Biotech industrialization of agriculture or new alternatives for family farmers? Paper presented at CEISAL July 3 to 6, 2002, Amsterdam. Available at http://www.agroeco.org/brasil/material/hisano.pdf (last accessed Jan 17, 2008).

Hund-Rinke, K., and Simon, M. (2006). Ecotoxic effect of photocatalytic active nanoparticles (TiO2) on algae and daphnids, *Environ. Sci. Poll. Res.*, 13(4), pp. 225-232.

Hussain, S., Hess, K., Gearhart, J., Geiss, K. and Schlager, J. (2005). *In vitro* toxicity of nanoparticles in BRL 3A rat liver cells, *Toxico.l In Vitro,* 19, pp. 975-983.

Hussain, S,. Javorina, A., Schrand, A., Duhart, H., Ali, S. and Schlager, J. (2006). The interaction of manganese nanoparticles with PC-12 cells induces dopamine depletion, *Toxicol. Sci.*, 92(2), pp. 456-463.

IAASTD (2008). *International Assessment of Agricultural Knowledge, Science and Technology for Development*. Available at http://www.agassessment.org (last accessed April 4, 2009).

IFAD (2008). *Soaring Food Prices and the Rural Poor: Feedback from the field.* Available at http://www.ifad.org/operations/food/food.htm (last accessed Dec 2, 2008).

IFRI (2008). *Nanotechnology, Food, Agriculture and Development*. IFPRI Policy Seminar, June 18, 2008. Available at http://www.ifpriblog.org/2008/06/24/nanotechseminar.aspx (last accessed Dec 2, 2008).

IFST (2006). *Information Statement: Nanotechnology*, Institute of Food Science & Technology Trust Fund, London. Available at http://www.ifst.org/uploadedfiles/cms/store/ATTACHMENTS/Nanotechnology.pdf (last accessed Jam 15, 2008).

Innovest (2006). *Nanotechnology: Non-traditional Methods for Valuation of Nanotechnology Producers*, Innovest, Strategic Value Advisers.

IRIN (2008). *Timor Leste: Rice Price Hike Leads to Government Subsidies*, Humanitarian news and analysis: UN Office for the Coordination of Humanitarian Affairs 20 May 2008. Available at http://www.irinnews.org/Report.aspx?ReportId=78310 (last accessed Dec 2, 2008)

ISO/TS 27687 (2008). Nanotechnologies — Terminology and definitions for nano-objects — Nanoparticle, nanofibre and nanoplate available at http://www.iso.org/iso/catalogue_detail?csnumber=44278.

Joseph, T. and Morrison, M. (2006). Nanotechnology in Agriculture and Food, *Nanoforum Report*. Available at http://www.nanoforum.org/nf06~modul~showmore~folder~99999~scid~377~.html?action=longview_publication& (last accessed Dec 2, 2008).

Khanna, V., Bakshi, B. and Lee, L. (2008). Carbon nanofiber production: Life cycle energy consumption and environmental impact, *J Indust Ecol*, 12(3), pp. 394-410.

La Via Campesina (2008). *An Answer to the Global Food Crisis: Peasants and Small Farmers can Feed the World!* 1 May 2008. Available at http://www.viacampesina.org/main_en/index.php?option=com_content&task=view&id=525&Itemid=1 (last accessed Dec 2, 2008).

Lagarón, J., Cabedo, L., Cava, D., Feijoo, J., Gavara, R. and Gimenez, E. (2005). Improved packaging food quality and safety. Part 2: Nano-composites, *Food Additives and Contaminants,* 22(10), pp. 994-998.

Li, N., Sioutas, C., Cho, A., Schmitz, D., Misra, C., Sempf, J., Wang, M., Oberley, T., Froines, J. and Nel, A. (2003). Ultrafine particulate pollutants induce oxidative stress andmitochondrial damage, *Environ Health Perspect.*, 111(4), pp. 455-460.

Linse, S., Cabaleiro-Lago, C., Xue, W. F., Lynch, I., ,Lindman. S., Thulin, E., Radford, S. and Dawson, K. (2007). Nucleation of protein fibrillation by nanoparticles, *Proceedings of the National Academy of Sciences*, 104(21), pp. 8691–8696.

Lomer, M., Harvey, R., Evans, S., Thompson, R. and Powell, P. (2001). Efficacy and tolerability of a low microparticle diet in a double blind, randomized, pilot study in Crohn's disease, *Eur. J. Gastroenterol. Hepatol.*, 13, pp. 101-106.

Long, T., Saleh, N., Tilton, R., Lowry, G. and Veronesi, B. (2006). Titanium dioxide (P25) produces reactive oxygen species in immortalized brain microglia (BV2): Implications for nanoparticle neurotoxicity, *Environ. Sci. Technol.*, 40(14), pp. 4346-4352.

Lovern, B. and Klaper, R. (2006). *Daphnia magna* mortality when exposed to titanium dioxide and fullerene (c60) nanoparticles, *Environ. Toxicol. Chem.*, 25(4), pp. 1132-1137.

Lovern, S., Strickler, J. and Klaper, R. (2007). Behavioral and physiological changes in *Daphnia magna* when exposed to nanoparticle suspensions (titanium dioxide, Nano-C60, and C60HxC70Hx), *Environ. Sci. Technol. 41,* pp. 4465-4470.

Lucarelli, M., Gatti, A., Savarino, G., Quattroni, P., Martinelli, L., Monari, E. and Boraschi, D. (2004). Innate defence functions of macrophages can be biased by nano-sized ceramic and metallic particles, *Eur. Cytok. Net,* 15(4), pp. 339-346.

Magrez, A,. Kasa, S., Salicio, V., Pasquier, N., Won Seo., J. Celio, M., Catsicas, S., Schwaller, B. and Forro, L. (2006). Cellular toxicity of carbon-based nanomaterials, *Nano Lett.*, 6(6), pp. 1121-1125.

Maynard, A. and Kuempel. E. (2005). Airborne nanostructured particles and occupational health, *J. Nanopart. Res.*, 7, pp. 587–614.

Maynard A. (2006). *Nanotechnology: A Research Strategy for Addressing Risk.* Woodrow Wilson International Center for Scholars Project on Emerging Nanotechnologies. Available at http://www.nanotechproject.org /file_download/files/PEN3_Risk.pdf (last accessed Nov 24, 2008).

Medina, C., Santos-Martinez, M., Radomski, A., Corrigan, O. and Radomski, M. (2007). Nanoparticles: Pharmacological and toxicological significance, *Brit. J. Pharmacol.*, 150, pp. 552–558.

Melhus, A. (2007). *Silver Threatens the Use of Antibiotics*. Unpublished manuscript, received by email 30 January 2007.

Miller, G. and Senjen, R. (2008a). Discussion paper on nanotechnology standardisation and nomenclature issues. Available at http://www.nano.foe.org.au (last accessed Nov 24, 2008).

Miller, G. and Senjen, R. (2008b). Out of the laboratory and on to our plates. Available at http://www.nano.foe.org.au (last accessed Nov 24, 2008).

Mitchell, D. (2008). A note on rising food prices. Leaked draft report available at http://image.guardian.co.uk/sysfiles/Environment/documents/2008/07/10/Biofuels. PDF (last accessed: Nov 24, 2008).

Moss, P. (2007). Future foods: Friend or foe? *BBC News,* 6 February 2007. Available at http://news.bbc.co.uk/2/hi/technology/6334613.stm (last accessed Dec 2, 2008).

Mozafari, M., Flanagan, J., Matia-Merino, L., Awati, A., Omri, A., Suntres, Z. and Singh, H. (2006). Recent trends in the lipid-based nanoencapsulation of antioxidants and their role in foods, *J. Sci. Food Ag.,* 86, pp. 2038-2045.

Nel, A., Xia, T. and Li, N. (2006). Toxic potential of materials at the nanolevel, *Science,* 311, pp. 622-627.

Nyéléni - Forum for Food Sovereignty (2007). *Peoples' Food Sovereignty Statement.* Available at http://www.nyeleni2007.org/IMG/pdf/Peoples_Food_Sovereignty_Statement.pdf (last accessed Sept 29, 2007).

Oberdörster, G., Oberdörster, E. and Oberdörster, J. (2005). Nanotoxicology: an emerging discipline from studies of ultrafine particles, *Environ. Health Perspect.,* 113(7), pp. 823-839.

Opara, L. (2004). Emerging technological innovation triad for agriculture in the 21st century. Part 1. Prospects and impacts of nanotechnology in agriculture, *Ag. Engineering Internat.: CIGR J. Ag. Engineering Internat., Vol 6.*

Parry, J. (2008). New technology breakthrough. *Farmers Guardian*, 09.05.08. Available at http://www.farmersguardian.com/story.asp?sectioncode=21&storycode=18376 (last accessed Nov 13, 2008).

Petrelli, G., Figà-Talamanca, I., Tropeano, R., Tangucci, M., Cini, C., Aquilani, S., Gasperini, L. and Meli, P. (2000). Reproductive male-mediated risk: Spontaneous abortion among wives of pesticide applicators, *Eur. J. Epidemiol.,* 16, pp. 391-393.

Pimental, D., Hepperly, P., Hanson, J., Douds, D. and Seidel, R. (2005). Environmental, energetic and economic comparisons of organic and conventional farming systems, *Bioscience,* 55(7), pp. 573-582.

Project on Emerging Nanotechnologies (2007). *Nanotechnology and Lifecycle assessment.* Available at http://www.nanotechproject:org/file_download/files/NanoLCA_3.07.pdf (last accessed 24.11.08).

Renton, A. (2006). Welcome to the world of nanofoods. *Guardian Unlimited UK,* 13 December 2006. Available at http://observer guardian co uk/foodmonthly/futureoffood/story/0,,1971266,00.html (last accessed Jan 17, 2008).

Reynolds, G. (2007). FDA recommends nanotechnology research, but not labelling. *FoodProductionDaily.com,* News, 26 July 2007. Available at http://www.foodproductiondaily-usa.com/news/ng.asp?n=78574-woodrow-wilsonnanotechnologyhazardous (last accessed Jan 24, 2008).

Roco, M. C. (2001). From vision to implementation of the US National Nanotechnology Initiative, *J. Nanoparticle Research,* 3, pp. 5-11.

Roco, M. C. and Bainbridge, W. (eds) (2002). *Converging Technologies for Improving Human Performance: Nanotechnology, Biotechnology, Information Technology and Cognitive Science.* NSF/DOC-sponsored report. Available at http://www.wtec.org/ConvergingTechnologies/ (last accessed Jan 24, 2008).

Salamanca-Buentello, F., Persad, D., Court, E., Martin, D., Daar, A. and Singer, P. (2005). Nanotechnology and the developing world, *PLoS Med.*, 2(5), e97.

Sanguansri, P., and Augustin, M. (2006). Nanoscale materials development – A food industry perspective. *Trends Food Sci. Technol.* 17, pp. 547-556.

Sayes, C. Wahi, R. Kurian, P. Liu, Y. West, J. Ausman, K. Warheit, D. and Colvin, V. (2006). Correlating nanoscale titania structure with toxicity: A cytotoxicity and inflammatory response study with human dermal fibroblasts and human lung epithelial cells, *Toxicol. Sci.*, 92(1), pp. 174-185.

SCENIHR (2006). *The Appropriateness of Existing Methodologies to Assess the Potential Risks Associated with Engineered and Adventitious Products of Nanotechnologies,* Brussels: European Commission.

Schneider, J. (2007). Can microparticles contribute to inflammatory bowel disease: Innocuous or inflammatory? *Exp. Biol. Med.* 232, pp. 1-2.

Scott-Fordsmand, J., Krogh, P., Schaefer, M. and Johansen, A. (2008). The toxicity testing of double-walled nanotubes-contaminated food to *Eisenia veneta* earthworms, *Ecotoxicol. Environ. Safety*. 71(3), pp. 616– 619.

Scrinis, G. and Lyons, K. (2007). The emerging nano-corporate paradigm: Nanotechnology and the transformation of nature, food and agri-food systems, *Internat. J. Sociol. Agric. and Food,* 15(2). Available at http://www.csafe.org.nz/ (last accessed Mar 2, 2008).

Şengül, H,, Theis, T. and Ghosh, S. (2008). Towards sustainable nanoproducts: An overview of nanomanufacturing methods, *J. Indust. Ecol.,* 12(3), pp. 329-359.

Shelke, K. (2006). Tiny, invisible ingredients. *Food Processing.com.* Available at http://www.foodprocessing.com/articles/2006/227.html (last accessed Aug 8, 2007).

Sorrentino, A., Gorrasi, G. and Vittoria, V. (2007). Potential perspectives of bio-nanocomposites for food packaging applications, *Trends Food Sci. Technol.,* 18, pp. 84-95.

Templeton, R., Ferguson, P., Washburn, K., Scrivens, W. and Chandler, G. (2006). Life-cycle effects of single-walled carbon nanotubes (SWNTs) on an estuarine meiobenthic copepod, *Environmental Science and Technology,* 40, pp. 7387-7393.

Throback, I., Johansson, M., Rosenquist, M., Pell, M., Hansson, M. and Hallin, S. (2007). Silver (Ag(+)) reduces denitrification and induces enrichment of novel nirK genotypes in soil, *FEMS Microbiol. Lett.,* 270(2), pp. 189-194.

Tran, C., Donaldson, K., Stones, V., Fernandez, T., Ford, A,. Christofi, N., Ayres, J., Steiner, M., Hurley, J., Aitken, R. and Seaton, A. (2005). A scoping study to identify hazard data needs for addressing the risks presented by nanoparticles and nanotubes, *Research Report, Instit. Occup. Med., Edinburgh.*

Tucker, J. and Zilinskas, R. (2006). The promise and the peril of synthetic biology, *New Atlantis,* 12, pp. 25-45.

U.K. HSE (2004). Nanoparticles: An occupational hygiene review. Available at http://www.hse.gov.uk/research/rrpdf/rr274.pdf (last accessed Jan 17, 2008).

U.K. DEFRA (2006). *UK Voluntary Reporting Scheme for Engineered Nanoscale Materials. September 2006,* London: Chemicals and Nanotechnologies Division Defra. Available at http://www.defra.gov.uk/environment/nanotech/policy/pdf/vrs-nanoscale.pdf (last accessed Jan 15, 2008).

U.K. RCEP (2008). *Novel Materials in the Environment : The Case of Nanotechnology,* Royal Commission on Environmental Pollution. Available at http://www.rcep.org.uk/novel%20materials/Novel%20Materials%20report.pdf (last accessed Nov 14, 2008).

U.K. RS/RAE (2004). *Nanoscience and Nanotechnologies: Opportunities and Uncertainties.* Available at http://www.nanotec.org.uk/finalReport.htm (last accessed Jan 17, 2008).

U.S. FDA (2006). *Nanoscale Materials [no specified CAS] Nomination and Review of Toxicological Literature,* December 8, 2006, Prepared by the Chemical Selection Working Group, U.S. Food & Drug Administration. Available at http://ntp.niehs.nih.gov/ntp/htdocs/Chem_Background/ExSumPdf/Nanoscale_materials.pdf (last accessed Jan 15, 2008).

U.S. FDA (2007). *Nanotechnology: A Report of the U.S. Food and Drug Administration Nanotechnology Task Force,* July 25, 2007. Available at http://www.fda.gov/nanotechnology/taskforce/report2007.html (last accessed Jan 15, 2008).

U.S. IOM (1998). *Dietary Reference Intakes for Thiamin, Riboflavin, Niacin, Vitamin B6, Folate, Vitamin B12, Pantothenic Acid, Biotin, and Choline. A Report of the Standing Committee on the Scientific Evaluation of Dietary Reference Intakes and its Panel on Folate, Other B Vitamins, and Choline and Subcommittee on Upper Reference Levels of Nutrients.* Food and Nutrition Board, Institute of Medicine, Washington D.C.: National Academies Press.

Van Balen, E., Font, R., Cavallé, N., Font, L., Garcia-Villanueva, M., Benavente, Y., Brennan, P. and de Sanjose, S. (2006). Exposure to non-arsenic pesticides is associated with lymphoma among farmers in Spain, *Occupation. Environ. Med.*, 63, pp. 663-668.

Vileno, B., Lekka, M., Sienkiewicz, A., Jeney, S., Stoessel, G., Lekki, J., Forró, L. and Stachura, Z. (2007). Stiffness alterations of single cells induced by UV in the presence of nanoTiO2, *Environ. Sci. Technol.,* 41(14), pp. 5149-5153.

Wang, B., Feng, W. Y., Wang, T. C., Jia, G., Wang, M., Shi, J. W., Zhang, F., Zhao, Y. L. and Chai, Z. F. (2006). Acute toxicity of nano- and micro-scale zinc powder in healthy adult mice, *Toxicol. Lett.,* 161, pp. 115-123.

Wang, B., Feng, W., Wang, M., Wang, T., Gu, Y., Zhu, M., Ouyang, H., Shi, J., Zhang, F., Zhao, Y., Chai, Z., Wang, H. and Wang, J. (2007a). Acute toxicological impact of nano- and submicro-scaled zinc oxide powder on healthy adult mice, *J. Nanopart. Res.,* 10(2), pp.263-276.

Wang, J., Zhou, G., Chen, C., Yu, H., Wang, T., Ma, Y., Jia, G., Gai, Y., Li, B., Sun, J., Li, Y., Jiao, F., Zhano, Y. and Chai, Z. (2007b). Acute toxicity and biodistribution of different sized titanium dioxide particles in mice after oral administration, *Toxicol. Lett.,* 168(2), pp.176-185.

Wolfe, J. (2005). Safer and guilt-free nanofoods, *Forbes.com US,* 10 August 2005. Available at http://www.forbes.com/2005/08/09/nanotechnology-kraft-hershey-cz_jw_0810soapbox_inl.html (last accessed Aug 8, 2007).

Wynne, B. and Felt, U. (2007). Taking European Knowledge Society Seriously. Available at http://ec.europa.eu/research/sciencesociety/document_library/pdf_ 06/european-knowledge-society_en.pdf (last accessed Nov 24, 2004).

Yang, L. and Watts, D. J. (2005). Particle surface characteristics may play an important role in phytotoxicity of alumina nanoparticles, *Toxicol. Lett.,* Volume 158(2), pp. 122-32.

Zhang, Y., Zhang, Y., Chen, J., Zhang, H., Zhang, Y., Kong, L., Pan, Y., Liu, J. and Wang, J. (2006). A novel gene delivery system: Chitosan-carbon nanoparticles, *Nanoscience,* 11(1), pp. 1-8.

Chapter 18

Treating Nanoparticles with Precaution: Recognising Qualitative Uncertainty in Scientific Risk Assessment

Fern Wickson, Frøydis Gillund & Anne Ingeborg Myhr

Scientific inventions and technologies, such as nanotechnology, interact with complex ecological and social systems at multiple levels and have the potential to cause novel, unprecedented consequences. This poses challenges to conventional risk assessment approaches, which presume both that the potential hazards of new technologies can be accurately predicted by scientific methods and that there is a clear distinction between a factual/objective expert-led risk assessment and normative/values-based risk management. In this chapter we describe various forms of qualitative uncertainty and seek to show how values, beliefs and interests are inevitably entangled in the science involved in risk assessment processes. We then present two available frameworks for exposing and analysing qualitative forms of uncertainty in science for policy and describe precautionary approaches as an alternative to risk based decision-making. We use nanoparticles as an illustrative case study and argue that adopting precautionary approaches to decision-making would necessarily involve recognising the importance of qualitative forms of uncertainty and that this would help promote more socially robust and transparent governance of this emerging field of technology development.

Nano Meets Macro - Social Perspectives on Nanoscale Sciences and Technologies
by K L Kjølberg & F Wickson

www.panstanford.com
978-981-4267-05-2

1. Introduction

Policymakers rely to a significant extent on expert scientific advice when evaluating and making decisions about emerging technologies. This privileged role for science in decision-making is generally based on a belief that science offers objective knowledge, free from the influence of particular values, interests or beliefs. Currently, the most prominent approach to incorporating science into political decision-making, particularly for new technologies, is based on the concept of 'risk' (Winner 1986). The risk-based approach to decision-making (or the process of risk analysis) is often described as consisting of three stages – risk assessment, risk management and risk communication. According to this approach, scientists first perform risk assessments, where potential adverse impacts associated with introducing a given technology are identified and their probabilities calculated. The risks identified by the scientists are then evaluated by policymakers, who decide the relative importance of the risks in question and how they will be managed. Finally, once decisions have been made, the general public is informed about both the risks and the chosen management initiatives. A crucial characteristic of this conventional approach is that it assumes a clear distinction between the stages of factual/objective expert-led risk assessment and normative/values-based risk management.

This risk-based approach to decision-making, and the clear separation it proposes between objective risk assessment and values-based risk management, have both been substantially challenged by a growing recognition of the existence and impact of various forms of uncertainty. However, conventional risk-based approaches to decision-making continue to treat uncertainty in a purely quantitative way, leaving other, more qualitative forms of uncertainty, unacknowledged. In this chapter, we will describe various types of qualitative uncertainty and seek to show how, even in the so-called objective stage of scientific risk assessment, the presence of these forms of uncertainties means that values, beliefs and interests are inevitably entangled in the science involved. We then outline two available frameworks for exposing and analysing these more qualitative forms of uncertainty - the uncertainty framework developed by Walker *et al.* (2003) and the notion of pedigree assessment inspired by Funtowicz & Ravetz (1990). After outlining these approaches we explore the notion of precautionary approaches as an alternative to risk-based decision-making. We use the case of nanoparticles to illustrate our arguments about the influence of values in scientific risk assessment, the importance of recognising various forms of uncertainty and the differences that a precautionary approach to governance might entail.

2. Risk and Science in Decision-Making

In a widely cited social science thesis, Ulrich Beck (1992) has suggested that risk is now the dominant organising principle of modern western societies. According to Beck's thesis, in modern western societies we have become increasingly aware of how the application of science and technology can be accompanied by unintended adverse effects and subsequently, increasingly concerned with how to handle the problems resulting from technological development. Beck (1992) describes this as representing a new phase of modernity, a phase in which the primary concern is no longer with the production and distribution of goods, but rather with the production and distribution of 'bads' (conceptualised as risks). Beck has therefore argued that the current social context is one in which new technological developments are increasingly scrutinised for their potential impacts on social and biological environments and that this scrutiny is increasingly structured around the notion of risk.

Conventional risk-based approaches to decision-making take the process of analysing risks as the primary way to inform policy-making in the attempt to promote benefits and avoid undesirable consequences. Technically, risk is often defined as the probability that a hazard (or undesirable event) will occur, multiplied by the magnitude of its impact. Importantly, conventional risk-based approaches presume that the potential hazards (and the probabilities associated with those hazards) can be accurately predicted and calculated using scientific methods. The general and widely shared image of science is of a process that produces verifiable, reproducible, and therefore trustworthy, objective, facts and theories about the material and biological world. Within both science in general and its approach to risk assessment, uncertainty is commonly characterized as quantitatively representable (e.g. in statistical terms) and reducible through further research (Stirling & Gee 2002; Stirling 2006; 2007). These understandings secure the currently predominant position that scientific advice and the process of risk assessment deserve a privileged position in decision-making processes.

Science has traditionally been dominated by reductionism. This is apparent in both the methodological belief that the best way to pursue an understanding of complex systems and processes is to reduce them to their smallest or most fundamental functional components, and in the ontological belief that the system itself is nothing more than the sum of these components. Recently, however, the experience of unexpected and undesired consequences from industrial and technological developments (such as the negative health and environmental impacts of chemical pesticides, asbestos, lead etc.) has created a growing recognition that a new approach is needed. The traditionally reductionist approach has arguably limited science's ability to see potential additive, cumulative, synergistic and indirect effects associated with new technologies (Stirling & Gee 2002). There are, however, emerging fields within science that aim to take complexity into account. Fields such as systems biology, resilience thinking, ecological economics, and sustainability science (see for instance Baumgärtner *et al.* 2008; Berkes, 2007; Kitano 2002; Kates *et al.* 2001), all emphasise the need to study 'systems as wholes' and investigate the dynamics, interactions, feedback loops, self organisation and responsiveness in these systems in a long term and broad scale perspective. This shift towards systems thinking in science is generating a greater sensitivity towards uncertainty and a greater awareness of the different forms it can take.

3. Types of Uncertainty

According to Prigogine (1980), unpredictability is a key feature of complex systems. Scientific inventions and technologies, such as nanotechnology, interfere with these complex systems at multiple levels and therefore have the potential to cause novel, unprecedented consequences. Accordingly, more research might not necessarily reduce uncertainty, but rather increase it and/or result in an increased awareness of the various forms of uncertainty characterising these systems. Several typologies characterising different types of uncertainty have recently emerged (e.g. Walker *et al.* 2003; Wynne 1992; Stirling 1999a&b; Stirling & Gee 2002; Faber *et al.* 1992; Felt & Wynne 2007; Funtowicz & Ravetz 1993) and these have important repercussions for the role of science in decision-making on emerging technologies. While the typologies differ in how they draw boundaries of distinction and define what constitutes the various forms of uncertainty, we believe some patterns can be extracted and developed into conceptually useful categories. The typology of different forms of uncertainty that we present below is therefore a synthesis that draws on the references listed above.

Firstly, the term risk always implies uncertainty to some extent. If we were certain that a particular impact would (or would not) occur, we would talk about it as a certainty, not as a 'risk'. According to recent typologies, however, the term risk is defined as specifically relevant to those situations where both the potential outcomes and the probabilities associated with those outcomes can be reasonably well characterised.

Uncertainty is then a term applied to those situations where there is some agreement about the potential outcomes or impacts of a technology or action, but the basis for assigning the relevant probabilities is not strong. This is because there is a lack of relevant information, a lack that can be reduced through further research. This understanding of risk and uncertainty is the one traditionally employed in processes of risk analysis. As the term 'uncertainty' has now been given a particular definition in our typology, we will follow Stirling (1999a&b) and use 'incertitude' as the collective term for the different forms. The types of incertitude that are more qualitative and that are not well addressed by conventional risk-based approaches to decision-making can be titled indeterminacy, ambiguity and ignorance.

Indeterminacy refers to a qualitative type of incertitude that exists because of the complexity associated with predicting outcomes (and probabilities) associated with the interaction of various open-ended social and natural systems (the situation we are in when any technology is commercially released outside the controlled conditions of a laboratory). This means that our knowledge will always be inherently incomplete, because science is simply unable to take every factor of a dynamic system into account. In other words, all scientific studies select frames of reference that are limited in their ability to include all factors of a complex and dynamic reality. While the various forms of systems science mentioned earlier are certainly more comprehensive and more favourable than reductionist approaches to understanding the social and biological impacts of new technologies, indeterminacy suggests that even these will be unable to take account of all relevant factors and interactions. Indeterminacy therefore implies that we should expect surprises and that we should acknowledge that our knowledge is always partial and conditional.

Ambiguity is a type of incertitude that results from contradictory information and/or the existence of divergent framing assumptions and values. By this we mean that there can be plural framings of an issue–different ways in which problems can be understood and approached and in which results can be interpreted. This potential for plural framings exists both in science (especially apparent in the way different disciplines approach particular issues) and in the socio-political arena (the way in which diverse interests, perspectives and value frameworks shape understandings of particular issues). This means that there will be different approaches to generating knowledge, different interpretations of the significance of the generated knowledge, different ways of evaluating the quality and strength of the knowledge and different understandings of how to act in light of the knowledge. Stirling (2007) provides a useful list of places where ambiguity can manifest itself in the framing of scientific risk assessments, starting from how problems are defined and hypotheses are formulated, to the choice of tools and methods for the study and its analysis, and the interpretation and communication of the results. Determining the levels of significance that are deemed appropriate is also a choice based on values. Ambiguity as a form of incertitude implies that we should acknowledge the diversity of possible framings, negotiate across different ones where possible, and at least be transparent about the particular frames we choose and the reasons for their selection.

Finally, *ignorance* can be described as our inability to conceptualise, articulate and consider the outcomes and causal relationships that lie beyond current frameworks of understanding. It has been described as the things 'we don't know that we don't know' and represents an inability to ask the right questions, rather than a failure to provide the right answers. The idea here is that there will be potential impacts that we have not yet thought about, not yet even imagined as possible. An illustrative example is our previous ignorance about the potential for chemicals to act as endocrine disrupters, or about the ozone depleting potential of chlorofluorocarbons. While it might be suggested that there is nothing we can do about the things we don't know we don't know, the best approach to this situation may be to not 'put all our eggs in one basket' and to try and pursue a range of diverse policy options so as to maintain flexibility, resilience and reversibility, as well as to consistently and vigilantly monitor for potential surprises.

Table 1: Typology of incertitude in science for policy

Type of Incertitude	Explanation
QUANTITATIVE FORMS	
Risk (Probability calculated)	We can imagine a possible impact and calculate the probability of that impact occurring, even though whether it will occur or not remains unknown.
Uncertainty (As yet uncalculated)	We can imagine a possible impact but we don't know the probability that it will occur. It is possible to calculate that probability, but we haven't enough knowledge to do so yet.
QUALITATIVE FORMS	
Indeterminacy (Unable to calculate completely)	For complex, open, interacting systems, it is impossible to include all the relevant factors and interactions in the calculations, therefore knowledge is conditional and fallible.
Ambiguity (Various ways to frame a calculation)	We can variously frame both the impacts we are interested in and the way we approach, interpret and understand the knowledge and calculations generated about them.
Ignorance (Not aware of what to calculate)	We can not imagine the possible impact. Not only have we not yet calculated the probability of the event, we are unaware of what we should make calculations for.

Through providing a way to conceptualise different forms of incertitude (see Table 1), this typology enables us to see the way in which risk assessments generally fail to take account of qualitative incertitude in the forms of ambiguity, indeterminacy, ignorance, and even uncertainty in some cases (Stirling & Gee 2002; Wynne 1992). This typology particularly emphasises that various plausible descriptions of complex systems and their processes are possible (Cilliers 2005) and that all are potentially fallible. This in turn highlights how scientists inevitably frame risk assessments in particular ways and how these framing choices reflect underlying assumptions, world-views, values and interests of the scientists' and scientific disciplines involved (Stirling 1999a; Stirling & Gee 2002; Stirling 2006; 2007).

4. Incertitude in the Case of Nanoparticles

Concerns about the potential adverse effects of nanoparticles can be seen as primarily related to the following factors (EC Sanco 2004; Nanoforum 2004; The Royal Society & Royal Academy of Engineering 2004; Royal Commission on Environmental Pollution 2008):

1. Their large surface area relative to volume. This characteristic enhances their reactivity and means that some manufactured nanoparticles may be more toxic per unit of mass than their bulk counterparts. This feature can also facilitate transport in the environment and in human/animal tissue and organs.
2. Their small size. Ultra-fine particles have different biological behaviour and mobility than larger particles, with no linear relationship between mass and effect. *Prima facie*, it is likely that nanoparticles will be absorbed and taken up by cells more readily than larger particles.
3. Their physical and methodological 'invisibility'. This characteristic means that such particles could accidentally be distributed to living systems through air, soil and water, with the potential to cause damage to plants, animals and humans without standardised and efficient methods for detection being available.

In the case of nanoparticles and concerns about their potential adverse effects on human and environmental health, one may argue that we are not in a position to assess risks at all. This is because there are serious knowledge gaps inhibiting our ability to carry out risk assessments on nanoparticles in terms of imagining the range of impacts, calculating probabilities for these, and multiplying these probabilities by predicted magnitudes (Royal Commission on Environmental Pollution 2008). This situation of uncertainty begins very basically with a lack of knowledge about how to accurately characterize various nanoparticles, as well as how to detect and measure them (EFSA 2008). There is also very limited information about likely exposure levels, dose-response relationships, modes of action, and fate in the environment – all of which are compounded by a lack of agreed and standardised test procedures and equipment (Myhr & Dalmo 2007; Wickson *et al.* in press, Royal Commission on Environmental Pollution 2008). All of this is, however, arguably uncertainty that can be reduced through further research.

Indeed, an increasing number of ecotoxicity studies are beginning to present empirical evidence that occupational and environmental exposure to nanoparticles may lead to adverse health effects (see for example the overview provided in chapter 17). Almost all present studies on potential adverse effects of nanoparticles have, however, focused on acute rather than chronic effects and have been health related, with almost no studies initiated on (a) nanoparticle mobility within air, water and sediments, (b) bioavailability and transfer between organisms, (c) ecotoxicology on organisms from fixed nanoparticles, and (d) the ability of nanoparticles to act as carriers for larger toxic contaminants. This means that these are all areas where unexpected effects may arise over either the short or long term. It has been suggested that decades of research could be required before we will be in a position to assess the risks to human and environmental health from nanoparticles (Royal Commission on Environmental Pollution 2008).

However, what we are arguing in this chapter (along with others, see for instance Renn 1998; Mayo & Hollander 1991; Wynne 1992), is that not only is uncertainty (the current lack of knowledge) a problem for risk-based approaches to decision-making, but that the other, more qualitative, types of incertitude described above also pose a serious challenge to policymakers. To help explore the idea and importance of these various forms of incertitude in the case of nanoparticles, we employ a hypothetical example below.

Box 1: Incertitude in Science for Policy: Hypothetical example for toxicology of nanoparticles

Susan is a policymaker who works for the Fijian government. She has the responsibility to decide whether to permit the commercial sale of products containing two different types of nanoparticles (x and y). To help her decide whether these products are safe for human health and the environment, she has appointed an expert scientific advisory committee. This committee has reviewed the scientific literature and found that a number of studies have been performed examining the effects inhaling nanoparticle x has on rats. Given this basis of scientific information, Susan is faced with different types of incertitude depending on the questions she is interested in and the types of studies that are available.

Risk

The scientific studies available mean that the expert committee arguably has a reasonable basis for calculating the probability that rats die after inhaling nanoparticle x. Therefore, if Susan is interested in the possibility that rats will die after inhaling nanoparticle x, the available scientific studies put her in a position where her committee can assess this risk.

Uncertainty

While the available science might mean that Susan can be well informed about the risk to rats of inhaling nanoparticle x, if she wanted to know whether rats would die after inhaling nanoparticle y, she would arguably find herself in a situation of uncertainty. This is because more experimental studies would need to be performed with nanoparticle y to provide a reasonable basis for calculating the risk involved.

Indeterminacy

Should Susan be interested in understanding the risk to rats under real world rather than controlled laboratory conditions, she would have to admit a degree of indeterminacy in the knowledge available to her. This is because of the way in which the currently available studies exclude a range of relevant factors and potential interactions, such as alternate routes of exposure (the rats eating the nanoparticles rather than inhaling them for example).

Ignorance

If Susan was interested in the risks nanoparticles x and y pose to rats' lives in a general sense, under the currently available knowledge she would arguably remain ignorant about the potential for the nanoparticles to kill rats through chronic (long term), synergistic (interactive) or multitrophic (through the food chain) effects, or for the nanoparticles to have non-fatal effects, such as negative impacts on immune system fitness.

Ambiguity

Despite the scientific studies available to help her make a decision, Susan still faces the problem of ambiguity. For example, she has to consider and weigh the range of answers available to questions such as: What weight should be given to different procedures and methods used in scientific studies? How should diverse results be interpreted? What is the relationship of results with rats for understanding risks to humans, bats or beetles? Who decides what organisms are of interest in the assessment? Should the focus only be on acute lethal effects or are impacts such as reduced reproductive rates also of interest? None of these questions have clear, objective and un-debatable scientific answers. The scientific studies that inform policy-making inevitably make a range of choices and assumptions (on these and a range of other questions) and therefore no matter how extensive the studies available to her are, Susan will inevitably face the challenge of ambiguity in science for policy.

5. Analysing Incertitude

Sarewitz (2004) has argued that incertitude is partly due to the fact that scientists from different disciplines conduct research that results in contradicting knowledge claims about the same problem (a form of what we have labelled ambiguity). Sarewitz uses the scientific controversy concerning genetically modified (GM) crops as an example to show how opposing views in the debate can be related to contrasting disciplinary views of nature. Ecologists are for instance primarily concerned with complexity, interconnectedness and lack of predictability, and consequently emphasise the possibility of unexpected side effects from GM crops. Molecular geneticists, on the other hand, are more concerned with controlling the attributes of specific organisms and tend to emphasise possible human benefits. Empirical support for this type of divergence can be found in Kvakkestad *et al.* (2007). Nanotechnology is a broadly cross-disciplinary endeavour, involving fields as diverse as physics, chemistry, biology, materials science, information science and engineering. How might these different disciplines approach the question of risk? What would they consider to be appropriate methods and models for identifying and reducing the risk involved? Is there any basis to assume that there will not be similar disagreements and debates between the different disciplines involved in nanotechnology as there has been for GM crops? If there is disagreement and debate between different disciplines, how should this be handled in political decision-making that aims to be based on 'sound scientific advice'?

Choices made by scientists when formulating and testing hypotheses give a good, simple, illustration of how values permeate science. Typically, one of two types of hypotheses will be tested in risk or toxicological types of research – H_0:'There is no adverse impact' and H_1: 'There is an adverse impact'. When testing these hypotheses and determining the statistical level of significance, more concern is traditionally given to avoiding Type I errors (false positives – situations where you reject H_0 and claim there will be an adverse effect, but in fact no adverse consequences manifest). What this focus does, however, is to increase the chance of Type II errors (false negatives – situations where you reject H_1 and claim there will be no adverse effect, but in fact adverse consequences do occur). The current scientific focus on avoiding Type I errors (false positives), means that strong evidence is required in order to claim that hazardous consequences may occur. This practice has been accused of favouring the developers of a new technology at the possible expense of human, animal and environmental health. Hence, the choice of which types of errors a scientist wants to avoid is a value judgement, and both choices have their own pitfalls.

Although seldom made explicit, the idea that science and risk assessment processes are inevitably shaped by values and interests, and that these vary across different disciplines, ultimately leads to the question: Should science (in the singular) have a privileged and authoritative role as knowledge provider for political decision-making? The active inclusion of diverse sources of knowledge, such as a range of disciplinary perspectives, as well as knowledge held by members of the public and specific interest groups and affected parties, would help to critically broaden the scope of risk assessment processes (Aslaksen & Myhr 2006) and enhance transparency around the incertitude involved. Science in this picture becomes a necessary, but not sufficient, condition for policy-making for new technologies. As a source of knowledge, it becomes subject to critical scrutiny and review from a range of alternative perspectives so as to reveal the impact and relevance of various forms of incertitude in its findings and in its approach to assessing risks. Indeed, many scholars have argued that the focus of decision-making has to shift away from a primary focus on scientific risk assessment, towards an increasing interest in negotiating and managing the more qualitative forms of incertitude involved (see for instance Funtowicz and Ravetz 1990; Stirling 1999a; Wynne 1992).

In order to cope with incertitude in science for policy, it is arguably beneficial that the various forms are recognised, identified and negotiated wherever possible. Identification and openness about incertitude in science used to inform decision-making can improve policies and help identify future research priorities. In this section we shall briefly outline two different frameworks available for recognising and negotiating incertitude – the uncertainty framework of Walker & Harremoës (Walker *et al.* 2003) and the concept of pedigree assessment inspired by Funtowicz & Ravetz (1990).

5.1 The Walker & Harremoës (W&H) uncertainty framework

The Walker & Harremoës (W&H) uncertainty framework is a conceptual framework for the systematic treatment of incertitude in model based decision support. It intends to give scientists an opportunity to express and reflect upon different forms of incertitude associated with their field of interest (Walker *et al.* 2003). Rather than incertitude, Walker *et al.* (2003) prefer to use 'uncertainty' as a collective term. They define uncertainty as "any departure from the unachievable ideal of complete determinism" (Walker *et al.* 2003, p. 8) and describe it as a three dimensional concept consisting of *location*, *level* and *nature* of uncertainty.

1. *Location* refers to where the uncertainty manifests itself within a given system or model. This relates to how the context of the study is defined; e.g. which parameters should be included, which should be left out, and which allow for alternative interpretations.
2. *Level* characterises the severity of the uncertainty identified, or how extensive a lack of knowledge is considered to be. This is generally understood as a gradual change from knowing for certain to not even knowing what you do not know.
3. *Nature* of uncertainty can either be epistemological or ontological. Epistemological uncertainty refers to a lack of knowledge or appropriate methodologies to properly investigate a scientific problem (i.e. reducible uncertainty). Ontological uncertainty describes uncertainty that is due to the inherent variability and complexity of the problem or system under investigation (i.e. irreducible uncertainty).

Krayer von Krauss *et al.* (2004; 2008) were the first to test the W&H framework and used it to identify scientists' judgement of incertitude relating to GM crops. Gillund *et al.* (2008a; 2008b) used the framework to investigate possible consequences and uncertainties associated with the use of DNA vaccines[1] for farmed fish. Using the framework,

[1] DNA vaccination is defined as the intentional transfer of genetic material to somatic cells for the purposes of influencing the immune system (Foss 2003)□

scientists are asked to identify uncertainties in a model that represents the system they are investigating. For instance, Gillund *et al.* (2008a; 2008b) presented twelve scientists involved in research on vaccines for farmed fish with a model portraying three stages of DNA vaccination: (i) the fish' immune response, (ii) tissue distribution and expression of the injected DNA vaccine and (iii) environmental release of the DNA vaccine. During individual interviews, the scientists were asked to comment on the specific model and identify uncertainties according to the dimensions given in the W&H framework. When identifying the 'location' dimension, the scientists were asked to identify the parameters in the model that could be described in several plausible ways. This dimension shares some of the characteristics of 'ambiguity' as described above, recognising that different scientific framings allow for several plausible ways to describe a system. Once the 'locations' are identified, the scientists were asked to indicate the severity of the uncertainty at each of the identified locations, using a scale ranging from a situation described as deterministic (complete knowledge) through to risk, uncertainty and ignorance (where we don't know what we don't know). This constitutes the 'level' dimension in the uncertainty framework and corresponds to our earlier characterisation of risk, uncertainty and ignorance. Finally the scientists were asked to describe the 'nature' of the uncertainty for each of the 'locations', i.e. whether they considered the uncertainty to be reducible with more science or not.

Walker *et al.* (2003) suggest that the data from such elicitations should be fed into an uncertainty matrix, which could serve as a map of the incertitude involved, indicating the 'level' and 'nature' of each of the 'locations' identified as uncertain. Presented in this way, policymakers could get an overview of the incertitude involved when deciding on management strategies for a given system or technology. None of the studies applying the W&H framework have, however, found it appropriate to present the data in such a matrix. Still, Krayer von Krauss *et al.* (2004; 2008) found the framework useful for the identification of different types of incertitude, as well as illustrating how the judgements differed depending of the background of the participating scientists. Gillund *et al.* (2008b) experienced that the framework can also function as a reflexive tool, stimulating scientists to reflect upon and express incertitude relevant to their work and possibly adjust their future practices accordingly.

5.2 Pedigree assessment

The concept of developing pedigrees of scientific knowledge used for policy was first proposed by Funtowicz & Ravetz (1990), with the idea said to refer to capturing 'the history behind a number'. The tool of pedigree assessment is based on the belief that qualitative forms of incertitude manifest through the different value-laden choices and assumptions involved in developing knowledge. Pedigrees aim to generate a broadly negotiated evaluative account of the crucial aspects where choices and assumptions have been made in the generation of knowledge (usually in the form of particular scientific studies, papers or assessments being used to inform policy). In this sense, pedigrees aim to provide "an evaluative account of the production process of information" (van der Sluijs *et al.* 2005, p. 482), with the assumption that a transparent and reflexive identification and negotiation of the various choices and assumptions involved will "enhance the quality and robustness of the knowledge input in policy-making" (Craye *et al.* 2005, p.216). Pedigree assessments have been performed in relation to cases as diverse as nuclear power production (Craye *et al.* in press), the health effects of waste incineration (Craye *et al.* 2005), non-target impacts of GM crops (Wickson 2009), and a particular energy model used by the Netherlands environmental assessment agency (van der Sluijs 2005).

In the first instance, pedigrees of knowledge are concerned with identifying the 'crucial aspects' in the production of scientific knowledge where incertitude is able to manifest through the availability of different choices and assumptions. These crucial aspects can include things like: how the scientific problem is defined, what method is chosen for the research, what endpoints and indicators are used, what statistical tools are applied, how the results are interpreted, what kind of review the study has gone through and how the findings have been communicated. Ideally, sets of critical questions are then developed for each crucial aspect to help encourage reflection and assessment. Following the development of critical questions, a qualitative scale is negotiated to help evaluate and assess the aspect in relation to the critical question. For example, if the crucial aspect is the choice of indicators (the thing you are going to measure to determine an effect), a critical question may be "How well does the selected indicator cover the effect one wants to have knowledge about?" For example, if you want to know whether there has been a negative health effect on an organism, you could select from a range of indicators, such as observed fatalities, number of tumours, white blood cell counts, strength of fingernails etc. The qualitative scale for the aspect of indicators might then progress from exact measure, good fit, well correlated, to weakly correlated, where strength of fingernails might be seen as weakly correlated and fatalities an exact measure. The pedigree is the resulting matrix (presentable in a range of forms) of the assessment of the knowledge, based on the qualitative ranking of the various crucial aspects. The issue of defining crucial aspects, critical questions, qualitative scales, and how the evaluation process producing the pedigree proceeds, remains open. Generally, however, emphasis is placed on these being developed through negotiation in workshops involving a range of stakeholders and particularly those with different perspectives, values and interests.

The development of a pedigree of knowledge will usually involve various levels of dissent and debate, and in many ways, this is part of its usefulness – the process enables the importance of value-laden choices and assumptions in science for policy to be made apparent through contestation, as well as the meaningfulness of these to be assessed and discussed from various perspectives. It should be noted, however, that this type of assessment process will almost always involve confrontations between technical facts and social issues that may be incommensurable (Giampietro *et al.* 2006). For instance, for new technologies, unavoidable value incommensurability often exists between potential costs (e.g. possible long-term adverse health effects and/or environmental degradation) and predicted benefits (e.g. economic growth and/or more efficient transport, medicine, and communication systems). Rather than providing an objective basis for decision-making in situations of incommensurability, what techniques like pedigree assessment do is to increase transparency about the role of various forms of incertitude in science for policy.

We have suggested that the challenge incertitude poses to the idea of objective scientific knowledge 'speaking truth to power', means that arguments can be made for decision-making to shift away from a sole focus on analysing risk, and towards a more direct engagement with the task of negotiating incertitude. While the two frameworks outlined here are useful for generating increased transparency about the incertitude involved in particular scientific studies, an even broader shift might be seen in what are called precautionary approaches to decision-making.

6. Precautionary Alternatives?

6.1 The precautionary principle

A classic formulation of the precautionary principle is stated in Principle 15 of the Rio Declaration on Environment and Development (Rio Declaration on Environment and Development 1992): "Where there are threats of serious or irreversible damage, lack of full scientific certainty shall not be used as a reason for postponing cost-effective measures to prevent environmental degradation". While a wide range of alternative formulations, with diverging legal status, exist (Foster *et al.* 2000), four central components are commonly associated with this principle: (i) taking preventive action in the face of uncertainty, (ii) shifting the burden of proof to the proponents of an activity, (iii) exploring a wide range of alternatives to possible harmful actions, (iv) increasing public participation in decision-making (Kriebel *et al.* 2001). The precautionary principle represents a clear recognition of the existence of uncertainty and suggests that it is important for decision-makers to actively consider the limitations of scientific knowledge when making decisions.

The extent to which the principle enables engagement with the full range of incertitude is, however, debatable. Wynne (1992) suggests that some of the existing interpretations of the precautionary principle, although acknowledging the importance of uncertainty for decision-making, do not necessarily dictate an engagement with ambiguity, indeterminacy and ignorance. This is because the understanding of 'a lack of scientific certainty' is often one in which the only type of incertitude involved is conceptually reducible through further research. Another criticism of the principle's approach to uncertainty from Levidow (2001) is that since full scientific certainty is rarely, if ever, claimed in judgments of safety (i.e. nothing is deemed to be safe with 100% certainty), the degree of uncertainty involved can be viewed as ambiguous. Additionally, Levidow (2001) questions the usefulness of the 'cost-effective' criterion because this necessarily implies that there is adequate knowledge to predict the degree of potential damage and therefore enable an assessment of what a 'cost-effective' measure for avoidance might be. Stirling (2002) also makes similar criticisms of the formulaic version of the principle suggesting that the idea of what counts as a threat, the criteria for judging seriousness or irreversibility, how the degree of uncertainty is gauged and the yardstick for judging what is cost effective, are all issues for which no objective or single rational answer exists. Additionally, the precautionary principle may represent a desire to shift the burden of proof, but the issue of how far along the axis of 'guilty until proven innocent' decision-makers need to slide is also problematic. The most outspoken opponents of the precautionary principle fear that it will reduce incentives for technological development and economic growth (Morris 2002). For example, does adoption of the principle as a policy guide mean that decision-makers are required to avoid the use of nanoparticles at the slightest hint of danger? In practice, applying the notion of precaution as a dogmatic principle or rule is problematic at least and paralysing at worst.

All the above described problems suggest that while the precautionary principle may be seen to represent admirable sentiments, important questions remain about how it could be practically applied in political decision-making. In fact, the combined weight of the criticisms is driving an emerging theoretical shift away from discussions of how and when a specific 'precautionary principle' might be applied, towards a description of what a 'precautionary approach' to decision-making might entail. In this sense, the notion of precaution moves away from being a formulaic decision-making rule towards representing a particular approach to the use of science in decision-making under conditions of incertitude.

6.2 Precautionary approaches to decision-making

Precautionary approaches are specifically focused on how to acknowledge and handle the problem of incertitude in decision-making (Foster *et al.* 2000; Sandin 2004). According to Stirling & Gee (2002) precautionary approaches imply (i) humility and a recognition of the limits of available knowledge, (ii) research and monitoring that goes beyond theoretical models and laboratory tests to cover a more complete array of indirect causal mechanisms for harm, (iii) participation of a full range of interested and affected parties and (iv) a consideration of both the pros and cons of a variety of alternative options as well as the more general features of technological commitments. This reflective approach can be undertaken by not only exposing particular scientific knowledge claims to the scrutiny of various other scientific disciplines, but also to stakeholders and the public more broadly (Stirling 2002) – i.e. to the type of 'extended peer review' described by Funtowicz & Ravetz (1990; 1993; 1994) or the 'negotiated science' approach presented by Carr & Levidow (1999). This means that a precautionary approach importantly broadens the notion of expertise and expands the evidence-base to include public views (Oreszczyn 2005).

In summary, the important elements of what represents a precautionary approach to decision-making are:

1. The use of scientific research that is broadly framed, interdisciplinary, able to consider indirect causal mechanisms, and contributory to a lifecycle approach to analysis.
2. A recognition of the limitations of this scientific knowledge and a willingness to expose the knowledge to critical reflection and 'extended peer review', particularly so as to create transparency about embedded choices and assumptions.
3. A commitment to reducing uncertainties and minimising surprises generated by ignorance through vigilance and ongoing research and monitoring.
4. A transparent handling of ambiguity and indeterminacy through interdisciplinary approaches and broad based public participation. This handling includes the consideration and implementation of a range of socio-technical alternatives and policy options.

As precautionary approaches are more about process than outcome, methodologies aiming to facilitate them will have to be continuously developed, modified and evaluated depending on the specific context or issue of concern and certainly further discussion and debate of the practical implications of these approaches is necessary.

6.3 Precautionary approaches to the governance of nanoparticles?

What would a precautionary approach to the governance of nanoparticles entail? We will now sketch some of the important features we see in correlation with the summary points provided above.

1. A precautionary approach to the governance of nanoparticles would place emphasis on science that examines potential adverse effects on human and environmental health in a broadly framed and interdisciplinary perspective. This would firstly mean dramatically enhanced funding of basic research into the characterization of different nanoparticles, methods to detect and measure their persistence and distribution in humans, animals and the environment, as well as their toxicological potential. It is worth noting that while there has been a consistent emphasis on this need for more fundamental and (eco)toxicological research, solid financial support from policymakers remains lacking (Editor 2008), with the budget for this kind of research rarely exceeding 3% of the total amount available for nanoscience and technology (Project on Emerging Technologies 2008; Royal Commission on Environmental Pollution 2008). According to a precautionary approach, toxicological impacts would not, however, only be narrowly conceptualized as direct impacts. Knowledge of issues such as the movement of nanoparticles in soil, air and water would be emphasized, as would their movement within and between biological organisms and food webs, considering a range of potential exposure routes. The potential role for nanoparticles to act as carriers of toxic compounds would also be an important field of research, as would their potential for synergistic effects with other particles and chemicals. The development of agreed standards and methods for testing would be considered a particularly important process for investment. Present research approaches would need to be supplemented with modern analytic tools

that study whole systems under various conditions. While the number, scope and time required for the necessary research under this notion of a precautionary approach may seem overwhelming, it is our opinion that without this research, any claim to a comprehensive process of risk assessment is baseless.

2. Under a precautionary approach, scientific knowledge that is given a role in political decision-making would ideally be exposed to some form of extended peer review, involving a wide range of stakeholders to reveal the influence and negotiate the importance of various forms of incertitude. This could proceed using frameworks such as the Walker & Harremoës uncertainty framework, the approach of pedigree assessment, or some other means for structuring critical reflection and broad-based discussion. This process would highlight the weaknesses, limitations and ambiguities in the science involved, and help begin the process of debating and negotiating the differential weight and importance that should be given to these.

3. Funding and importance would be placed on ongoing research and monitoring, particularly for any products that were approved for commercial production. It is crucial that methods for detection and monitoring are initiated with the purpose to follow up and map actual health and environmental effects and to identify unanticipated effects. While this would of course involve the work of scientists, importance may also be place on a notion of collective monitoring through the cultivation and support of a perceptive public. This is relevant because in the past it has often been members of the public that alert authorities to negative impacts and potential hazards from new technologies. For collective public monitoring of nanoparticles, however, the public would need to be informed about where and in what products these particles were being used, which is not currently the case.

4. Finally there would continue to be broad-based public engagement around the use of nanoparticles in various technologies and commercial applications. Importantly, however, this public participation should not be restricted to particular 'invited' events nor confined to discussing the potential risks associated with the use of nanoparticles. Discussion and debate should be encouraged and embraced around the various alternatives to these uses that might be available, the overall trajectory of

the technology in question, the social expectations and values this trajectory represents, and what kind of a future is ultimately desirable. In this way, it is important that the discussions move away from a focus on governing risk, to a focus on governing innovation processes more broadly (Felt & Wynne 2007).

7. Conclusion

In this chapter, we have described various kinds of qualitative uncertainty and argued that the presence of these forms of incertitude mean that values, beliefs and interests are inevitably entangled in the science involved in decision-making. We have argued that qualitative forms of incertitude affecting any risk assessment of nanoparticles need to be acknowledged, and that good strategies for the management of these need to be developed. This involves a shift away from a singular focus on risk assessment processes towards more precautionary approaches, where an important step is to identify and characterise the relevant incertitude apparent in research on nanoparticles and in the science being employed in policy-making. Furthermore, since we emphasise that values and interests influence the choices made by scientists, when science is used to inform policy decisions, these choices should be made as transparent as possible and negotiated in terms of their public and political support. Uncertainty and complexity challenge the traditional notion of science and its privileged role in decision-making, so broad-based public debate must be seen as having a greater role to play in innovation and technological decision-making. A precautionary approach to the governance of nanoparticles may lead to a slow down in the commercialisation of certain innovation streams, but it may also make new technologies more socially robust and environmentally sound in the long term. According to our assumptions and values, this is a desirable goal to pursue.

Questions for Reflection:

1. What does 'increased transparency in research' imply in practical terms?
2. How would various disciplines (for example chemistry, quantum physics, ecology, microbiology, epidemiology etc) investigate the potential risks posed by nanoparticles differently?
3. Should science have a privileged role in decision-making? Why/Why not?

Bibliography

Aslaksen, I. and Myhr, A. I. (2006). The worth of a wildflower: Precautionary perspectives on the environmental risk of GMOs, *Ecol. Econ.*, 60, pp. 489-497.

Baumgärter, S., Becker, C., Frank, K., Müller, B. and Quaas, M. (2008). Relating the philosophy and practices of ecological economics: The role of concepts, and case studies in inter-and transdisciplinary sustainability research, *Ecol. Econ.*, 67, pp. 284-393.

Beck, U. (1992). *Risk Society: Towards a New Modernity*, London: Sage.

Berkes, F. (2007). Understanding uncertainty and reducing vulnerability: Lessons from resilience thinking, *Nat. Hazards*, 41, pp. 283-295.

Carr, S. and Levidow, L. (1999). Negotiated Science - The case of agricultural biotechnology regulation in Europe. In Collier, U., Orhan, G. and Wissenburg, M. (eds) *European Discourses on Environmental Policy,* Aldershot: Ashgate Publishers, pp. 159-72.

Cilliers, P. (2005). Complexity, Deconstruction and Relativism, *Theor. Cult. Soc.*, 22, pp. 255-267.

Covello, V. T. and Merkhofer, M. W. (1993). *Risk Assessment Methods; Approaches for Assessing Health and Environmental Risks*, New York: Plenum Press.

Craye, M., Funtowicz, S. and Van Der Sluijs, J. P. (2005). A reflexive approach to dealing with uncertainties in environmental health risk science and policy, *Int. J. Risk Assessment Manage.*, 5, pp. 216-236.

Craye, M., Laes, E, and van der Sluijs, J. Re-negotiating the role of external cost calculations in the Belgian nuclear and sustainable energy debate. In Pereira, A., and Funtowicz, S. (eds) *Science for Policy*, Oxford: Oxford University Press, in press.

Sanco E. C.(2004). Nanotechnologies: A Preliminary Risk Analysis on the Basis of a Workshop Organized in Brussels on 1-2 March 2004 by the Health and Consumer Protection Directorate General of the European Commission, European Commission Community Health and Consumer Protection.

Editor (2008). The same old story, *Nature Nanotechnology*, 3, pp. 697.

European Food Safety Authority (2008). *Draft Opinion of the Scientific Committee on the Potential Risks Arising from Nanoscience and Nanotechnologies of Food and Feed Safety.* http://www.efsa.europa.eu/EFSA/efsa_locale1178620753812_ ScientificOpinionPublicationReport.htm (last accessed Nov 15, 2008).

Faber, M., Manstetten, R. and Proops J. L. R. (1992). Humankind and the environment: An anatomy of surprise and ignorance, *Environ. Value*, 1, pp. 217-242.

Felt, U. and Wynne, B. (2007). *Taking European Knowledge Society Seriously.* Report on the Expert Group on Science and Governance to the Science, Economy and Society Directorate, Directorate – General for Research, European Commission.

Foss, G. S. (2003). *Regulation of DNA Vaccines and Gene Therapy on Animals,* The Norwegian Biotechnology Advisory Board. http://www.bion.no/publikasjoner/regulation_of DNA_vaccines.pdf (last accessed Mar 6, 2009).

Foster, K. R., Vecchia, P. and Repacholi, M. H. (2000). Science and the precautionary principle, *Science*, 288, pp. 979-981.

Funtowicz, S. O. and Ravetz, J. R. (1990). *Uncertainty and Quality in Science for Policy*, Dordrecht: Kluwer.

Funtowicz, S. O. and Ravetz, J. R. (1993). Science for the post normal age, *Futures*, 25, pp. 739-755.

Funtowicz, S. O. and Ravetz, J. R. (1994). The worth of a songbird: Ecological economics as a post-normal science, *Ecol. Econ.*, 10, pp. 197-207.

Giampetro, M., Mayumi, K. and Munda, G. (2006). Integrated assessment and energy analysis: Quality assurance in multi- criteria analysis of sustainability, *Energy*, 31, pp. 59-68.

Gillund, F., Dalmo, R., Tonheim, T. C., Seternes, T. and Myhr A. I. (2008a). DNA vaccination in aquaculture – Expert judgements of impacts on environment and fish health, *Aquaculture*, 284, pp. 25-34.

Gillund, F., Kjølberg, K. A., Krayer von Krauss, M., Myhr, A. I. (2008b). Do uncertainty analyses reveal uncertainties? Using the introduction of DNA vaccines to aquaculture as a case, *Sci. Total Environ.*, 407, pp. 185-196.

Kates, R. W., Clark, W. C., Corell, R., Hall, J.M., Jaeger. C. C., Lowe, I., McCarthy, J. J., Schellnhuber, H. J., Bolin, B., Dickson, N. M., Faucheux, S., Gallopin, G. C., Grübler, A., Huntley, B., Jäger, J., Jodha, N. S., Kasperson, R. E., Mabogunje, A., Matson, P., Mooney, H., Moore III, B., O'Riordan, T., and Svedin, U. (2001). Sustainability science, *Science*, 292, pp. 641-642.

Kitano, H. (2002). System biology: A brief overview, *Science*, 295, pp. 1662-1664.

Krayer von Krauss, M. P., Casman, E. A. and Small, M. J. (2004). Elicitation of Experts judgements of uncertainty in the risk assessment of herbicide-tolerant oilseed crops, *Risk Anal.*, 24, pp. 1515-1527.

Krayer von Krauss, M. P., Kaiser M., Almaas V., van der Sluijs, J. and Kloprogge P. (2008). Diagnosing and prioritizing uncertainties according to their relevance for policy: The case of transgene silencing, *Sci. Total Environ.*, 390, pp. 23-34.

Kriebel, D., Tickner, J., Epstein, P., Lemons, J., Levins, R., Loechler, E. L., Quinn, M., Rudel, R., Schettler, T. and Stoto, M. (2001). The precautionary principle in environmental science, *Environ. Health Persp.*, 109, pp. 871-876.

Kvakkestad, V. Gillund, F., Kjølberg, K. A. and Vatn, A. (2007). Scientists' perspectives on the deliberate release of GM crops, *Environ. Values*, 16, pp. 79-104.

Lam, C. W., James, J. T., McCluskey, R. and Hunter, R. L. (2004). Pulmonary Toxicity of single-wall carbon nanotubes in mice 7 and 90 days after intratracheal instillation, *Toxicol. Sci.*, 77, pp. 126-134.

Levidow, L. (2001). Precautionary uncertainty: Regulating GM crops in Europe, *Soc. Stud. Sci.*, 31, pp. 842-74.

Mayo, D. G. and Hollander, R. D. (1991). *Acceptable Evidence: Science and Values in Risk Management,* Oxford: Oxford University Press.

Morris, J. (2002). The relationships between risk analysis and the precautionary principle, *Toxicology*, 181-182, pp. 127-130.

Myhr, A. I. and Dalmo, R. A. (2007). Nanotechnology and risk: What are the issues? In Allhoff, F., Lin, P., Moor, J., Weckert, J. (eds) *Nanoethics, Examining the Societal Impact of Nanotechnology*, John Wiley & Sons Inc.

Nanoforum (2004). *Benefits, Risks, Ethical, Legal and Social Aspects of Nanotechnology. European Nanotechnology Gateway*. Available at www.nanoforum.org (Accessed 2005 Sept 5).

Oberdörster, G., Sharp, Z., Atudorei, V., Elder, A., Gelein, R., Kreyling, W. and Cox C. (2004). Translocation of Inhaled Ultrafine Particles to the Brain, *Inhal. Toxicol.*, 16, pp. 437-446.

Oberdörster, E. (2004). Manufactured Nanomaterials (fullerenes, C_{60}) Induce Oxidative Stress in Juvenile Largemouth Bass, *Environ. Health Persp.* , 112, pp. 1058-1062.

Oreszczyn, S. (2005). GM crops in the United Kingdom: Precaution as process, *Sci. Public Policy*, 32, pp. 317-324.

Pellizzoni, L. (2004). Responsibility in environmental governance, *Environ. Politics*, 13, pp. 541-565.

Peterson, M. (2007). The precautionary principle should not be used as a basis for decision making, *EMBO J.*, 8, pp. 305-308.

Prigogine, I. (1980). *From Being to Becoming: Time and Complexity in the Physical Sciences,* San Fransisco: W.H. Freeman.

Project on Emerging Technologies (2008). *Europe Spends Nearly Twice as Much as U.S. on Nanotech Risk Research.* http://www.nanotechproject.org/news/archive/ehs update/ (last accessed Dec 18, 2008).

Renn, O. (1998). Three decades of risk research: accomplishments and new challenges, *J. Risk Res.*, 1, pp. 44-71.

Rio Declaration on Environment and Development (1992). Un.Doc/CpNF.151/5/Rev.1.

Royal Commission on Environmental Pollution (2008). *Novel Materials in the Environment: The Case of Nanotechnology. Twenty-seventh report.* The Stationary Office, Norwich. http://www.rcep.org.uk/novelmaterials.htm (last accessed Dec 18, 2008).

Royal Society and Royal Academy of Engineering (2004). *Nanoscience and Nanotechnologies: Opportunities and Uncertainties,* London: Royal Academy of Engineering, 29 July 2004. http://www.nanotec.org.uk/finalReport.htm.

Sandin, P. (2004). The precautionary principle and the concept of precaution, *Environ. Value.*, 13, pp. 461-475.

Sarewitz, D. (2004). How science makes environmental controversies worse, *Environ. Sci. Policy*, 7, pp. 385-403.

Stirling, A. (1999a). *On Science and Precaution on Risk Management of Technological Risk*, An Esto project prepared for the European Commission - JRC Institute for Prospective Technological Studies, Seville.

Stirling, A. (1999b). Risk at a turning point? *J. Environ. Med.*, 1, pp. 119-126.

Stirling, A. (2002) Risk, uncertainty and precaution: Some instrumental implications from the social sciences. In Berkhout, F., Leach, M. and Scoones, I. (eds) *Negotiating Environmental Change*, Cheltenham, UK: Edward Elgar, pp. 33-76.

Stirling, A. (2006). Uncertainty, precaution and sustainability: Towards a more reflective governance of technology. In Voss, J. P., Mauknecht, D. and Kemp, T. (eds) *Reflexive Governance for Sustainable Development*, Cheltenham, UK: Edward Elgar, pp. 225-272.

Stirling, A. (2007). Risk, precaution and science: Towards a more constructive policy debate, *EMBO J.* , 8, pp. 309-312.

Stirling, A. and Gee, D. (2002). Science precaution and practice, *Public Health Rep.*, 117, pp. 521-533.

Van der Sluijs, J., Craye, M., Funtowicz, S., Kloprogge, P., Ravetz, J. and Risbey, J. (2005) Combining quantitative and qualitative measures of uncertainty in model-based environmental assessment: The NUSAP system, *Risk Anal.,* 25, pp. 481-492.

Walker, W. E., Harremoöes, P., Rotmans, J., van der Sliuijs, J. P., van Asselt, M. B. A., Janssen, P. and Krayer von Krauss, M. P. (2003). Defining uncertainty; A conceptual basis for uncertainty management in model based decision support, *Integr. Assessment* , 4, pp. 5-17.

Warheit, D. B., Laurence, B. R., Reed, K. L., Roach, D. H., Reynolds, G. A. M. and Webb, T. R. (2004). Comparative pulmonary toxicity assessment of single-wall carbon nanotubes in rats, *Toxicol Sci.*, 77, pp. 117-125.

Wickson, F. (2009) Reliability rating and reflective questioning: A case study of extended review on Australia's risk assessment of Bt Cotton, *J. Risk Res.,* 12, pp. 749-770.

Wickson, F., Grieger, K. D. and Baun, A. Nature and nanotechnology: Science, Ideology and policy. In Torres, R. and Gould, K. (eds) *Nanotechnology, Social Change and the Environment*, Lanham: Rowman & Littlefield, forthcoming.

Winner, L. (1986). *The Whale and the Reactor: A Search for Limits in an Age of High Technology,* Chicago, London: University of Chicago Press.

Wynne B. (1992). Uncertainty and environmental learning: Reconceiving science and policy in the preventive paradigm, *Global Environ. Change*, 2, pp. 111-127.

Chapter 19

Nanotechnology and Public Engagement: A New Kind of (Social) Science?

Sarah R. Davies, Matthew Kearnes & Phil Macnaghten

Nanotechnology is often framed as revolutionary, both within science and in the effects it will have on our lives. Increasingly, the relationship between nanotechnology and society is cast as a hallmark of this distinctiveness. In current debates concerning its governance and regulation nanotechnology is cast as an opportunity to 'get things right' and 'avoid the mistakes of the past'. Recent emphasis on the responsible development of nanotechnology, for example, demonstrates the increasing incorporation of both ethics and public participation initiatives into nanotechnology research programmes. In this chapter we examine this trend, reviewing current research on the relations between laypeople and nanotechnology. We reflect on both quantitative and qualitative literatures as well as the broader difficulties of engaging with an emergent technology such as nano, ending by suggesting issues, problematics and challenges that future social science on public engagement with nanotechnology should engage with. Not least of these is the question of how to research public concerns about a technology that is 'in-the-making'.

Nano Meets Macro - Social Perspectives on Nanoscale Sciences and Technologies
by K L Kjølberg & F Wickson

www.panstanford.com
978-981-4267-05-2

1. Introduction

Nanotechnology, we are told, will change all of our lives. Enthusiasts argue that the technology will usher in a 'new industrial revolution' that will include breakthroughs in computer efficiency, pharmaceuticals, nerve and tissue repair, catalysts, sensors, telecommunications and pollution control. The US's National Nanotechnology Initiative envisions a future "in which the ability to understand and control matter on the nanoscale leads to a revolution in technology and industry",[1] while a 2007 report by the insurer Lloyds suggests that "current and potential areas of application include transport, manufacturing, biomedicine, sensors, environmental management, food technology, information and communications technology, materials, textiles, sports equipment, cosmetics, skin care and defence" (Lloyds 2007, p.1-2) – and notes that the list is not exhaustive. For some, these visions culminate in the suggestion that nanotechnology represents a new kind of science, one that is both substantively distinct from existing disciplinary approaches and that has the potential for radical transformation. Thus a number of commentators have suggested that one of the defining features of nanotechnology is an attempt to unify an otherwise disparate range of scientific and technical approaches and disciplines. Similarly, the philosopher Alfred Nordmann sees nanotechnology as paradigmatic of a new kind of 'noumenal technology' which exists beyond human perception and control (Nordmann 2006).

Increasingly, the relationship between nanotechnology and society is cast as a hallmark of this distinctiveness. In current debates concerning its governance and regulation, nanotechnology is commonly cast as an opportunity to 'get things right' and 'avoid the mistakes of the past' (Krupp & Holliday 2005; Rip 2006; Randles 2008). Nanotechnology is thus represented as a unique opportunity to learn from previous scientific controversies and mishaps, including that of genetically modified crops and foods (Thompson 2008) and to create a new, socially robust science

---

[1] See the 'Frequently asked questions' on the initiative's website, www.nano.gov/html/facts/faqs.html (accessed March 2009).

(Kearnes *et al.* 2006). Building on the model established by the ELSI programme of the Human Genome Project, in which a proportion of genetics research funding was reserved for identifying the ethical, legal and social implications of human genetics research, early nanotechnology policy documents spoke of nanotechnology as a "rare opportunity to integrate the societal studies and dialogues from the very beginning and to include societal studies as a core part of the [nanotechnology] investment strategy" (Roco & Bainbridge 2001, p. 2). As a culmination of this now international consensus, a new discourse has emerged, speaking of the "responsible development of science and technology" (see for example: European Commission 2008; Meridian Institute and National Science Foundation 2004; NNI no date; Tomellini and Giordani 2008). It is suggested that the responsible development of nanotechnology might be achiveved through the incentivisation of responsible behaviour on the part of the nano scientists and developed together with the incorporation of social science and ethics into nanotechnology research programmes (Kearnes & Wynne 2007; Kearnes & Rip 2009). One practical effect of this discourse has been a proliferation of ethical codes on responsible practice in nanoscience and the increasingly strategic role of programmes of public engagement (Kearnes & Rip 2009).

Thus public engagement, through the incorporation of public participation and deliberation into nanotechnology research programmes, is seen as a key enabler of responsible development (Royal Society and Royal Academy of Engineering 2004), and as such operates as an epistemic definition of the field. This emphasis can be viewed as the endpoint of a distinct history. Macnaghten and colleagues (Macnaghten *et al.* 2005) trace the development of attitudes to science-public relations from the communication-focussed deficit model (Irwin & Wynne 1996), through a new emphasis on dialogue (House of Lords 2000), to the notion of upstream engagement (Wilsdon & Willis 2004). Beyond this, however, they call not only for a new mode of doing science but for a corresponding new role for the social sciences in becoming "an actor in these changes and [providing] insights that are simultaneous with scientific, technological, and social changes" (Macnaghten *et al.* 2005, p.

269). Nanotechnology, they suggest, presents an opportunity for robust social insight to be built into scientific innovation at an early stage.

In this chapter we consider what this might mean for studies of nascent and emerging public responses to nanotechnology. In particular we focus on public engagement activities and processes, discussing the current state of the art before moving on to set out a future agenda. We begin by briefly reviewing what we know already about the relations between laypeople and nanotechnology. How visible is nanotechnology? What do people think of it? How do they think it should be developed? We start to answer these questions from both quantitative and qualitative literatures, and, in conclusion, point to the challenges for a future social science fit for researching the social dimensions of a technology that is 'in-the-making'.

2. Risks and Benefits: Governance, Trust, and Survey Research

The advent of interest in nanotechnology, then, has brought with it an intense debate about public participation and the nature of 'responsibility'. A number of scholars have responded to this debate by suggesting that we are, in fact, currently witnessing the emergence of a new set of governmental techniques[2]—principally public deliberation, ethics and foresight—which are increasingly being incorporated into the formal governance of nanotechnology (Kearnes & Wynne 2007; Kearnes *et al.* 2006; Kearnes & Rip 2009; Macnaghten *et al.* 2005). Though there is considerable debate about the precise role that these techniques might play in the governance of science, Macnaghten *et al.* (2005) suggest that these deliberative and anticipatory techniques are increasingly cast as central to the successful development of nanotechnology. For example, UK policy on nanotechnology now indicates an official commitment:

---

[2] This terminology is drawn from Rose and Miller's (1992) distinction between the 'problematics of government'—that is the rationalities that constitute the logic of government—and the technologies of government—defined as the 'the complex of mundane programmes, calculations, techniques, apparatuses, documents and procedures through which authorities seek to embody and give effect to governmental ambitions' (p. 175). Authors have explored the embodiment of governmental rationality in a range of

> *to enable [public] debate to take place 'upstream' in the scientific and technological development process, and not 'downstream' where technologies are waiting to be exploited but may be held back by public scepticism brought about through poor engagement and dialogue on issues of concern* (Department of Trade and Industry/Department for Education and Skills/HM Treasury 2004, p.105).

Significantly, the initiation of forms of participatory and deliberative approaches is set in the context of what has been described as a deficit in public trust concerning science and technology (House of Lords Select Committee on Science and Technology 2000). Public engagement is here represented as a mechanism through which public trust can be restored by increasing the transparency and accountability of scientific governance and policy development. Though UK policy increasingly speaks of a commitment to forms of upstream public engagement, the rationale of this policy development tends to be framed as ensuring that technologies are not 'held back' by public scepticism (Kearnes & Wynne 2007). Public concern, set against a broad lack of public trust in regulatory institutions and a succession of well-known technological controversies, can therefore be identified as the assumed backdrop to the current proliferation of studies on public perceptions of nanotechnology: the implicit assumption is that by measuring public opinion and perceptions of nanotechnology, and by actively engaging the lay citizenry in the development of nanotechnology, public trust can be restored and nanotechnology 'successfully' developed. These aims appear explicitly, for example, in a recent UK government document (Department for Innovation Universities and Skills 2008) in which the Department for Innovation, Universities and Skills outlines the broad ambitions of its science and society policy as producing:

1. A society excited by and valuing science;
2. A society that is confident in the use of science; and

---

different technical areas—for example in techniques of auditing and accounting (Porter 1995; Power 1997), in psychiatric practice (Rose 1991) and in techniques of public participation and consultation (Cruikshank 1999; Lezaun and Soneryd 2007).

3. A society with a representative, well-qualified scientific workforce.

If current governmental techniques are fundamentally driven by a desire to measure and mould citizens who trust in and are excited by science, it is perhaps not surprising that this has been made manifest in a swathe of surveys of public opinion. Over the last six years there have been several key studies which have examined different aspects of public perceptions of nanotechnology, starting with an early, internet-based survey by William Bainbridge (2002) which suggested high levels of enthusiasm and expectation of future social benefit for nanotechnology and little concern about possible dangers. Two years later Michael Cobb and Jane Macoubrie conducted the first national phone survey of Americans' perceptions of nanotechnology, set up to measure public knowledge, levels of familiarity, sources of information, perceptions of risks and benefits, and levels of trust (Cobb & Macoubrie 2004). Critically, and as expected, the survey found that most citizens of the United States were unfamiliar with nanotechnology, with 80% of survey respondents reporting that they had heard 'little' or 'nothing' about nanotechnology, and with only one in three correctly answering questions designed to measure factual knowledge. Notwithstanding this low awareness, the respondents nevertheless anticipated the greater probability of benefits over risks, with 40% agreeing that benefits would outweigh risks, compared to 22% agreeing that risks would outweigh benefits. However, the survey found respondents expressing low levels of trust in the nanotechnology industry, with 60% of respondents stating that they had 'not much trust' in business leaders' ability or willingness to minimise the risks of nanotechnology to human health. The survey was interpreted to suggest that Americans are basically positive towards nanotechnology (even when it is presented within negative frames) but that trust in elites is low.

A more elaborate follow-up study in 2005 aimed at providing an in-depth look at 'informed public perceptions of nanotechnology and trust in government' (Macoubrie 2005; 2006). Funded as part of the Woodrow Wilson's 'Project on Emerging Nanotechnologies', this research differs from most other studies by focusing on informed lay publics and by incorporating qualitative aspects into its design. In many respects, though, its findings echo those of previous work. Awareness of nanotechnology was low (the media did not appear to be a significant source of information); general attitudes towards nanotechnology were enthusiastic (50% being positive rather than neutral or negative); 71% thought that benefits would equal or exceed risks; and there was little support for any ban on the technology. Reported concerns included uncertainty as to impacts, regulation and risks, and the impacts on human health and the environment. As with previous studies, there was a deep distrust of government, industry and regulatory authorities – largely ascribed to prior experiences of these bodies. Finally, the study reports a widespread desire for more information and openness and to be included in decision-making processes.

These studies all focus on the United States. In contrast, a 2004 report commissioned by the Royal Society and the Royal Academy of Engineering's Nanotechnology Working Group (BMRB Social Research 2004) provides a UK perspective. The study found limited awareness of nanotechnology (just 29% of respondents said they were aware of the term), although awareness was higher among men (40%) than women (19%), and was slightly lower for older respondents. The majority (68%) of those who were able to give a definition of the word felt that it would improve life in the future, compared to only 4% who thought it would make things worse, depending on how it was used. Use of the Eurobarometer survey tool also provides a US-Europe comparison, revealing some key differences as well as similarities (Gaskell *et al.* 2004; 2005). When asked whether nanotechnology will improve our way of life, 50% of the US sample agreed against only 29% of Europeans. The authors suggest that "people in the US assimilate nanotechnology within a set of pro-technology cultural values" (2005, p. 81) and are thus more positive about science and technology generally. By contrast, in Europe there is "more concern about the impact of technology on the environment, less commitment to economic progress and less confidence in regulation" (Gaskell *et al.* 2005, p. 81).

The overall picture from these studies, then, is of low public awareness and a cautious enthusiasm. As our earlier discussion would suggest, however, these findings need to be understood in the context of their production. Given the performative nature of all social science, and the assumptions driving the desire to measure citizens, we should be aware of the limitations of the social science which has 'found' and reported them. Significantly, surveys tend to utilise a framing in which risk is the assumed key point of interest for publics with regard to new technologies: public attitudes are thus understood to be focussed around issues of safety and to involve assessments of the possible risks of nanotechnologies (see Bowman & Hodge 2007; Peter D Hart Research Associates 2007). Benefits, similarly, tend to be either assumed or framed in economic terms with little effort devoted to examining how the promised benefits relate to social values. Broader framings, concerns and meanings are thus either ignored or under-represented, with minimal scope for meanings and understandings to be expressed in participants' own terms. This limitation has potentially profound implications in that surveys may be unwittingly imposing pre-defined categories, questions and issues that reflect the researcher's own assumptions, often in close alignment with regulators and corporate interests, and possibly at odds with wider public sentiment. Some recent work, for example, focuses on public knowledge of particular 'facts' relevant to nanotechnology under the explicit belief that public "understanding of nanotechnology will be an important challenge to avoid a backlash by a less than informed public" (Waldron *et al.* 2006; see also Castellini *et al.* 2007). Further limitations arising from the specific character of the technology include the highly questionable assumption that nanotechnology exists as a unified research programme to which it is possible to have a single, stable response or 'attitude'; the fact that most nanotechnologies remain at an early or pre-market stage of development, existing largely in terms of their promise; and the reality that most people are unfamiliar with the term, and so presumably do not have pre-existing attitudes as traditionally conceived. These challenges—which will apply to any engagements with a technology as emergent as nano—will be discussed further in the conclusion.

3. Beyond Risk: Findings from Qualitative Studies and Deliberative Processes

Survey research on public attitudes, then, runs the risk of slipping back into understanding how publics should relate to nanotechnology in terms of a deficit model, whether of knowledge or of trust (Wynne 2006): the tacit implication of this research is that the key challenge for public policy on nanotechnology is one of improving public knowledge about, and thereby trust in, nanotechnological development. In this section we turn to examine how findings from qualitative research and deliberative or dialogue processes might both broaden our understanding of public responses to nanotechnology and challenge this assumption.

The Royal Society and Royal Academy of Engineering working group commissioned the market research group BMRB to undertake both qualitative and quantitative research as part of its study activity (BMRB Social Research 2004). The qualitative aspects of this aimed to examine public awareness and attitudes and public views on potential environment, health and safety impacts, and social and ethical dimensions. Perhaps not surprisingly – given the scope this aspect of the research provided for more indepth discussion – the research found considerable ambivalence towards the technology. While enthusiasm and excitement was expressed towards prospective applications, notably in the medical domain, and in its potential to improve quality of life, concerns were also expressed as to its impending transformative impacts in restructuring social and economic life, coupled with unease on possible long-term and unforeseen effects. The report concluded that considerable 'public engagement' initiatives were required to ensure that constructive and proactive debate about the future of the technology developed before deeply entrenched or polarised positions appeared. Clearly influenced by their recent bruising experience of genetically modified foods, where public attitudes were seen to have played a formative role in the development of the controversy, the UK Government and associated funders launched a series of initiatives aimed at proactive or 'upstream' public engagement (Bowman & Hodge 2007).

A report by the Nanotechnology Engagement Group (NEG) has summarised the findings of key UK activities which resulted from this funding emphasis, including the NanoJury UK (a citizen's jury); the Small Talk programme (which sought to coordinate science communication-based dialogue activities); Democs (a conversation game designed to enable small groups of people to engage with complex public policy issues); and the Nanodialogues project (a series of practical experiments in four different institutional contexts to explore how the public can meaningfully inform decision-making processes related to emerging technologies) (Gavelin *et al.* 2007). The NEG report discusses each project's findings in detail, as well as synthesising these in the form of recommendations for science policy and for public engagement, and including the suggestion that there are three key areas which are consistently raised by lay publics deliberating nanotechnology.

The first of these relates to the fact that public attitudes are not formed in relation only to the technical aspects of particular technologies. Rather, public responses to new technologies are embedded in what we might call the 'political economy' of technological innovation: the trajectory of scientific and technological development as it is shaped by a host of culturally embedded 'imaginaries' of possible social transformations. Such imaginaries often take the form of unquestioned assumptions that implicitly shape technoscientific goals and priorities (Kearnes, *et al.* 2006; Marcus 1995). Public responses to new technologies might therefore be considered as a reaction to these inbuilt values and assumptions together with the political and policy conditions that enable technological innovation. Qualitative studies of latent public concerns demonstrate that public participants are not only concerned with the potential benefits and risks of nanotechnologies, but also with broader questions concerning who these benefits and risks are most likely to affect, why this technology and not another, and what this will mean for questions of control.

The second observation concerns the institutional dimensions of risk perceptions. Public attitudes to risk, uncertainty, and regulation were found to be interconnected with the perceived ability of regulation and regulatory authorities to manage complex risks. Perceptions of risks were thus mediated by public perceptions of those institutions charged with oversight – their honesty, independence, competence and so on – all of which influenced people's reception of current claims (see also Wynne

1982; 1992). And, thirdly, there was a consistent demand for more open discussion and public involvement in policymaking relating to the management of nanotechnology policy, invoking the sense that such matters were too important to be left to 'experts' but needed instead to become part of public discourse and civic life.

The reports from the individual projects discussed in the NEG report flesh out these findings in more detail and with more specific emphases (see Kearnes *et al.* 2006; NanoJury UK 2005; Smallman & Nieman 2006; Stilgoe 2007). For example, the ESRC-funded *Nanotechnology, Risk and Sustainability* project[3] (Kearnes *et al.* 2006) incorporated a focus group phase where laypeople were introduced to and discussed nanotechnology in the context of their experience of other technologies. The authors identify key themes of enthusiasm and ambivalence and gradually evolving concerns around risks, but also questions of control, power, inequality and the kind of 'utopian' futures being promised. As with most other studies, participants had little knowledge of what nanotechnology was. The authors note that "when pressed, people tended to define it as something that was scientific, clever, small, possibly medical, futuristic and associated with science fiction" (Kearnes *et al.* 2006, p.47-8). Further analysis from this research suggests that there are, in fact, several broad areas of concern which are key in shaping lay responses to nanotechnology. These patterns of concern include: their potential for harm, mishap and potential irretrievability; the inevitability of technological innovation as being double-edged; the likelihood that the technology would reduce autonomy, choice and personal control; the ability of technology to transgress limits and to 'play God'; and, finally, the speed of technological innovation as beyond the control of governance (see Macnaghten 2009).

More recently, a number of deliberative initiatives have been commissioned concerning nanotechnology. The consumer organisation *Which?* funded a deliberative process similar to a citizen's jury (see Wakeford 2002), designed to look at how nanotechnology would "affect consumers" (Opinion Leader 2008). Again, the jury identified key opportunities: medical applications, increased consumer choice, the

[3] Phil Macnaghten and Matthew Kearnes were directly involved in this research.

potential to help the environment and developing countries. The process brought up safety as a key concern, along with the current lack of effective regulation and labelling and, accordingly, recommendations focussed around the need for better regulation and information. While this process might be considered problematic in its strong focus on participants as consumers and corresponding emphasis on risk (the report's introduction notes that the organisers were "keen that consumers should be able to make educated choices about the extent to which they use nanotechnologies ... being aware of the areas in which uncertainty remains"; Opinion Leader 2008, p. 3), it is striking that despite these framings broader issues still emerged. The report notes, for example, that some participants were concerned about relying on 'high-tech' rather than currently available 'low-tech' solutions, or about whether nanotechnology was simply a money-making opportunity for big business.

The Engineering and Physical Sciences Research Council (EPSRC) funded a process with a rather different emphasis, using deliberative workshops to help define priority areas for research in nanotechnology for healthcare (Bhattachary *et al.* 2008; see also Corbyn 2008). One striking finding from this process was a concern about nanotechnological devices removing control or agency: those developments which would empower people, rather than taking control of healthcare from them, were seen as more likely to be beneficial. Other concerns and enthusiasms were familiar from previous processes. Diagnostics and treating serious illnesses such as cancer were seen as priorities, while there were concerns about issues of privacy, surveillance, safety and personal autonomy.

Internationally, there are few divergences from the key findings the NEG reported (Gavelin *et al.* 2007), with engagement processes in France (Ile-de-France 2007) and Switzerland (Rey 2006) producing similar recommendations to UK processes. Indeed, a key finding of the Swiss *publifocus* process was simply a marked ambivalence towards nanotechnology (Burri & Belluci 2008). Public engagement activity in the United States has been more limited, although the investment of a NSF funded Centre for Nanotechnology in Society has created a context for deliberative research which is rapidly being translated into initiatives, the most notable of which is an integrated set of consensus conferences on human enhancement set within a National Citizens' Technology Forum. Loosely based on the Danish Consensus Conference practice, and conducted across six sites in the United States, the research was set up to present the informed, deliberative opinions of ordinary, non-expert people for the consideration of policy makers who are responsible for managing these technologies before those technologies are deployed. The process itself was extensive, involving parallel panels of approximately 15 individuals undertaking a guided process of learning and deliberating in order to create a set of recommendations arrived upon by consensus. The final reports show common themes: the call for regulation, the need for a new and dedicated policy commission, concerns over access and equity, the need to prioritise remediation over enhancement, and the requirement for wise and judicious oversight (see National Citizens' Technology Forum 2008).

The literature we have examined so far—whether survey-based, drawn from qualitative research projects, or reporting deliberative processes—has focused on publics and their perceptions of nanotechnology. While this has—as we described above—been the emphasis of much research, social science studies have also examined more general public discourse on nanotechnology in the media, policy communities or fiction. There is, for example, a small literature examining media coverage of nanotechnology (for a brief review, see Kjølberg & Wickson 2007; also Anderson *et al.* 2005; Faber 2006; Gaskell *et al.* 2005; Kulinowski 2004; Stephens 2005). Scheufele *et al.* (2007) have, for example, demonstrated that public discourse concerning the risk of nanotechnology has tended to emanate from with the scientific community itself. Toumey's work explicitly relates media coverage to public perceptions. As well as tracking nanotechnology's 'creation myth' through the scientific and popular press (Toumey 2005), he has argued that the narratives surrounding nanotechnology will help anticipate public reactions to it (Toumey 2004). Drawing on the histories of recombinant DNA and cold fusion research, he suggests that if certain conditions are met—including polarised and hyperbolic discourse and exacerbated differences in power and wealth—then negative stories about nanotechnology may rapidly become dominant. As he notes, a "little bit of recklessness or disdain will easily be magnified and transmuted into a compelling story about amoral scientists arrogantly producing terribly dangerous threats to our health and our environment" (Toumey 2004, p. 108). Similarly, Schummer (2005) attempts to understand public interactions with nanotechnology through examining patterns of book buying. Using a complex network analysis based on data from Amazon.com, he argues that there is high public interest in nanotechnology, with many purchasers of 'nanobooks' being new to science and technology literature, and that this interest is focused in books about forecasting and investment. He also suggests that interest in fiction and non-fiction about nanotechnology remains mostly separate, but that links between them are growing—as are connections to the business world—through 'border-crossing' authors.

A related literature has focused on the visions or imaginaries that are manifest in nanotechnology policy and discourse and their role in constructing future-oriented promises and expectations. Informed by wider social science interest on the role of expectations in constructing socio-technical futures (Brown & Michael 2003; Selin 2007; van Lente 1993), and on the master narratives of technoscience that drive and frame current science and technology policy (Felt & Wynne 2007), research has begun to explore the multiple ways in which scenarios, foresight or vision assessment techniques can be deployed to help anticipate the likely social and ethical implications of nanotechnology. For example, van Merkerk and van Lente explored the concept of 'emerging irreversibilities' underlying the dynamics of on-going technological development of nanotubes, with the aim of rendering the technology more socially accountable (van Merkerk & van Lente 2005); the European Framework 6 project Nanologue has developed three scenarios aimed at setting out three possible futures in the development of nanotechnology with the aim of structuring the debate around 'responsible innovation' (Nanologue 2007); while scenarios have been incorporated into research projects around upstream public engagement (Kearnes *et al.* 2006), and green technology foresight (Jørgensen *et al.* 2006). Other studies have examined the role of science fiction in the development of nanotechnology policy (Milburn 2004), and in shaping the moral imagination of practitioners (Nerlich 2008; Berne 2006).

4. Conclusion: Nanotechnology and Public Engagement

Qualitative research, and that derived from dialogic or deliberative processes, then, has certainly broadened our understanding of the ways in which lay publics relate to and negotiate nanotechnologies beyond a framing of 'risks and benefits'. There is in fact remarkable consistency across different studies: we find optimism—particularly about the social benefits of new technology—mingled with concern, particularly around the motivations and trustworthiness of those driving the technology, and combined with a desire for increased openness, information, and public deliberation. In addition, not only has this consistency been expressed as a research outcome, but also as a deliberative finding from engagement processes seeking to involve publics in nanotechnology's 'upstream' direction. Can we assume, then, that social science is doing its job, and innovating in a way which involves genuine intervention in nanotechnology's development, making it increasingly socially robust (cf. Rip 1986)?

Despite considerable advances in the sophistication of both research engagements and policy-oriented activities, we would suggest that there is more work to be done in building robust social research into the "complex and difficult terrain" (Macnaghten *et al.* 2005, p. 278) of emerging technologies. To celebrate not insignificant successes—not least in bringing the language of public participation into, in many countries, the policy mainstream—without acknowledging the limitations of these is to ignore at least two important challenges that continue to face public engagement with nanotechnology.

The first of these relates to nanotechnology's emergent nature. The difficulties of negotiating a technology which largely exists in promises and speculation create a number of problems not just for survey research —as discussed above—but for any kind of engagement (see Macnaghten 2009). At this stage, in other words, nanotechnology is effectively invisible to most people. In particular, we can point to two related aspects of this 'invisibility' which are key challenges for public engagement processes but which remain largely unresolved: *problem definition* and *stakeholder definition*. The first draws on an understanding of engagement and deliberation as the negotiation of difference around a shared question or problem (Benhabib 1996); without a commonly acknowledged problematic issue, in other words, there is nothing on which to focus deliberation (see also Andersen & Jaeger 1999). With downstream or already-controversial technologies, the shared problem is frequently obvious (often coming down to the question: how can we mutually resolve this — problematic — situation?). But with nanotechnology, problem definition, and therefore the structure and outcomes of public engagement, is by no means clear. Nanotechnologies are generally felt to lack a material presence in everyday life. The fact that the majority o people are unaware of the term's meaning, coupled with disjunctions between technically expert and lay concerns, can make identifying shared problems around which to engage deeply challenging and result in a lack of focus within engagement processes.

Stakeholder definition is, of course, related to this. Where there is a commonly acknowledged problem it is usually clear who is a stakeholder in resolving this; to use Collins and Evans' (2002) language, there are a range of different forms of interactive and contributory expertise which have relevance to the issue. In the case of nanotechnology, there are currently only relatively few groups who would consider themselves as stakeholders in the technology and its development (we might suggest: nanoscientists, those who make nano-related policy, representatives from nanotechnological industry, and a small number of NGOs). Most laypeople— as our review of public engagement literature has highlighted —are unaware of the technology and do not see themselves as having a stake in it; even when their awareness of it is raised, they often feel disempowered to the extent that they cannot view themselves as active participants in its coproduction (Kearnes, *et al.* 2006). At the very least, this raises the practical point that public engagement on nanotechnology inevitably must involve a phase of awareness or consciousness raising in which lay participants can develop their own concerns around the technology (similar, in fact, to the qualitative focus groups described above). At a more fundamental level, this and the question of problem definition highlight the essential performativity of public engagement on emerging technologies. While all engagement processes shape their participants into certain forms of citizens mobilised around certain kinds of issues (see, for example, Irwin 2001; Goven 2006), those focussed around upstream technologies are exemplary in doing this in that they take what to many participants is an invisible issue—a non-issue—and create a problem, *ex nihilo*. In doing so they not only construct a problem to be solved but a group of citizens who are to perform their concerns in certain ways. This feature of public engagement with nanotechnology—and the impact that this creative process has on resulting public debate and wider discourses—remains, if not exactly a challenge to be overcome, a key aspect of the role of social science to be reflected upon.

A second challenge relates to the wider context into which the findings of public engagement are received. As a number of authors have pointed out, while the language of dialogue, public engagement and participation may have become a reality in policy, it is far less clear how such activities relate to the everyday practice of governing nanotechnology (Irwin 2006; Joly & Kaufman 2008; Rogers-Hayden & Pidgeon 2007; Wynne 2006). Assumptions about right relations between science, laypeople, and policy are often deeply embedded, and it has always been optimistic to assume that because language and, to some extent, practice have altered, such assumptions will also change automatically (Joly & Kaufman 2008; Jones & Irwin 2009). Despite repeated calls to the contrary, direct and obvious outcomes such as those arising from the 2008 EPSRC deliberative process (see above), which clearly shaped research priorities, remain unusual. In addition it should be noted that understanding this wider policy context is not simply a matter of checking off whether policy commitments relating to public engagement have been put into practice; whether, in other words, public recommendations have been acted on (although this would certainly be a start). Rather, the imperative is also towards finding where it is possible to have an impact, where irreversabilities are already and are becoming fixed, and in what ways public engagement should be structured in order to best shape these (Joly & Kaufman 2008). Without greater analysis of these questions, and the development of more effective participation mechanisms and institutional structures, public engagement with nanotechnology runs the risk of becoming what Alfred Nordmann and

Astrid Schwarz have characterized as an unending buzz of conversation, acting merely as a soundtrack as, by design or default, decisions are made elsewhere (Wordmann and Schwarz 2009). Accordingly the authors of the *Reconfiguring Responsinility* report have called for public debate on nanotechnology to move from 'conversation' to 'deliberation'. They highlight the danger of public engagement remaining simply in 'conversational mode' suggesting that "the broader point remains: public debate, whatever its exact from, must take place in a space that takes the issues that arise seriously, and acts upon them" (Davies *et al.* 2009).

Public engagement with nanotechnology therefore remains a fertile field for analysis and for methodological innovation, offering the opportunity to develop thinking on deliberation in a way that enables it to cope better with the challenges of emerging technologies. As we conclude we wish to emphasise one concept which, we suggest, might help develop a social science that is equipped to do this. A key feature of our discussion has been that nanotechnology is distinctive in a number of ways (to the extent that it has been framed as a new kind of science). Not the least part of its uniqueness as an object for social science study is its emergence: it presents particular challenges due to its current upstream position and public invisibility. In order to deal with these challenges we might suggest that we need a social science which is also emergent; that is, in development, experimental, exploratory—and therefore also multidisiplinary and 'messy' (cf. Law 2004). Taking this concept as a framework will allow methods—of both engagement practice and analysis—that are flexible, growing with the technology and with lay agency, and open to shaping by a range of actors (cf. Rip 1986). It justifies innovation and experimentation, enabling us to retreat, if necessary, from a range of tried and tested practices that may no longer be appropriate. It will allow us to reflect on and acknowledge the work being done by our own methods and processes, and the citizens, sciences, and futures being performed by them. Finally, we would hope that as these new methods of social science unfold alongside nanotechnology's own development, they would enable this development to be more robust, socially valid, and resilient.[4]

---

[4] This work has been supported by the DEEPEN (Deepening Ethical Engagement and Participation with Emerging Nanotechnologies) project, an EU Sixth Framework Programme funded project and Europe's leading partnership for integrated understanding of the ethical challenges posed by nanotechnologies (see www.geography.dur.ac.uk/projects/deepen). We would like to acknowledge productive discussions around this topic with the other DEEPEN partners.

Questions for Reflection:

1. What visions and imaginaries are shaping developments in nanotechnology?
2. How can the development of nanotechnology become the subject of greater public discussion?
3. What do you think the role of social science should be in the development of new technologies?

Bibliography

Andersen, I. E. and Jaeger, B. (1999). Danish participatory models. Scenario workshops and consensus conferences: Towards more democratic decision-making, *Science and Public Policy,* 26(5), pp. 331-340.

Anderson, A., Allan, S., Petersen, A. and Wilkinson, C. (2005). The framing of nanotechnologies in the British newspaper press, *Science Communication,* 27(2), 200-20.

Bainbridge, W. S. (2002). Public attitudes toward nanotechnology, *Journal of Nanoparticle Research,* 4(6), 561-270.

Bainbridge, W. S. (2004). Sociocultural meanings of nanotechnology: Research methodologies, *Journal of Nanoparticle Research,* 6(2-3), 285-299.

Benhabib, S. (ed) (1996). *Democracy and Difference: Contesting the Boundaries of the Political*, Princeton: Princeton University Press.

Berne, R. (2006). *Nanotalk: Conversations with Scientists and Engineers about Ethics, Meaning, and Belief in the Development of Nanotechnology*, Nahwah N.J.: Lawrence Erlbaun.

Bhattachary, D., Stockley, R. and Hunter, A. (2008). *Nanotechnology for Healthcare*, London: BMRB.

BMRB Social Research (2004). *Nanotechnology: Views of the General Public. Quantitative and Qualitative Research Carried out as Part of the Nanotechnology Study*, London: The Royal Society and Royal Academy of Engineering Nanotechnology Working Group.

Bowman, D. and Hodge, G. A. (2007). Nanotechnology and public interest dialogue: Some international observations, *Bulletin of Science, Technology and Society,* 27(2), 118-132.

Brown, N. and Michael, M. (2003). A sociology of expectations: Retrospecting prospects and prospecting retrospects, *Technology Analysis and Strategic Management,* 15(1), 4-18.

Burri, R. V. and Bellucci, S. (2008). Public perception of nanotechnology. *Journal of Nanoparticle Research,* 10(3), 387-391.

Castellini, O., Walejko, G., Holladay, C., Theim, T., Zenner, G. and Crone, W. (2007). Nanotechnology and the public: Effectively communicating nanoscale science and engineering concepts, *Journal of Nanoparticle Research,* 9(2), 183-189.

Cobb, M. and Macoubrie, J. (2004). Public perceptions about nanotechnology: Risks, benefits and trust, *Journal of Nanoparticle Research,* 6(4), 395-405.

Collins, H. and Evans, R. (2002). The third wave of science studies: Studies of expertise and experience, *Social Studies of Science*, 32(2), 235-296.

Cook, A. J. and Fairweather, J. R. (2007). Intentions of New Zealanders to purchase lamb or beef made using nanotechnology, *British Food Journal,* 109(9), 675-688.

Corbyn, Z. (2008). Nanotech research funding halted by a thumbs-down from the public, *Times Higher Education Supplement*, 7 August, p. 11.

Cruikshank, B. (1999). *The will to empower: Democratic Citizens and Other Subjects*, Ithaca, N.Y.: Cornell University Press.

Davies, S., Macnaghten, M. and Kearnes, M (eds.) (2009). *Reconfiguring Responsibility: Lessons for Public Policy.* (Part 1 of the report on Deepening Debate on Nanotechnology), Durham: Durham University.

Department for Innovation Universities and Skills (2008). *A Vision for Science and Society: A Consultation on Developing a New Strategy for the UK,* London: Department for Innovation Universities and Skills.

Department of Trade and Industry (2000). *Excellence and Opportunity – A Science and Innovation Policy for the 21st Century*, London: DTI.

Department of Trade and Industry/Department for Education and Skills/HM Treasury (2004). *Science and Innovation Investment Framework 2004–2014*, London: HM Treasury.

European Commission (2004). *Towards a European Strategy for Nanotechnology*, Luxembourg: Commission of the European Communities.

European Commission (2008). *Commission recommendation of 07/02/2008 on a code of conduct for responsible nanosciences and nanotechnologies research,* 07/02/2008, C(2008) 424 final, Brussels: E. C.

Faber, B. (2006). Popularizing nanoscience: The public rhetoric of nanotechnology, *Technical Communication Quarterly,* 15(2), 141-169.

Felt, U. and Wynne, B. (2007). *Science and Governance: Taking European Knowledge Society Seriously*, Report of the Expert Group on Science and Governance to the

Science, Economy and Society Directorate, Directorate-General for Research, Brussels: E. C.

Fisher, E., Mahajan, R. L. and Mictham, C. (2006). Midstream modulation of technology: Governance from within, *Bulletin of Science and Technology,* 26(6), 485-496.

Gaskell, G., Eyck, T. T., Jackson, J. and Veltri, G. (2004). From our readers: Public attitudes to nanotech in Europe and the United States, *Nature Materials,* 3(8), 496.

Gaskell, G., Eyck, T. T., Jackson, J. and Veltri, G. (2005). Imagining nanotechnology: Cultural support for technological innovation in Europe and the United States, *Public Understanding of Science,* 14, 81-90.

Gavelin, K., Wilson, R. and Doubleday, R. (2007). *Democratic Technologies? The Final Report of the Nanotechnology Engagement group*, London: Involve.

Goven, J. (2006). Processes of inclusion, cultures of calculation, structures of power: Scientific citizenship and the royal commission on genetic modification, *Science, Technology & Human Values,* 31(5), 565-598.

Grove-White, R., Macnaghten, P., Mayer, S. and Wynne, B. (1997). *Uncertain World: Genetically Modified Organisms, Food and Public Attitudes in Britain*, Lancaster: Lancaster University, in association with Unilever.

Grove-White, R., Macnaghten, P. and Wynne, B. (2000). *Wising Up: The Public and New Technologies*, Lancaster: Lancaster University.

Hagendijk, R. P. and Irwin, A. (2006). Public deliberation and governance: Engaging with science and technology in contemporary Europe, *Minerva,* 44, 167-184.

Hill, A. and Michael, M. (1998). Engineering acceptance: Representations of 'the public' in debates on biotechnology. In Wheale, P., von Schomberg, R. and Glasner, P. (eds), *The Social Management of Genetic Engineering*, Aldershot: Ashgate, pp. 201-218.

HM Government (2005). *The Government's Outline Programme for Public Engagement on Nanotechnologies*, London: Department of Trade and Industry.

House of Lords Select Committee on Science and Technology (2000). *Science and Society*, London: HMSO.

Ile-de-France (2007) *Citizens Recommendations on Nanotechnology*. Paris: Espace Projets.

Irwin, A. (2001). Constructing the scientific citizen: Science and democracy in the biosciences, *Public Understanding of Science,* 10(1), 1-18.

Irwin, A. (2006). The politics of talk: Coming to terms with the 'new' scientific governance, *Social Studies of Science,* 36(2), 299-320.

Irwin, A. and Wynne, B. (1996). *Misunderstanding Science? The Public Reconstruction of Science and Technology*, Cambridge and New York: Cambridge University Press.

Joly, P. B. and Kaufmann, A. (2008). Lost in translation? The need for upstream engagement with nanotechnology on trial, *Science as Culture,* 17(3), 225-47.

Joly, P. B. and Rip, A. (2007). A timely harvest, *Nature,* 450(7167), 174.

Jones, K. E. and Irwin, A. (2009 forthcoming). Creating space for engagement? Lay membership in contemporary risk governance. In Hutter, B. (ed) *Anticipating Risks and Organizing Risk Regulation in the 21st Century*, Cambridge: Cambridge University Press.

Jørgensen, M. S., Hansen, A. G., Anderson, M. M., Jørgensen, U., Falch, M., Pedersen, T. T., Wenzel, H., Rasmussen, B., Olsen, S. I. and Willum, O. (2006). *Green Technology Foresight about Environmentally Friendly Products and Materials – Challenges from Nanotechnology, Biotechnology and ICT*, Copenhagen: Danish Environmental Protection Agency.

Kearnes, M. B., Grove-White, R., Macnaghten, P. M., Wilsdon, J. and Wynne, B. (2006). From bio to nano: Learning lessons from the agriculture biotechnology controversy in the UK, *Science as Culture,* 15(4), 291–307.

Kearnes, M. B. and Wynne, B. (2007). On nanotechnology and ambivalence: The politics of enthusiasm, *Nanoethics,* 1(2), 131-142.

Kearnes, M. B., Macnaghten, P. M. and Wilsdon, J. (2006). *Governing at the Nanoscale: People, Policies and Emerging Technologies*, London: Demos.

Kearnes, M. B. and Rip, A. (2009). The emerging governance landscape of nanotechnology. In Gammel, S., Lösch, A. and Nordmann, A. (eds) *Jenseits von regulierung: Zum politischen umgang mit der nanotechnologie*, Berlin: Akademische Verlagsgesellschaft, in press.

Kjølberg, K. and Wickson, F. (2007). Social and ethical interactions with nano: Mapping the early literature, *NanoEthics,* 1(2), 89-104.

Krupp, F. and Holliday, C. (2005). 'Let's get nanotech right', *Wall Street Journal*, Tuesday, June 14, 2005, Management Supplement, B2.

Kulinowski, K. (2004). Nanotechnology: From "Wow" To "Yuck"? *Bulletin of Science, Technology & Society,* 24(1), 13-20.

Law, J. (2004). *After Method: Mess in Ssocial Science Research*, London: Routledge.

Lee, C. J., Scheufele, D. A. and Lewenstein, B. V. (2005). Public attitudes toward emerging technologies — examining the interactive effects of cognitions and affect on public attitudes toward nanotechnology, *Science Communication,* 27(2), 240-267.

Lezaun, J. and Soneryd, L. (2007). Consulting citizens: Technologies of elicitation and the mobility of publics, *Public Understanding of Science,* 16(3), 279–297.

Lloyds (2007). *Nanotechnology, Recent Developments, Risks and Opportunities,* London: Lloyds Emerging Risks Team Report.

Macnaghten, P. M. (2009). 'Researching technoscientific concerns in-the-making: narrative structures, public responses and emerging nanotechnologies, *Environment & Planning A,* in press.

Macnaghten, P. M., Kearnes, M. B. and Wynne, B. (2005). Nanotechnology, governance and public deliberation: What role for the social sciences? *Science Communication,* 27(2), 268-287.

Macoubrie, J. (2005) *Informed Public Perceptions of Nanotechnology and Trust in Government,* Washington: D.C.: Woodrow Wilson International Center for Scholars.

Macoubrie, J. (2006). Nanotechnology: Public concerns, reasoning and trust in government, *Public Understanding of Science,* 15(2), 221-241.

Marcus, G. E. (ed) (1995). *Technoscientific Imaginaries: Conversations, Profiles and Memoirs*, Chicago: University of Chicago Press.

Meridian Institute and National Science Foundation (2004). *Report: International Dialogue on Responsible Development of Nanotechnology*, Washington, D.C.: Meridian Institute.

Milburn, C. (2004). Nanotechnology in the age of posthuman engineering: Science fiction as science. In Hayles, N. K. (ed) *Nanoculture: Implications of the New Technoscience*, Bristol: Intellect Books, pp. 109-28.

NanoJury U.K. (2005) *NanoJury UK: Provisional Recommendations*. Newcastle upon Tyne: PEALS.

Nanologue (2007). *The Future of Nanotechnology: We Need to Talk*, Döppersberg: Wuppertal Institute.

National Citizens' Technology Forum (2008). *National Citizens' Technology Forum on Technologies of Human Enhancement*, Arizona State University: Centre for Nanotechnology in Society.

Nerlich, B. (2008). Powered by imagination: Nanobots at the Science Photo Library, *Science as Culture*, 17(3), pp. 269-292.

NNI (no date). *International Cooperation on Responsible Development of Nanotechnology.* www.nano.gov/html/society/Responsible_Development.html (last accessed Mar 23, 2009).

NNI/Nanoscale Science Engineering and Technology Subcommittee (2004). *National Nanotechnology Initiative Strategic Plan*, Arlington, V. A. : National Nanotechnology Coordination Office.

Nordmann, A. (2006). Noumenal technology: Reflections on the incredible tininess of nano. In Schummer, J. and Baird, D. (eds) *Nanotechnology Challenges: Implications for Philosophy, Ethics and Society*, Singapore: World Scientific.

Nordmann. A. and Schwarz, A. E. (2009). The lure of the "yes": Reflections on the power of technoscience. In: S Maasen, M. Kaiser, M Kurath and C. Rehmannsutter (eds.) *Governing Future Technologies: Nanotechnology and the Rise of an Assessment Regime,* Heidelberg: Springer (Sociology of the Sciences Yearbook).

Opinion Leader (2008). *Report on the citizens' panel examining nanotechnologies*, London: Opinion Leader.

Peter D. Hart Research Associates (2007). *Awareness of and Attitudes Toward Nanotechnology and Federal Regulatory Agencies*, Washington: The Woodrow Wilson International Center For Scholars.

Porter, T. M. (1995). *Trust in Numbers: The Pursuit of Objectivity in Science and Public Life*, Princeton: Princeton University Press.

Power, M. (1997). *The Audit Society: Rituals of Verification*, Oxford: Oxford University Press.

Project on Emerging Nanotechnologies (2008). *An Inventory of Nanotechnology-Based Consumer Products Currently on the Market.* www.nanotechproject.org/inventories/consumer/ (last accessed Sept 9, 2008).

Randles, S. (2008). From nano-ethicswash to real-time regulation, *Journal of Industrial Ecology*, 12(3), 270-274.

Rey, L. (2006). *Public Reactions to Nanotechnology in Switzerland: Report on Publifocus Discussion forum 'Nanotechnology, Health and the Environment',*

Bern: Centre for Technology Assessment at the Swiss Science and Technology Council.

Rip, A. (1986). Controversies as informal technology assessment, *Knowledge: Creation, Diffusion, Utilization,* 8(2), 349-371.

Rip, A. (2006). Folk theories about nanotechnologists, *Science as Culture,* 15(4), 349-365.

Roco, M. and Bainbridge, W. S. (eds) (2001). *Societal Implications of Nanoscience and Nanotechnology*, Boston: Kluwer Academic Publishers.

Rogers-Hayden, T. and Pidgeon, N. (2007). Moving engagement "Upstream"? Nanotechnologies and the Royal Society and Royal Academy of Engineering's inquiry, *Public Understanding of Science,* 16(3), 345-364.

Rose, N. (1991). *Governing the Soul: The Shaping of the Private Self,* London: Routledge.

Rose, N. and Miller, P. (1992). Political power beyond the state: Problematics of government, *British Journal of Sociology* 43(2), 173-205.

Royal Society and Royal Academy of Engineering (2004). *Nanoscience and Nanotechnologies: Opportunities and Uncertainties*, London: Royal Society and Royal Academy of Engineering.

Scheufele, D. A., Corley, E. A., Dunwoody, S., Shih, T. J., Hillback, E. and Guston, D. H. (2007). Scientists worry about some risks more than the public, *Nature Nanotechnology,* 2, 732-734.

Scheufele, D. A. and Lewenstein, B. V. (2005). The public and nanotechnology: How citizens make sense of emerging technologies, *Journal of Nanoparticle Research,* 7(6), 659-667.

Schummer, J. (2005). Reading nano: The public interest in nanotechnology as reflected in purchase patterns of books, *Public Understanding of Science,* 14(2), 163-183.

Selin, C. (2007). Expectations and the emergence of nanotechnology, *Science, Technology & Human Values,* 32(2), 1-25.

Smallman, M. and Nieman, A. (2006). *Discussing Nanotechnologies*, London: Think-Lab.

Stephens, L. F. (2005). News narratives about nano S&T in major U.S. And non-U.S. newspapers, *Science Communication,* 27(2), 175-199.

Stilgoe, J. (2007). *Nanodialogues: Experiments with Public Engagement*, London: Demos.

Thompson, P. (ed) (2008). *What can Nano Learn from Bio? Social and Ethical Lessons for Nanoscience from the Debate over Agrifood Biotechnology and GMOs*, San Diego, CA: Academic Press.

Tomellini, R. and Giordani, J. (2008). *Report: Third International Dialogue on Responsible Research and Development of Nanotechnology*, Brussels: European Commission.

Toumey, C. (2004). Narratives for nanotech: Anticipating public reactions to nanotechnology, *Techne,* 8(2), 88-116.

Toumey, C. (2005). Apostolic succession, *Engineering and Science,* (1/2), 16-23.

Van Lente, H. (1993). *Promising Technology: The Dynamics of Expectations in Technological Development*, Enschede: Twente University.

Van Merkerk, R. and Van Lente, H. (2005). Tracing emerging irreversibilities in emerging nanotechnologies: The case of nanotubes, *Technological Forecasting and Social Change,* 72, 1094-111.

Wakeford, T. (2002). Citizens juries: A radical alternative for social research. *Social Research Update,* (37), Available at: http://sru.soc.surrey.ac.uk/SRU37.pdf.

Waldron, A., Spencer, D. and Batt, C. (2006). The current state of public understanding of nanotechnology, *Journal of Nanoparticle Research*, 8(5), 569-575.

Wilsdon, J. and Willis, R. (2004). *See-Through Science: Why Public Engagement Needs to Move Upstream*, London: Demos.

Wynne, B. (1982). *Rationality and Ritual: The Windscale Inquiry and Nuclear Decisions in Britain*, Chalfont St Giles: British Society for the History of Science.

Wynne, B. (1992). Public understanding of science research: New horizons or hall of mirrors? *Public Understanding of Science,* 1(1), 37-43.

Wynne, B. (2006). Afterword. In Kearnes, M. B., Macnaghten, P. M. and Wilsdon, J., *Governing at the Nanoscale*, London: Demos, pp. 70-78.

Chapter 20

Civil Society and the Politics of Nano-Scale Converging Technologies

Hope Shand

In the final decades of the 20th century, civil society and consumer campaigns against genetically modified organisms (GMOs) demonstrated that science and technology are no longer the private preserve of white-coated scientists. In the first decade of the 21st century, civil society and social movements have played a key role in education, advocacy and critical analysis of nano-scale converging technologies. Not only were civil society organisations (CSOs) among the first to question the potential health and safety risks of engineered nanoparticles, they have also brought to light broader concerns related to justice, corporate power, livelihoods, human rights and impacts on marginalised populations. Although nano-scale converging technologies are frequently hyped as the greatest and greenest industrial revolution ever, the full societal impacts of these technologies are not yet known. CSOs are diverse, and their positions on nano-scale converging technologies are not monolithic. However, the work and activities of civil society have influenced the development of nanotech policies, attracted considerable media attention and continue to build support for democratic governance and social control of new technologies. This chapter reviews some of the key developments in the civil society critique of nano-scale converging technologies and its role in building a broader movement for democratic science and social control of technology.

Nano Meets Macro - Social Perspectives on Nanoscale Sciences and Technologies
by K L Kjølberg & F Wickson

www.panstanford.com
978-981-4267-05-2

1. What is BANG and Why is it Important to Understand Nanotechnology in the Larger Context of Technological Convergence?

Many civil society advocacy organisations, especially those that have campaigned against genetic engineering in agriculture, stress the need to understand new developments in nano-scale science in a much wider context. Nanotech is not a discreet technology or an industry sector. It simply refers to a range of technologies that operate at the nano-scale. The concept of technological convergence is important to keep in mind—that is, the use of nano-scale science to manipulate and exploit **bits** (the basic unit of information in digital computing), **atoms**, **neurons** and **genes.** International civil society advocacy organisation, ETC Group, uses the acronym BANG (**b**its, **a**toms, **n**eurons and **g**enes) to describe the power of converging technologies. In short, BANG refers to a broad technological platform that seeks to extend control over all matter, life and knowledge—from the bottom up. The U.S. government uses the acronym NBIC (Nanotechnology, Biotechnology, Information technology, Cognitive neurosciences) to describe technological convergence.

The power of technological convergence stems from the fact that nano-scale science involves the manipulation of the building blocks of all matter. At the nano-scale, atoms combine to form molecules and molecules combine to form all larger structures. In the words of one social scientist: "The big deal here is that we're domesticating atoms. We're trying to make the basic building blocks of our world do our bidding" (Beck 2007). At the nano-scale, scientific disciplines are converging (physics, chemistry, molecular biology, material sciences, bioinformatics, etc.). There is no difference between living and non-living matter at the nano-scale.

"Nanobiotechnology" refers to the application of nano-scale science to the life sciences. In some cases, nanobiotechnology involves the merging of living and non-living matter to make hybrid materials and organisms. Chemists, for example, are entering the realm of biology by creating electronic components out of viruses and bacteria (Leo 2002). With the new, nano-scale science of synthetic biology (often a synonym for nanobiotechnology) scientists aim to create designer organisms built from synthetic DNA. They are re-engineering living organisms to do things they can't do in nature, creating new organisms that have never existed before, and also manipulating living organisms to perform mechanical functions (ETC Group 2007a). Nano-scale technologies potentially involve all of these areas.

2. Civil Society Concerns Related to BANG

For many civil society organisations, there are three broad areas of concern related to the power of BANG:

1. Nano-scale technologies will have enormous social, economic and ethical impacts, especially for marginalised communities, for which society is not prepared.
2. Nano-scale technologies offer new opportunities for corporations to gain sweeping monopoly control over the building blocks of nature.
3. Nano-scale technologies have the potential to introduce new hazards to human health and the environment.

Each of these is briefly elaborated below.

2.1 Societal impacts for marginalised communities

Today, governments, corporations and international institutions are promoting nano-scale technologies as the magic bullet for cheap energy, clean water, arresting climate change, curing cancer and addressing poverty. Although nanotech enthusiasts insist that nano-scale technologies will address the global South's most pressing needs, far less attention has focused on the potential *disruptive* impacts of nanotech. As industry analysts point out, nanotechnology has the potential to "ultimately displace market shares, supply chains, and jobs in nearly every industry" (Lux Research 2004). The introduction of powerful new technologies can result in sudden economic upheavals, and, it is usually the most vulnerable populations that are hardest hit. If researchers succeed in engineering the metabolic pathways of microbes, for example to secrete a natural rubber substitute (a current R&D project supported by the U.S. Department of Agriculture[1]), what will this mean for the six million farmers who produce rubber in Thailand? If textile companies continue to develop synthetic fabrics with nano-scale molecular coatings that mimic the feel and properties of cotton, how will this affect the world's 100 million families engaged in cotton production? Will nanotech companies like QuantumSphere (http://www.qsinano.com/) and Nanostellar (http://www.nanostellar.com/) succeed in developing nano-scale materials that replace the need for platinum? And if so, what impact will this have on South Africa that accounts for over 87% of the world's platinum reserves? Will carbon nanotubes (large molecules of pure carbon) replace copper? Worker displacement brought on by commodity obsolescence will hurt the poorest and most vulnerable, particularly workers in the global South who do not have the economic flexibility to respond to sudden demands for new skills or different raw materials (ETC Group 2005a). In the face of perennially low and volatile prices for primary export commodities, and the persistent poverty experienced by many workers who produce commodities in the South, few would argue in favor of preserving the *status quo*. Preservation of

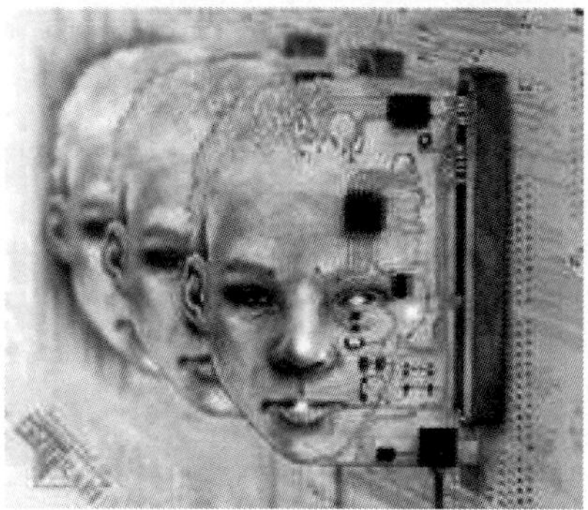

Figure 1. Artist: Reymond Pagé.

[1] See http://www.ars.usda.gov/research/projects/projects.htm?ACCN_NO=408518

the status quo is not the issue. The immediate and most pressing issue is that nanotechnologies are likely to bring huge socio-economic disruptions for which society is not prepared.

Societal impacts of nanotechnology go beyond issues of livelihood and justice for marginalized communities in the global South. The explicit goal of using nano-scale technologies for enhancing human performance also raises broad social and ethical concerns. According to a seminal workshop on technology convergence hosted by the U.S. National Science Foundation in December 2001, the ultimate goal of manipulating bits, atoms, neurons and genes is to enhance the physical and cognitive capabilities of humans, both individually and collectively. For example, research on mapping the human brain is accelerating rapidly. The Blue Brain Project at the Swiss Federal Institute of Technology is attempting to use computer-based simulations to reverse-engineer the brain at the molecular and cellular level (Stix 2008). Neural implants are being developed to treat tremors, paralysis, psychiatric disorders and memory loss. In 1997 the U.S. Food and Drug Administration (FDA) approved a neural implant for clinical use that replaces neurons damaged by Parkinson's Disease[2]. In February 2009 the FDA approved a brain implant for treatment of obsessive-compulsive disorders (Vedantam 2009). According to one estimate, more than 35,000 people have already received neural implants around the world (Harding 2009). For now, these devices are being used to treat people who are suffering from disease. With future advances in brain-machine interface technology, however, it will likely become harder to distinguish what is deemed a therapeutic use, and what is "less-than-optimal performance" that needs to be corrected. The potential use of neural technologies to control thought, emotions and memory could have profound implications for the future of democracy and dissent. Biochemist and disability rights activist, Gregor Wolbring, warns that human performance enhancement technologies will create an "ability divide" between the enhanced and unenhanced—resulting in new groups of marginalised people (Wolbring 2006).

---

[2] See http://www.fda.gov/bbs/topics/ANSWERS/2002/ANS01130.html

> *As much as human enhancement technology will become an enabling technology for the few, it will become a disabling technology for the many...if we go on as we are today we will see the appearance of a new underclass of people—the unenhanced* (Wolbring 2006, p. 126).

Historians often refer to major technology introductions as 'technology waves'. In the case of BANG technologies, however, it may be more of a 'high-tech tsunami.'

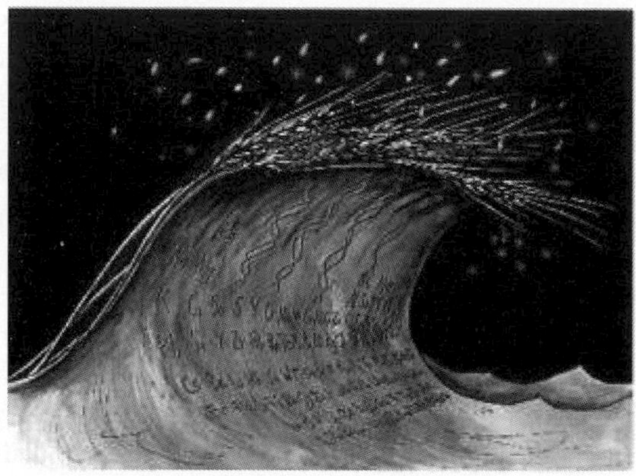

Figure 2. Society is not prepared for the technological tsunami approaching. If current trends continue, nanotech threatens to widen the gap between rich and poor and further consolidate power in the hands of multinational corporations. Artist: Reymond Pagé.

2.2 The threat of monopoly control over the building blocks of nature

Worldwide, governments and industry invested US$13,500 million in nanotech R&D in 2007 (Lux Research 2008). Global corporate and venture capital spending on nanotech R&D surpassed government spending for the first time (Lux Research 2008). According to nanotech industry analysts, the market for nanotechnology-based products is expected to reach US$3.1 trillion ($3,100,000 million) by 2015, up from

US$147,000 million in 2007 (Katz 2008). Although more than 60 countries have established national programs to support nanotech R&D, half of them are in Europe (ETC Group 2005a). The largest players are the EU, USA and Asia (primarily China, Korea and Japan). In 2005, the remaining countries accounted for only 4% of R&D investment (Lux Research 2005).

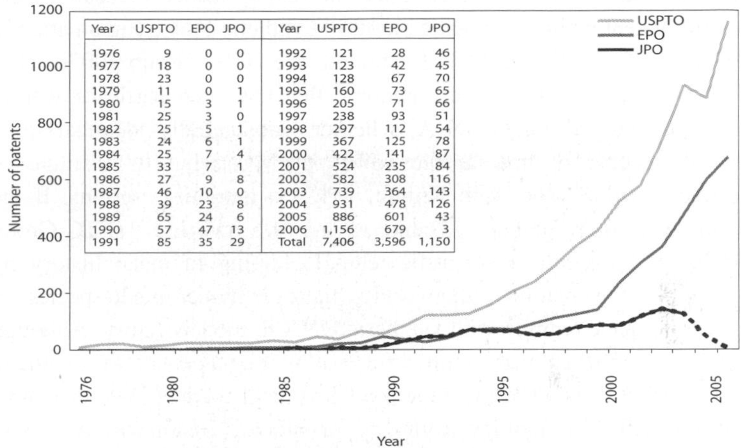

Year	USPTO	EPO	JPO	Year	USPTO	EPO	JPO
1976	9	0	0	1992	121	28	46
1977	18	0	0	1993	123	42	45
1978	23	0	0	1994	128	67	70
1979	11	0	1	1995	160	73	65
1980	15	3	1	1996	205	71	66
1981	25	3	0	1997	238	93	51
1982	25	5	1	1998	297	112	54
1983	24	6	1	1999	367	125	78
1984	25	7	4	2000	422	141	87
1985	33	7	1	2001	524	235	84
1986	27	9	8	2002	582	308	116
1987	46	10	0	2003	739	364	143
1988	39	23	8	2004	931	478	126
1989	65	24	6	2005	886	601	43
1990	57	47	13	2006	1,156	679	3
1991	85	35	29	Total	7,406	3,596	1,150

Figure 3 Number of nanotechnology patents published by the USPTO, EPO and JPO according to publication date. The drop in the number of USPTO patents in 2005 in due to the USPTO enforcing a striker definitions of nanotechnology. The decline in the number of JPO patents for 2005 and 2006 is due to the delay between the publication and granting of patents at the JPO. For colour reference turn to page 566.

Source: H. Chen., *et al.*, "Trends in Nanotechnology Patents," *Nature Nanotechnology*, Vol 3, March 2008, p. 123.

The world's largest transnational corporations, university labs and nanotech start-up companies are racing to win monopoly control of tiny tech's colossal market by aggressively seeking patents on nanotech's novel materials, devices and manufacturing processes. With nanotech, the reach of exclusive monopoly patents is not just on life, but all of nature. The current nanotech patent grab is reminiscent of the early days of biotech. Whereas biotech patents make claims on biological products and processes, nanotechnology patents may literally stake claims on chemical elements, as well as the compounds and the devices that incorporate them. Molecular-level manufacturing provides new opportunities for sweeping monopoly control over both animate and inanimate matter. Just as we've seen with biotech and other new technologies - intellectual property will play a major role in determining who will gain access to nano-scale technologies, and what price they will pay.

Concerns about 'extreme monopoly' extend to synthetic biology. In December 2007 ETC Group revealed that Craig Venter's research team is filing patent claims on a wide swath of synthetic biology in an attempt to become the 'Microbesoft' of synthetic life (ETC Group 2007b). The claims extend to virtually any genome that has been partly or wholly modified using synthetic DNA, whether 'substantially identical' to a natural genome or not. In the words of Massachusetts Institute of Technology professor, Tom Knight, "This is extremely serious. If the claims were to be granted, it's like saying 'we own life'" (ETC Group 2007b, p. 1). Venter's scientific team is hoping to make history by announcing the creation of the world's first-ever human-made species – a bacterium made entirely with synthetic DNA in the laboratory. Although Venter's team has already applied for worldwide patents on the synthetic bacterium, it remains a theoretical achievement to date. Venter claims that his synthetic biology company, Synthetic Genomics, Inc., will

Figure 4. Artist: Reymond Pagé.

ultimately create commercial microbes that produce drugs, chemicals and fuels. In 2007 he told *Business Week*, "If we made an organism that produced fuel, that could be the first billion - or trillion-dollar organism. We would definitely patent that whole process" (Sheridan 2007). The World Intellectual Property Organization published the Venter Institute's patent application claiming a hydrogen-producing bacterium in late November 2008.

2.3 Hazards to human health, safety and the environment

Despite the fact that hundreds of consumer products containing engineered nanoparticles are already on the market, there are no labeling requirements, no worker protection protocols and no government in the world has adopted comprehensive, mandatory regulations governing nanotechnology. Today, it is widely accepted that there are major gaps in research related to toxicology, health and environmental effects of engineered nanomaterials. In the early years of the 21st century, however, industry, scientists and governments were reluctant to validate any critique that might slow the race to commercialise nanotech-based products.

3. Civil Society Sounds the Alarm

It was neither government regulators nor industry that first blew the whistle on the potential health and environmental hazards of nanoparticles. In 2002, ETC Group published a report on possible toxicity of engineered nanomaterials and called for a moratorium on release of manufactured nanoparticles until laboratory protocols were established to protect workers and until mandatory regulations were in place to protect consumers and the environment (ETC Group 2002). A number of other civil society organisations have endorsed the call for a moratorium, including Greenpeace, the International Center for Technology Assessment and Friends of the Earth Australia.

Figure 5. Artist: Reymond Pagé.

In response to ETC Group's call for a moratorium, the nanotech industry's U.S. trade journal, *SmallTimes*, featured several editorial articles attacking ETC Group as a "merry band of miscreants" with "avowed Maoist sympathies." (Modzelewski 2003, p. 8). ETC Group staff were described as a "militant" group that attended the 2002 World Summit on Sustainable Development in Johannesburg "to distribute fliers and fear about the potential unintended consequences" of nano-scale technologies (Herrera 2003, p. 16). The then-executive director of the NanoBusiness Alliance, emphasised that ETC Group "is not run by a scientist... Their bizarre beliefs seem to be driving their attacks on legitimate science and social advances to the detriment of all of us" (Modzelewski 2003, p. 8).

4. Scientific Evidence Grows

The body of scientific evidence supporting health and environmental risks associated with engineered nanomaterials continues to grow. In 2004, for instance, the UK's Royal Society and Royal Academy of Engineering advised: "Until more is known about their environmental impact we are keen that the release of nanoparticles and nanotubes to the environment is avoided as far as possible" (Royal Society 2004, p. 46). Also in 2004, a report prepared for the European Parliament concluded: "Release of nano-particles should be restricted due to the potential effects on environment and human health" (Haum *et al.* 2004, p. 38). In 2006 the Quebec-based scientific research organisation, Institut de recherche Robert-Sauvé en santé et en sécurité du travail (IRSST), published an in-depth review of health effects of nanoparticles, concluding that "documented toxic effects on animals and the physicochemical characteristics of nanoparticles justify immediate application of all necessary measures to limit exposure and protect the health of potentially exposed individuals" (Ostiguy *et al.* 2006, p. 43).

Despite high-profile warnings, most governments continue to act as enthusiastic cheerleaders – not regulators – for nanotech R&D, and over 800 nanotech consumer products have come to market[3] in the absence of toxicological research and government oversight. In addition to increasing public awareness and highlighting the pressing need for a larger societal debate, the early call for a moratorium by civil society organizations was (and remains) a legitimate response to the regulatory vacuum that continues to put the public, researchers and workers at risk.

[3] See http://www.nanotechproject.org/inventories/

The growing body of scientific evidence about potential risks of engineered nanoparticles is prompting renewed calls for regulation and oversight. Two studies published in 2008 confirm that some forms of carbon nanotubes, with characteristics similar to asbestos fibres, are dangerous and can potentially cause lung disease if inhaled (Poland *et al.* 2008; Shvedova *et al.* 2008). In September 2008, 70 governments, 12 intergovernmental organisations, and 39 nongovernmental organisations meeting at the VI Intergovernmental Forum on Chemical Safety in Dakar, Senegal agreed that countries should have the right to refuse imports of manufactured nanomaterials and that these materials should be labeled (International Forum on Chemical Safety 2008). In November 2008 the UK's Royal Commission on Environmental Pollution warned, "the pace at which new nanomaterials are being developed and marketed is beyond the capacity of existing testing regulatory arrangements to control the potential environmental impacts adequately" (Royal Commission on Environmental Pollution 2008, p. 1).

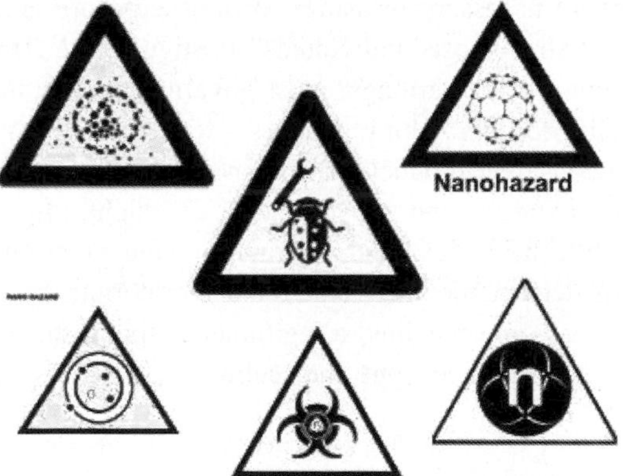

Figure 6. In September 2006 ETC Group launched an *International Nano-hazard Symbol Design Competition* to capture public attention and spark debate on the potential impacts of nanotechnology. Six sample entries appear above. For colour reference turn to page 566.

Concerns about nano-scale technologies are not limited to engineered nano-particles. In the rapidly advancing field of synthetic biology, for example, scientists have already built complete working viruses using synthetic DNA – including dangerous pathogens such as the poliovirus (Pollack 2002), the 1918 flu virus (Kolata 2005) and the SARS virus (Keim 2008). Whether by deliberate misuse or as a result of unintended and unpredictable consequences, will scientists be able to control or contain these new life forms? Synthetic biology has the potential to introduce new and potentially catastrophic societal risks, yet it is barely on the public radar.

5. Techno-Optimists Respond

As growing concerns about environmental health and safety risks associated with nano-scale science appear more prominently in the media, the nanotech industry has largely abandoned its early, clumsy and defensive (and factually inaccurate) attacks on nanotech naysayers. In response to heightened concerns, some industry and governments have adopted more sophisticated strategies designed to "manage public acceptance" and preempt concerns voiced by civil society critics[4]. These strategies include, for example:

1. Image make-over.
2. Promoting voluntary, self-governance
3. Fostering public engagement to manage public acceptance

5.1 Image make-over (green, clean and pro-poor)

Although nano-scale technologies are heralded as new and revolutionary, the way in which they are being introduced is reminiscent of earlier technologies. Beginning in the late 1980s, for example, the biotechnology industry promised that genetically engineered crops would feed hungry people and clean up the environment. Today, governments,

[4] It should be noted that just as civil society is not a monolithic entity, it is also true that there are many different views represented within industry and governments

corporations and international institutions advocate for environmental and "pro-poor" nanotech applications such as solar power, water clean-up or cheap vaccines. European governments have identified nanotechnology as an important tool for achieving the United Nations Millennium Development Goals. In the hyperbolic words of one industry CEO:

> *Around the world, [ ] the brightest minds of this generation are focusing on nanotechnology as the solution to our very biggest global problems. The goal is nothing less than using nanotechnology to bring the basic necessities of life to every corner of the globe, raising the standard of living for every citizen of the world, and helping facilitate world peace with more universal abundance* (Rickert 2008).

Five years ago, the 'nano' prefix was popular and trendy, and manufacturers were eager to capitalise on the 'nano' label. With increased media coverage on the possible toxicological risks, however, 'nano' has suddenly become a liability for some companies (Singer 2008). In response, many nano-enthusiasts adopted the motto of 'responsible nanotechnology' and have tried to re-introduce nano-scale technologies as 'clean tech'. The 'clean tech' and 'pro-poor' labels are designed to win moral legitimacy for the technologies and promote public acceptance.

5.2 Push for voluntary self-governance

Industry and governments have pushed for voluntary or self-governance of nano-scale technologies as a way to pre-empt public debate, avoid government scrutiny, or postpone calls for rigorous, mandatory regulations. In recent years industry and governments have supported various initiatives such as 'codes of conduct', voluntary reporting schemes and self-governance proposals for nano-scale technologies. The UK government, for example, adopted a voluntary scheme to encourage companies to report the use of nanomaterials and to make available risk assessment information. Few companies are participating. In the US,

DuPont and the Environmental Defense Fund developed a voluntary "Nano Risk Framework" that was flatly rejected by a coalition of civil society and labour organisations as 'fundamentally flawed' because it reflects voluntary standards dictated by industry, rather than government regulatory systems to protect the public (International Center for Technology Assessment 2007).

In May 2006, 38 civil society organisations signed an Open Letter urging the synthetic biology community to reject its proposals for self-governance (ETC Group 2006). The letter from civil society urged the scientists to participate in a process of open and inclusive oversight. In the words of one signatory, Dr. Sue Mayer of GeneWatch UK, "Scientists creating new life forms cannot be allowed to act as judge and jury. The possible social, environmental and bio-weapons implications are all too serious to be left to well-meaning but self-interested scientists. Proper public debate, regulation and policing is needed" (ETC Group 2006, p. 1).

5.3 Fostering public engagement to manage public acceptance

A growing number of governments and industry advocates recognise that successful commercialization of controversial nano-scale technologies will ultimately require regulatory oversight of nano-scale technologies, which they view, first and foremost, in the framework of science-based risk assessment (that is, action is only taken when there are scientific reasons for concern). Unfortunately, the emphasis on health and safety issues and science-based risk assessment also narrows the discourse on new technologies to an elite group of science 'experts' and usually avoids discussion of larger societal impacts (control and ownership). In recent years there has been considerable interest in public dialogue and the concept of 'upstream engagement' involving, for example, citizen juries or 'nano cafés'. However, the focus on public dialogue and engagement is often driven by the goal of protecting commercial interests and insuring national competitiveness. Instead of protecting the public interest and fostering a genuine debate, public engagement exercises are used to manage public acceptance for the technology. Often

Examples of NanoActions

In recent years, civil society and social movements have launched a number of campaigns to encourage public debate, draw attention to potential risks associated with nano-scale technologies, reject voluntary regulations and demand government action. The following are just a few examples:

- June 2003 – ETC Group, in collaboration with The Greens/EFA in the European Parliament, Dag Hammarskjold Foundation, Greenpeace, Genewatch UK, Clean production Action and the Ecologist Magazine held a day-long seminar in the European Parliament on the impacts of nano-scale technologies. http://www.etcgroup.org/en/materials/publications.html?pub_id=156
- May 2006 – 38 civil society organisations signed an Open Letter urging the synthetic biology (nanobiotechnology) community to reject proposals for self-governance. http://www.etcgroup.org/en/materials/publications.html?pub_id=8
- September 2006 – ETC Group launched an *International Nano-hazard Symbol Design Competition* to capture public attention and spark debate on the potential impacts of nanotechnology. The competition netted 482 unique designs from 24 countries. The winners of the competition were announced at the World Social Forum in Nairobi in January 2007. http://www.etcgroup.org/nanohazard
- March 2007 – the International Union of Food, Agricultural, Hotel, Restaurant, Catering, Tobacco and Allied Workers' Associations (IUF) became the first trade union to call for a moratorium on nanotech products in food and agriculture. www.iuf.org
- July 2007, a broad international coalition of civil society, public interest, consumer, environmental and labor organisations called for strong precautionary oversight for nanotechnologies which, as yet, are not subject to mandatory regulation anywhere in the world. The coalition's oversight principles have been endorsed by over seventy organisations worldwide. http://www.icta.org/press/release.cfm?news_id=26
- January 2008 – the UK-based Soil Association – one of the world's pioneers of organic agriculture – announced that it had banned human-made nanomaterials from the organic cosmetics, foods and textiles that it certifies. http://www.soilassociation.org/web/sa/saweb.nsf/librarytitles/1EC7A.HTMl/$file/Nanotechnology.pdf
- May 2008 – the International Center for Technology Assessment filed a legal petition challenging the U.S. Environmental Protection Agency's failure to regulate nanomaterials in pesticides. http://www.icta.org/nanoaction/doc/CTA_nano-silver_press_release_5_1_08.pdf
- June 2008 – the European Trade Union Confederation (ETUC) adopted a resolution demanding that the precautionary principle be applied to nanotechnologies and nanomaterials. http://www.etuc.org/a/5159?var_recherche=nanotechnology

the starting place for public dialogue/discussion is that the potential benefits of nano-scale technologies outweigh the risks. Therefore, the question of whether products of nano-scale technologies are necessary or desirable, or the possibility of rejecting the technology outright is seldom addressed.

6. Conclusion: Towards a New Era of Technology Politics/Governance

Given the wide-ranging scope of nano-scale converging technologies and the speed with which they are being developed, it is neither practical nor possible to respond to each emerging technology (biotech, nanotech, synthetic biology) in a piecemeal fashion as if they were discreet technologies. Many civil society activists are discovering that the same issues they faced when campaigning against GMOs, for example, are resurfacing with BANG technologies. In a globalised world, nano-scale converging technologies are highly decentralised and their impacts are trans-boundary. Governance options must therefore be debated in an international framework. To overcome the cycles of crises that accompany each new technology introduction, the international community needs an independent body dedicated to assessing major new technologies on an ongoing basis that can provide an early warning/early listening system. One possibility put forward by ETC Group is the establishment of an intergovernmental facility that could be called the International Convention on the Evaluation of New Technologies – ICENT (ETC Group, 2005b). The objective of ICENT would be to create a socio-political and scientific environment for the timely evaluation of new technologies in a participatory and transparent process that supports societal understanding, encourages social and scientific innovation, and facilitates equitable benefit-sharing. Further, the inter-governmental framework would also ensure the conservation of useful, conventional or culturally distinct technologies and also promote technological diversification and decentralization. Most importantly, the process of United Nations negotiations to develop an international agreement such as ICENT would stimulate high-level and broad societal discussion, encourage national and regional legislative and institutional initiatives that would complement an international agreement for assessing new technologies.

Ultimately, ICENT is only one part of much-needed innovation and reform in technology politics and governance. Civil society is now exploring broader, more inclusive mechanisms for promoting social control of emerging technologies. Concerns about technology governance go beyond the desire to 'regulate' a new and powerful technology. The issue is not simply how to regulate a risky technology, but how to decide whether it is socially acceptable or desirable. Should its development be publicly supported? Should the technology proceed at all? Are other technological approaches preferable? Civil society organisations and social movements are advancing ideas to fundamentally transform and democratise science and technological research, and to promote decentralised production and people-oriented technologies.

Nano-scale converging technologies are being unleashed in the context of converging crises. In the midst of collapsing ecosystems, climate chaos and global financial meltdown, the pressure to embrace techno-fix solutions has never been greater. For example, U.S. President Barack Obama has pledged to establish a US$150 billion "Apollo Project" for energy independence, and his new Energy Secretary, Nobel Prize laureate Steven Chu, believes that synthetic biology is key to achieving that goal. At the University of California Berkeley, Chu was responsible for overseeing the creation of the world's largest public-private partnership, a US$500 million R&D deal with oil giant BP, to promote energy production via biotechnology and synthetic biology. The partnership envisions a future where engineered microbes will be used to extract and ferment the sugars in biological feedstocks (agricultural crops, grasses, forest residues, plant oils, algae, etc.) and convert them into fuels. Corporate agribusiness, energy and chemical giants are also investing in synthetic biology platforms to produce high-value fuels, pharmaceuticals, chemicals and plastics. Theoretically, any chemical made from the carbon in oil could be made from the carbon in plants. It may sound green and clean, but it will be the catalyst for a corporate grab on all plant matter – and destruction of biodiversity on a massive scale

(ETC Group 2008). Promoting nano-scale technologies as a 'painless' techno-solution for addressing the challenges of climate change and peak oil may be politically expedient, but it is dangerous and counter-productive. It threatens to shift the debate away from the most urgently needed policy measures – such as cutting greenhouse gas emissions and promoting radical changes in consumption patterns (especially in the OECD countries).

Figure 7. ETC Group uses this cartoon to illustrate the post-petroleum sugar economy as envisioned by synthetic biologists and corporate investors. Artist: Stig. For colour reference turn to page 567.

Civil society organisations from five continents convened in late November 2008 outside of Montpellier, France to discuss long-term strategic planning on nano-scale converging technologies—BANG—in the context of converging crises. They concluded that nano-scale converging technologies are being introduced in blatant violation of the Precautionary Principle and rejected the infallibility of science-based decision making – seeing it as a limited, easily manipulated and exclusionary form of knowledge that seeks to focus on safety issues and risk assessment – while side-stepping the issues of control and ownership of new technologies. Converging technologies are being propelled by the pursuit of profits, not human development needs and social justice.

Although participants at the meeting came from very diverse backgrounds, there was agreement that new technologies are never a 'silver bullet' for resolving old injustices. Hunger, poverty and environmental degradation are the consequences of inequitable systems – not inadequate technologies. Furthermore, technologies are *not* neutral. They are inextricably linked to the power and ideologies of those who develop and control them. Although they may be promoted in the name of fighting poverty, arresting climate change and curing disease, nano-scale technologies that reinforce corporate power threaten to deepen existing inequalities, accelerate environmental degradation and introduce new societal risks. Civil society discussions will continue on a larger scale at the 2009 World Social Forum in Belem, Brazil. It is the first time that the World Social Forum will dedicate a full day to discussion and debates on science, society and democracy—but it is just the beginning of a new era in technology politics[5].

Questions for Reflection:

1. Why might it be useful to understand nanoscale technologies within the larger context of BANG?
2. Imagine that to protect public health and the environment from potential risks, a moratorium was placed on the commercial release of nanotechnology products. What would it take for this to be lifted?
3. What role do civil society organisations have in the governance of nanoscale technologies?

---

[5] I gratefully acknowledge the work and contributions of ETC Group staff to this chapter.

Bibliography

Beck, A. (2007). Unknown health impact of nanotech worries some, *Reuters*, November 14. http://www.reuters.com/article/scienceNews/idUSN1221261620071114.

ETC Group (2002). No Small Matter! Nanotech Particles Penetrate Living Cells and Accumulate in Animal Organs, *Communiqué* # 76, May/June 2002. http://www.etcgroup.org/en/materials/publications.html?pub_id=192 (last accessed May 28, 2009).

ETC Group (2005a). *The Potential Impacts of Nano-Scale Technologies on Commodity Markets: The Implications for Commodity Dependent Developing Countries*, Research Paper published by the South Centre, November 2005.

ETC Group (2005b), *Special Report – NanoGeoPolitics: ETC Group Surveys the Political Landscape NanoGeoPolitics*, July/August 2005. http://www.etcgroup.org/en/materials/publications.html?pub_id=51.

ETC Group (2006). *News Release, Global Coalition Sounds the Alarm on Synthetic Biology, Demands Oversight and Societal Debate*, May 19. http://www.etcgroup.org/upload/publication/8/01/nr_synthetic_bio_19th_may_2006.pdf.

ETC Group (2007a). *Extreme Genetic Engineering: An Introduction to Synthetic Biology*.

ETC Group (2007b). *News Release, Extreme Monopoly: Venter's Team Makes Vast Patent Grab on Synthetic Genomes*, December 8, 2007. http://www.etcgroup.org/en/materials/publications.html?pub_id=665 .

ETC Group (2008). Who Owns Nature? Corporate Power and the Final Frontier in the Commodification of Life, *Communique*, #100, November 2008. http://www.etcgroup.org/en/materials/publications.html?pub_id=707 (last accessed May 28, 2009)

Harding, A. (2009). Brain Implant Better than Meds for Parkinson's Disease, *CNN online*, January 6. http://www.cnn.com/2009/HEALTH/01/06/parkinsons.deep.brain.stimulation/index.html

Haum, R., Petschow, U. and Steinfeldt, M. (2004). *Nanotechnology and Regulation within the Framework of the Precautionary Principle*, Final Report for ITRE Committee of the European Parliament, Institut für ökologische Wirstschaftforschung (IÖW) gGmbH, Berlin, February 24.

Herrera, S. (2003). Swatting at gadlies: Anti-nano celebs succeed in made-for-media provocation, *SmallTimes*, Vol. 3, No. 2, March/April, p. 16.

International Center for Technology Assessment (2007). *News Release, Civil Society-Labor Coalition Opposes Voluntary NANO Risk Framework of Environmental Defense and DuPont*, April 12, 2007. http://www.icta.org/detail/news.cfm?news_id=120&id=218.

International Forum on Chemical Safety (2008). Forum VI was held 15-19 September 2008 in Dakar, Senegal. http://www.who.int/ifcs/forums/six/ en/index.html.

Katz, J. (2008). Nanotechnology Boom Expected by 2015, *Industry Week,* July 22. http://www.industryweek.com/ReadArticle.aspx?ArticleID=16884&SectionID=4 (last accessed May 28, 2009)

Keim, B. (2008). Synthetic Viruses Could Explain Animal-to-Human Jumps, *Wired Science Blog*, November 24. http://blog.wired.com/wiredscience/2008/11/synthetic-virus.html#more.

Kolata, G. (2005). Experts Unlock Clues to Spread of 1918 Flu Virus, *New York Times*, October 6. http://www.nytimes.com/2005/10/06/health/06flu.html?scp=1&sq=1918%20flu%20virus%20synthesized&st=cse (last accessed May 28, 2009)

Leo, A. (2002). The State of Nanotechnology; Coming Soon: Nanoelectronics for infotech and medicine, *Technology Review*, June. http://www.technologyreview.com/computing/12862/?a=f.

Lux Research, Inc. (2004). *The Nanotech Report, 2004*, Vol. 1, New York.

Lux Research, Inc. (2008). *Overhyped Technology Starts to Reach Potential: Nanotech to Impact $3.1 Trillion in Manufactured Goods in 2015*, Press Release, July 22. http://www.luxresearchinc.com/press/RELEASE_Nano-SMR_7_22_08.pdf.

Modzelewski, M. (2003). Militant group's anti-nano manifesto a bizarre blend, *SmallTimes*, Vol. 3, No. 2.March/April, p. 8

Ostiguy, C., Soucy, B., Lapointe, G., Woods, C. and Ménard, L. (2006). *Health Effects of Nanoparticles, Robert-Sauvé en santé et en sécurité du travail (IRSST), Studies and Research Projects / Report R-469*, Montréal. http://www.irsst.qc.ca/files/documents/PubIRSST/R-469.pdf, http://www.irsst.qc.ca/en/_publicationirsst_100209.html (last accessed May 28, 2009).

Poland, C., Duffin R., Kinloch, I., Maynard, A., Wallace ,W., Seaton, A, Stone, V., Brown, S., MacNee, W. and Donaldson, K. (2008). Carbon nanotubes introduced into the abdominal cavity of mice show asbestos-like pathogenicity in a pilot study, *Nature Nanotechnology* 3, 423-428.

Pollack, A. (2002). Traces of Terror: The Science; Scientists Create a Live Polio Virus, *New York Times*, July 12.

Rickert, S. (2008). Taking the NanoPulse – Nanotechnology. It Makes a World of Difference, *Industry Week*, November 20.

http://www.industryweek.com/ReadArticle.aspx?ArticleID=17834 (last accessed May 28, 2009)

Royal Commission on Environmental Pollution (2008). *News Release, Urgent Action Needed on Testing and Regulation of Nanomaterials*, November 12. http://www.rcep.org.uk/novel%20materials/Novel%20materials%20press%20release.pdf.

Royal Society and Royal Academy of Engineering (2004). *Nanoscience and Nanotechnologies: Opportunities and Uncertainties*, Science Policy Section, July. http://www.nanotec.org.uk/finalReport.htm.

Sheridan, B. (2007). *Newsweek International*, 4 June, J. Craig Venter quoted in interview with Sheridan. http://www.newsweek.com/id/34352/page/2.

Shvedova, A., Kisin, E., Murray, A. Johnson, V., Gorelik, O., Arepalli, S., Hubbs, A., Mercer, R., Keohavong, P., Sussman, N., Jin ,J., Yin, J., Stone, S., Chen, B., Deye, G., Maynard, A., Castranova, V., Baron, P. and V. Kagan (2008). Inhalation vs. aspiration of single-walled carbon nanotubes in C57BL/6 mice: inflammation, fibrosis, oxidative stress, and mutagenesis, *American Journal of Physiology - Lung Cellular and Molecular Physiology,* 295: L552-L565, 2008. Published July 25, 2008. http://ajplung.physiology.org/cgi/content/abstract/295/4/L552.

Singer, N. (2008). New Products Bring Side Effect: Nanophobia, *New York Times*, December 3.

Stix, G. (2008). Jacking into the Brain – Is the Brain the Ultimate Computer Interface? *Scientific Amerian*, November. http://www.sciam.com/article.cfm?id=jacking-into-the-brain.

Vedantam, S. (2009). FDA Allows Brain Implants for Obsessions, *Washington Post*, February 24, p. HEO2.

Wolbring, G. (2006). The unenhanced underclass, Chapter 12. In Miller, P. and Wilsdon, J. (eds) *Better Humans: The Politics of Human Enhancement and Life Extension,* U.K.: Demos Collection 21, Demos Institute. [For more information about Wolbring's work, visit the website of the International Center for Bioethics, Culture and Disability: http://www.bioethicsanddisability.org/start.html]

Chapter 21

Moving In

Lachlan Atcliffe

Timothy Drennan is working from his hospital bed, an intravenous delivery implant in one arm and a stack of paperwork in the other.

He is a young man in his early thirties. The skin on his hands and around his eyes has a faint, glinting tracery of nano-lithographed tattoo-circuitry, only visible when he passes directly under light. His eyes have long since lost their natural colouring, now mandelbrot-patterned with quantum dots that transmit images directly to his retina. Wireless monitor stickers are just visible on his chest and arms, passing his vital signs to the little monitor box on the wall, and one is stuck to his forehead to check his brain activity. His active styling gel has moved his thin black hair out of its way to prevent interference.

It is not a very dignified place to work from, his modesty protected only by a thin sheet and a cordon of swirling, cloudlike curtains. But if he dwells too long on the silvery-red fluid running into him from a large tank sitting next to his bed, he tends to panic. So he is sitting up and trying not to think about it too much, his lap overflowing with rolls of little plastic flimsies—sheet computers, with the screen and touchpad on one side and a glittering pattern of circuitry on the other. He is taking his mind off his treatment by ploughing through cases ahead of next week's court hearings.

Outside of the porous windows with their built-in air conditioning, the sun is setting. The fading light leaks through the cloudy display curtains and the clinical white of the back wall of the hospital ward is bathed in a faint sheen of red and yellow light. By contrast, all the

Nano Meets Macro - Social Perspectives on Nanoscale Sciences and Technologies
by K L Kjølberg & F Wickson

www.panstanford.com
978-981-4267-05-2

material on Timothy's flimsies is black on white. Occasionally he reaches for the pair of glasses that sit on the metal tank next to him. Not to read better, but to use their net connection to check some point of law.

He sets his current case aside and picks up a new one from the pile in his lap, tagged *Personal Injury*. The flimsy's surface blinks into life as he unrolls it and lays it out flat. The claim form swims into focus, alleging Mister Anthony Goddard, of 59 Bishopgate Street negligently allowed his nanomotile fence paint to migrate into his neighbour's garden. There it allegedly overran René Theissen's prize-winning hedge and lawn. It also supposedly triggered allergic reactions in his wife and daughter when they breathed in the resulting dust.

Tim flicks his eyes at the 'OK' button to confirm the automatic address check, to see if 59 Bishopgate Street knows it is owned by someone of that name. Then he clicks the hyperlinks to the attached medical data. Yes, 57 Bishopsgate's medical systems dispensed atmospheric medication on the claimed date and time. Several pages of medical reports are attached. The claimant is demanding compensation for damage to his garden and the drugs dispensed by his house, as well as the cost of the disposable aerostat nanites used to deliver them. Oh, and pain and suffering of course.

First step, liability for substances escaping from one's property is trite law, dating back to the 19th century. Tim clears his throat to start dictating and launches into an explanation of modern tort law that his implanted throat-mike transcribes as he speaks. He dismisses the paint company's disclaimers with a scathing critique, noting they need only display the ashen remains of the garden and the logs of the soil nutrient-release system to disprove it. There is no evidence of intervening bacteria or mutation among the other active systems in the garden to cause damage on such a scale. Mr Goddard was clearly negligent in failing to paint a restrictive strip at the bottom of his fence panels to prevent just such an accident. Mr Goddard might claim an outside signal or interference caused his paint to go haywire, but if so it would be for him to prove that it existed. Since Mr Theissen himself has a nano-active household, however, he was entirely capable of intercepting any harmful dust or gases released when the plants were swarmed. The attached log files contain no mention of any attempts to order the out-of-control paint to stop or to deploy the house's immune system against it. Regretfully it is Timothy's opinion as counsel that any judge will take this into account when assessing damages and make deductions accordingly. Finally some closing remarks about valuation, and the assignment of an autoclerk to trawl the Theissen household accounts to produce proof the plants were worth the amount claimed - provided Mr Theissen consents, of course. If Mr Theissen wishes to claim for pain, suffering, and the emotional loss of seeing his garden destroyed, he will need to produce logs from his memories to prove the impact was as great as he claims. Sincerely yours *et cetera et cetera*, a wiggle of a fingertip to sign it and so much for that one. On to the next!

There is an email from Vivienne waiting in his glasses, sent yesterday before she packed herself up for transit. It is a few quick lines of txtspk, with a recorded message at the end. He presses the play button and for a moment his cheek throbs. The remnants of her lipstick play back their parting kiss, a romantic moment that makes him blush nostalgically and close his eyes. He glances at the tank next to his bed and smiles.

The next case is criminal defence work, a charge of theft of processing time on a tank of yoghurt. The tank's bacteria were allegedly hijacked by the disgruntled personality of a swarm-entity that used to work in the same building. Fifty megabytes of conflicting log files to sift through. Tim takes a deep breathe, this one is going to take some time…

It is well into the evening, when Tim is disturbed by a warble. It takes him a second to realise it is his visitor alert. Tim says "Come in!" cheerily, putting down his current casefile and turning it off in case his visitor is someone nosy. At his command, the curtain that surrounds him swirls opens. The hovering swarm of aerostats keeping themselves aloft on internal fans and tiny vacuum bubbles double as anti-contamination watchdogs, lest the chemicals in his intravenous drip get out and cause compensation claims. In this case he is willing to take the risk.

His visitor is a tall young woman in a business suit, with dark skin, even darker hair, and a French accent.

"Hi Tim."

"Hi Ally." Alison sits two desks away at *Fuller & Partners*, and has obviously been sent by the department's well-wishers. She has the obligatory bunch of flowers and congratulation cards tucked under one besuited arm and she looks around for a flat surface to put them on.

"Try over there, maybe?"

"Okay..." Alison gingerly sets the bouquet down on the trolley next to the hospital bed, trying not to knock any of the wires. The flowers are nice, a vivid electric blue with sacs of perfume in the base emitting a faintly soporific smell. The cards are less subtle and blast their messages into Tim's glasses as he looks at them. They loudly play back recordings of his colleagues toasting his health or wishing him luck. He thinks "Pause" at them, and frustratedly grinds his teeth when their cheap circuitry doesn't respond. Alison, meanwhile, is heading straight for his pile of completed flimsies. Ahh priorities…

"Everybody wanted to drop by, but you weren't answering the conference calls." She smirks at the curtain. "Guess that's why, huh?"

"Could be." Tim specifically asked for a room with something that could block phone signals. "I'm getting network reception, but maybe it blocks calls? They don't want to have patients disturbed, I guess."

Alison nods knowingly, one junior caseworker to another, as she rolls up the finished flimsies and tucks them into a spare pocket in her suit jacket. Their charge lights come on as her suit notices their batteries are low.

"Viv been by yet? How is she?"

"Vivienne's already..." There is a nurse-like cough from the curtain, a polite hint from wherever the building's staff hang out, which these days, could be anywhere. Alison glances around in search of the speaker, and represses her embarrassed giggle at being caught.

"Sorry, I can't stay. It's not visiting hours, I only got five minutes to drop off personal items. They wanted me to template them and get them reassembled in here, but I put my foot down." Tim laughs at the thought of a bouquet suddenly forming mid-curtain, although he knows it's nonsense. There must be a printer somewhere on the ward.

"Thanks for coming Ally. It was nice to see somebody. See you next week."

"Don't hurry back! You've got time off, use it. And tell Viv I said hello!"

Tim sighs as he is left alone again, except for the substrate being pumped into him. The drug dispenser is starting to ache, and he can't turn the pain off because his immune system's boosts are shut down for the treatment to take effect. Should he whistle for a nurse's attention? No, no, he'll be determined and ignore it.

He looks at the curtains with his glasses and tells them to close and become transparent. Obediently, the aerostats shiver slightly in mid-air, and ripple from cloudy to clear, giving Tim a view through to the windows and outside. Night has fallen and the ward's walls are glowing softly, their paint rich with impregnated glowdust. The view in daylight, from here on the 215th floor, is a magnificent panorama of huge aerofoil towers poking out through the cloud layer. At night the towers are dark but the lights from the city below bloom within the clouds, giving the impression of flying above a huge glowing blanket. Occasional lights blink from the tower's warning beacons, or microgliding commuters flitting from tower to tower.

The young lawyer glances at his wrist, where the active inks in his watch tattoo tell him it is past 8 and the nurses will be enforcing the data curfew soon. Viv should be able to chat to him after that, but a hospital bed does not a romantic evening make. Even if she does project better scenery onto his eyeballs, it's still hard to forget his location. Oh well. Time for one more case.

The next file is a referral from the civil claims team. Arbitration of a commercial dispute between two companies blaming each other for the loss of an extremely valuable industrial manufacturing printer. It seems no-one remembered to refill the substrate reservoirs over a weekend. Of course nobody bothered to consider that nanites also obey the laws of thermodynamics and need some form of fuel.

The glasses blink the "Incoming Call" icon. How – oh yes, he'd told the curtains to go transparent. He'd have to remember that opened the phone lines as well. The caller ID tells him it's Max, the guy one desk away back at *Fuller & Partners*.

"Hi Max, c'est Tim."

"Ah, Timothy! Your French is as fake as ever, I see."

Tim's glasses project tiny lasers to the quantum dots in his iris, fooling his optic nerves into thinking Max is standing roughly where the curtains are in front of him. The entire back wall seems to silently fade into Max's office. He must be working late. Tim waves twice with his free hand – once to Max, once to Max's security plant, a small potted rosebush he keeps on his desk ever since that unfortunate business with the angry defendant who smuggled bits of himself past security inside a water cooler tank. If threatened, the plant blooms and fills the air with a cloud of disinfectant. Viv, needless to say, hates the thing.

"Just heard, and wanted to say congrats and all. When are you getting the ring?"

Tim fights back a blush and musters up a smile.

"Thanks. We were going with my parents' one, they kept the template"

"Ah, the old reconstructed heirloom trick....how traditional. I'll spare you the diamond jokes and just say best wishes."

"Heh. Thanks. Listen, Viv'll be able to talk in a bit if you want to say hi."

"No, no, don't let me get in the way. I'll get one of your agents to sign any letters you've left."

"Might be one on my desk…"

"Good, good." The main part of his call finished, Max nods to his friend's saline drip, switching from friendliness to concern.

"Your medical stuff there. Nothing too bad, I hope? Hang-gliding accident?"

"Hmm, this?" Tim smiles at the tank's inspection port, where the occupant bubbles back at him affectionately. The thick, syrupy substrate mix in his intravenous drip slurps into his bloodstream a little faster, sped on by the millions of tiny machines animating it.

"No, this is all Vivienne's. What did you think I meant when I said she was moving in with me?"

Questions for Reflection:

1. What aspects of our daily life will be changed by nanotechnological development and what will always remain the same?
2. How might the use of nanotechnology consumer products create legal issues?
3. To what extent can your identity be separated from your biological body?

Quoiazander

Leda C. Sivak

This painting speaks to the relationship between humans, technologies and environments. Although the painting is in no way directly 'about' Indigenous peoples, it is influenced by a deep respect for the sustainable relationships with technologies that are possible when these are developed and performed within contexts shaped by awareness of universal interconnection, interdependence and proportional consequence. The transformative serpent – be it imagination, adaptation, destruction, intervention, displacement, creation – necessarily holds potential for disruption.

Leda Sivak is an anthropologist who presently works in Central Australia. Her Masters thesis explored relationships between sciences and cultures within the context of the public debate about genetic modification in New Zealand.

Question for Reflection:

What is the relationship between humans, technologies and environments in your cultural context, and how may nanotechnologies affect this?

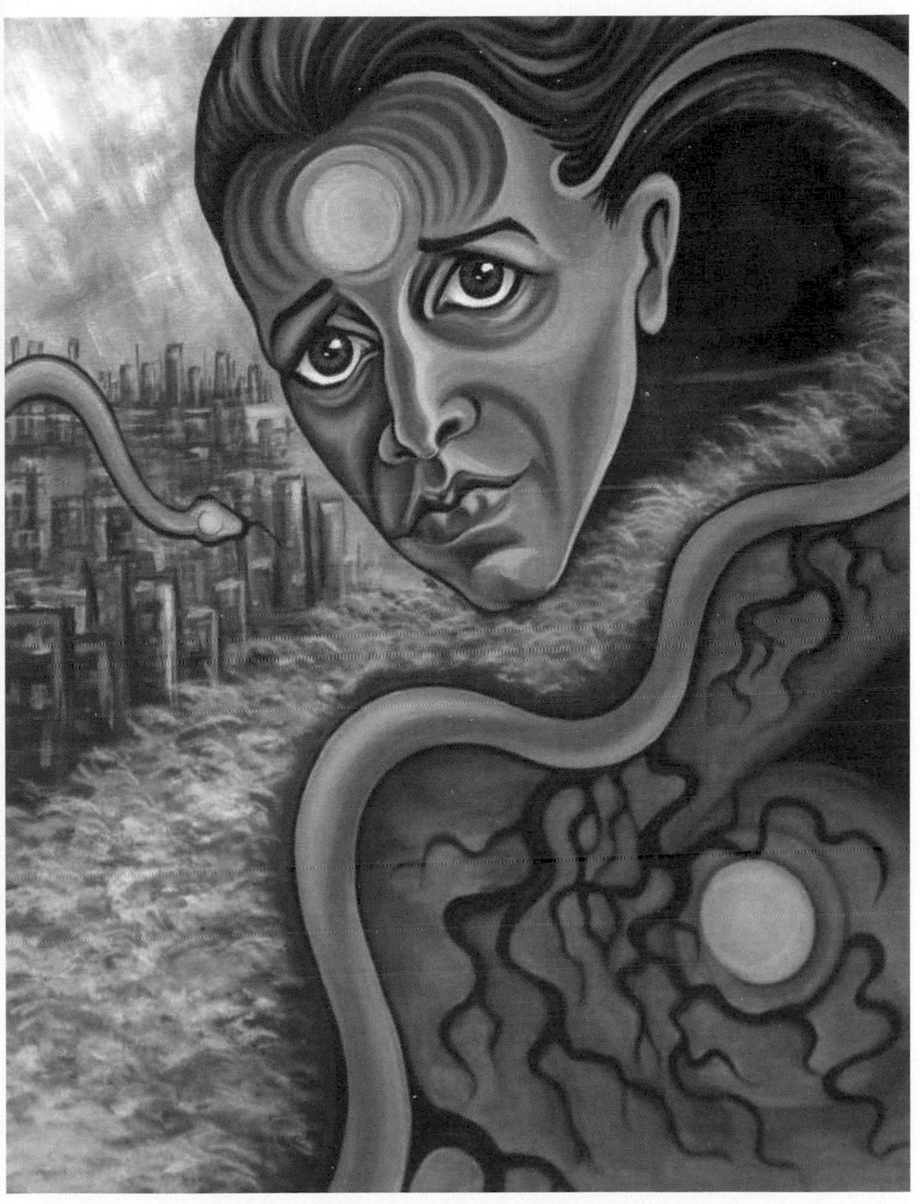

For colour reference turn to page 555.

Discovery,

David Hylton

I find nanotechnology fascinating as it is on the verge of completely reshaping our world. It also offers artists new insights on an aspect of nature that would otherwise be left unseen. My image, "Discovery," provides the viewer with a metaphorical glimpse of the myriad possibilities inherent in the exploration of nanotechnology and its applications. Through the translucent egg we see a lush, mysterious world shimmering in early-morning light. A winged creature is stirring and slowly rising toward a small fissure in the side of the shell. Intellectual curiosity and the desire for discovery compel us to anticipate the opening of the pod, even while we are fully aware of risks inherent in the exploration process. This piece was originally featured as part of NanoArt 21 2008.

David Hylton is an award-winning artist and his artwork has been featured in numerous international, national and regional exhibitions. For example, in the 'Siggraph Traveling Art Show' (France, USA); 'The Second International Festival for NanoArt' (Germany); 'Field of Vision: Extremes' (Germany); 'InterFaces New Media Art Exhibit' (Malaysia); 'The Melbourne Digital Fringe Festival' (Australia). In addition, his artwork has been included in *The History of Computer Graphics and Digital Art Project.* Hylton is an Associate Professor at California Polytechnic State University, Pomona.

Question for Reflection:

What is the balance between fear and curiosity in your approach to the advance of nanotechnology? Do you want to break the egg?

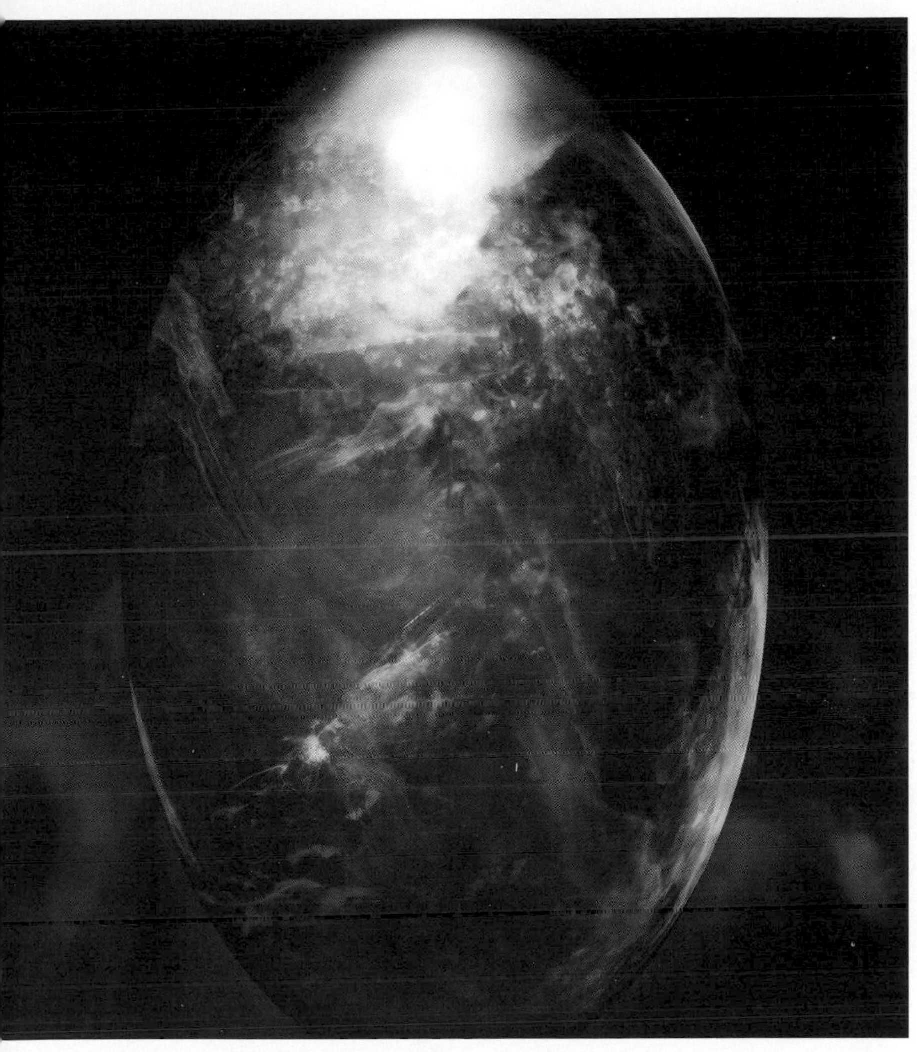

For colour reference turn to page 556.

Index

Colour Index

Artworks

Eigler's Eyes 2

Triangular Masterpiece No. 5

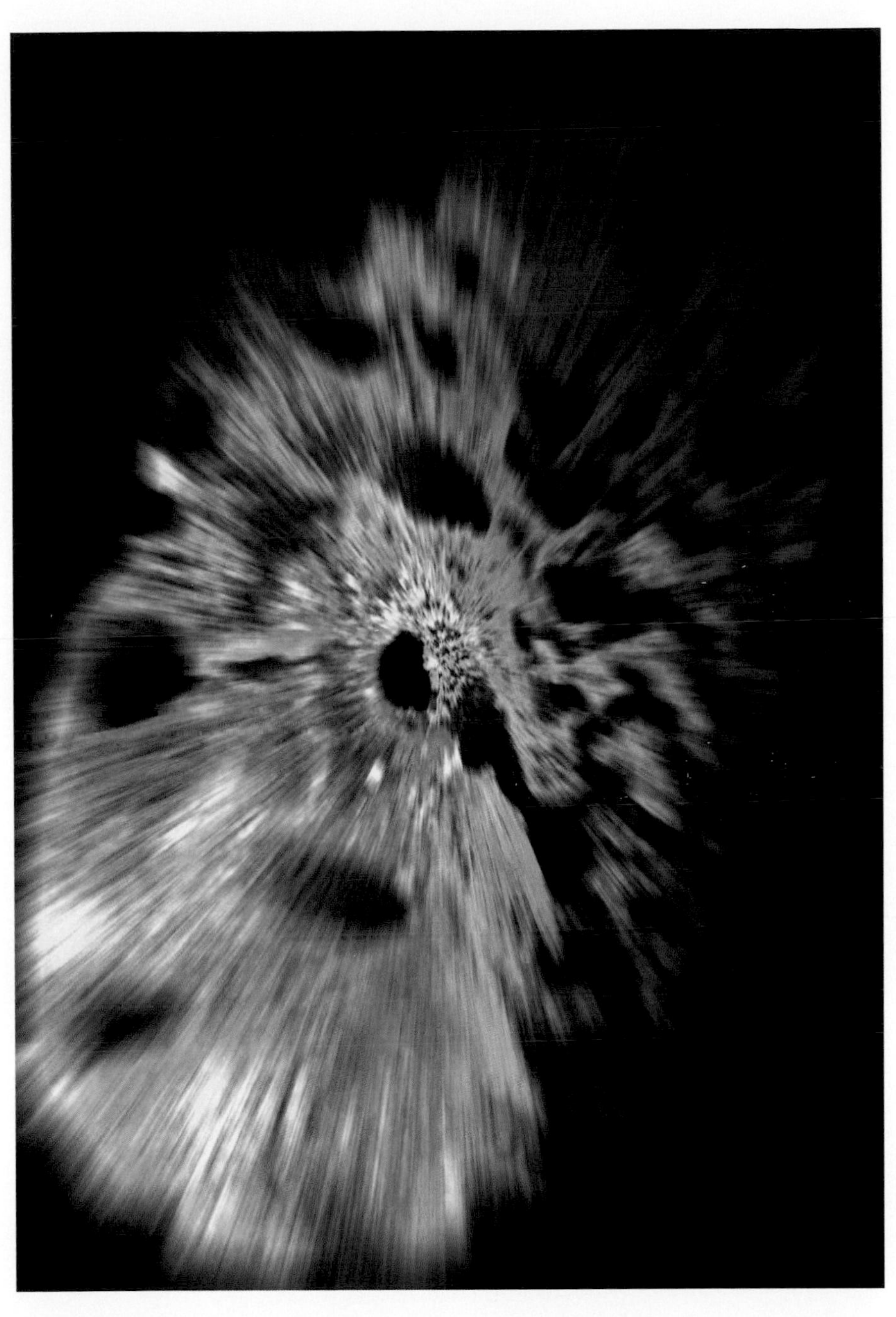

NancFireball

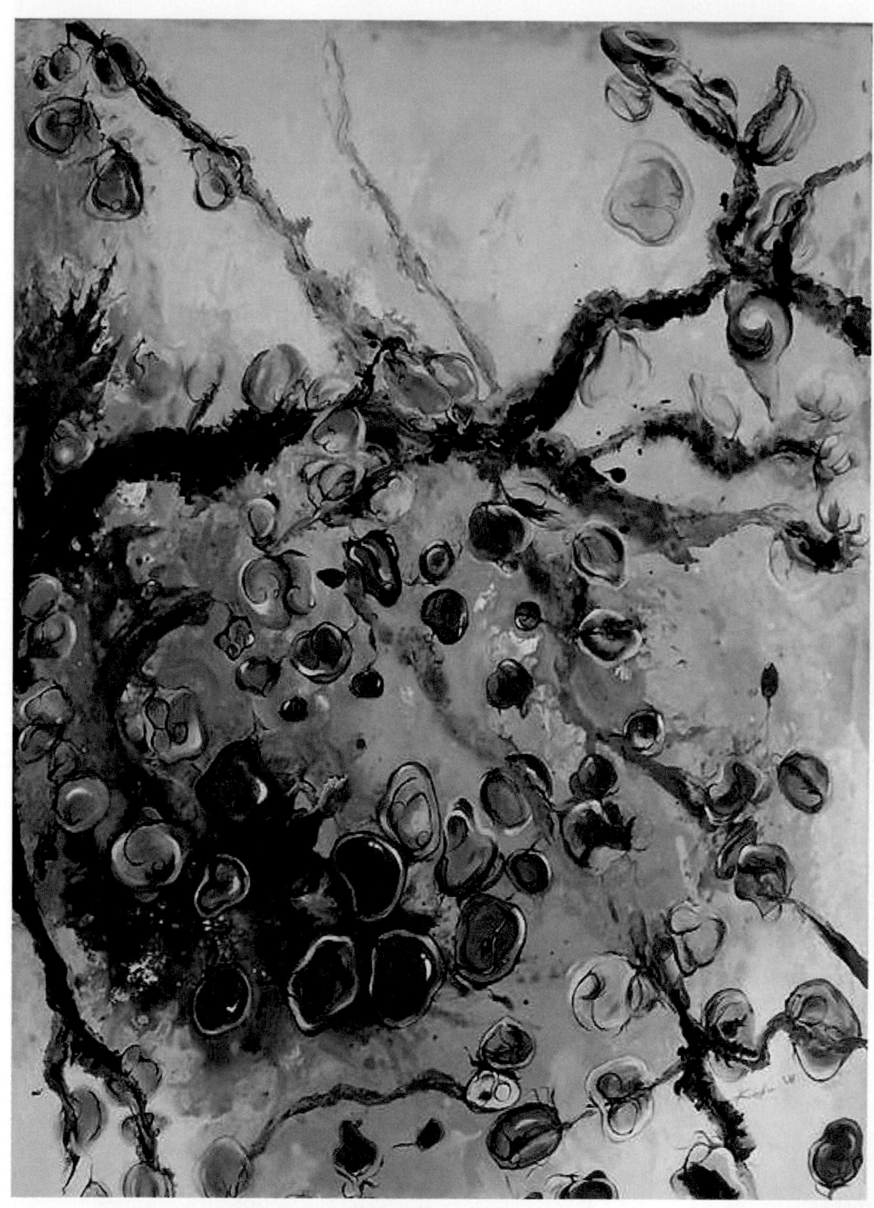

Periodic Table

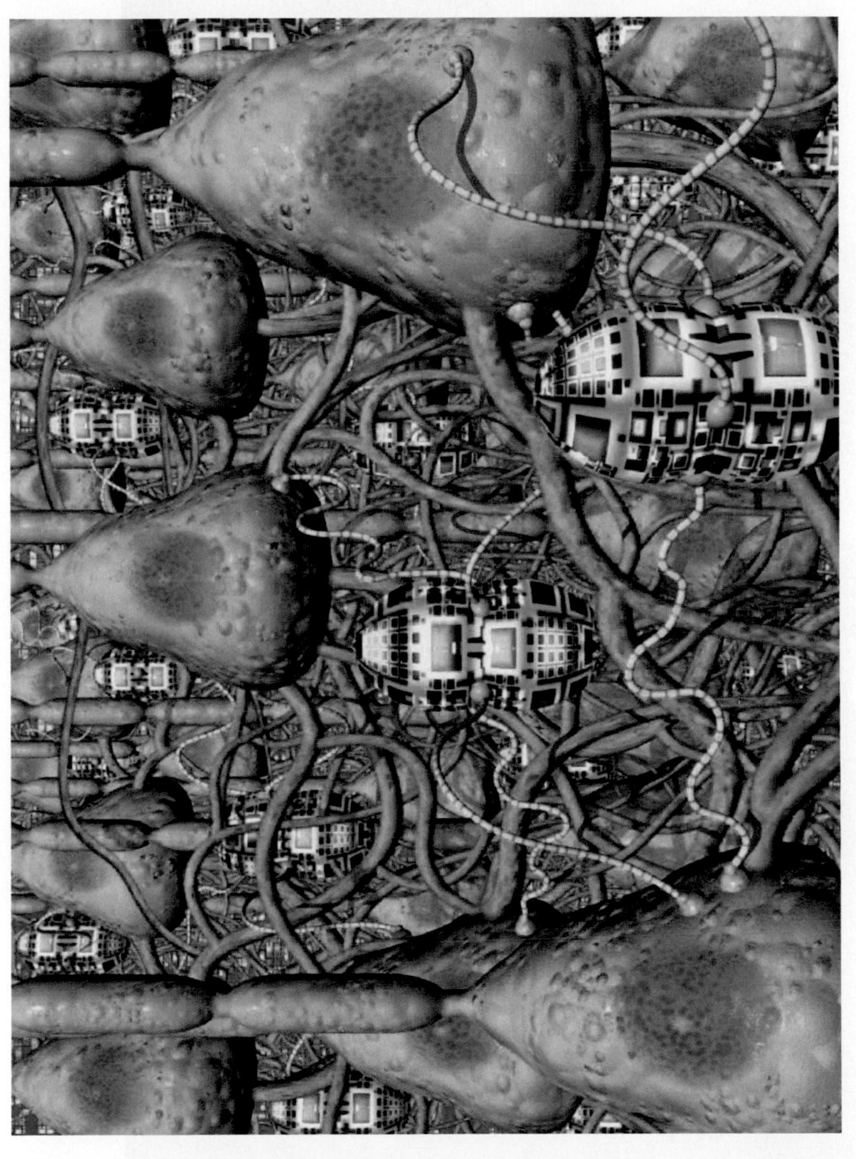

Brainbots

The Metamorphosis of Forms

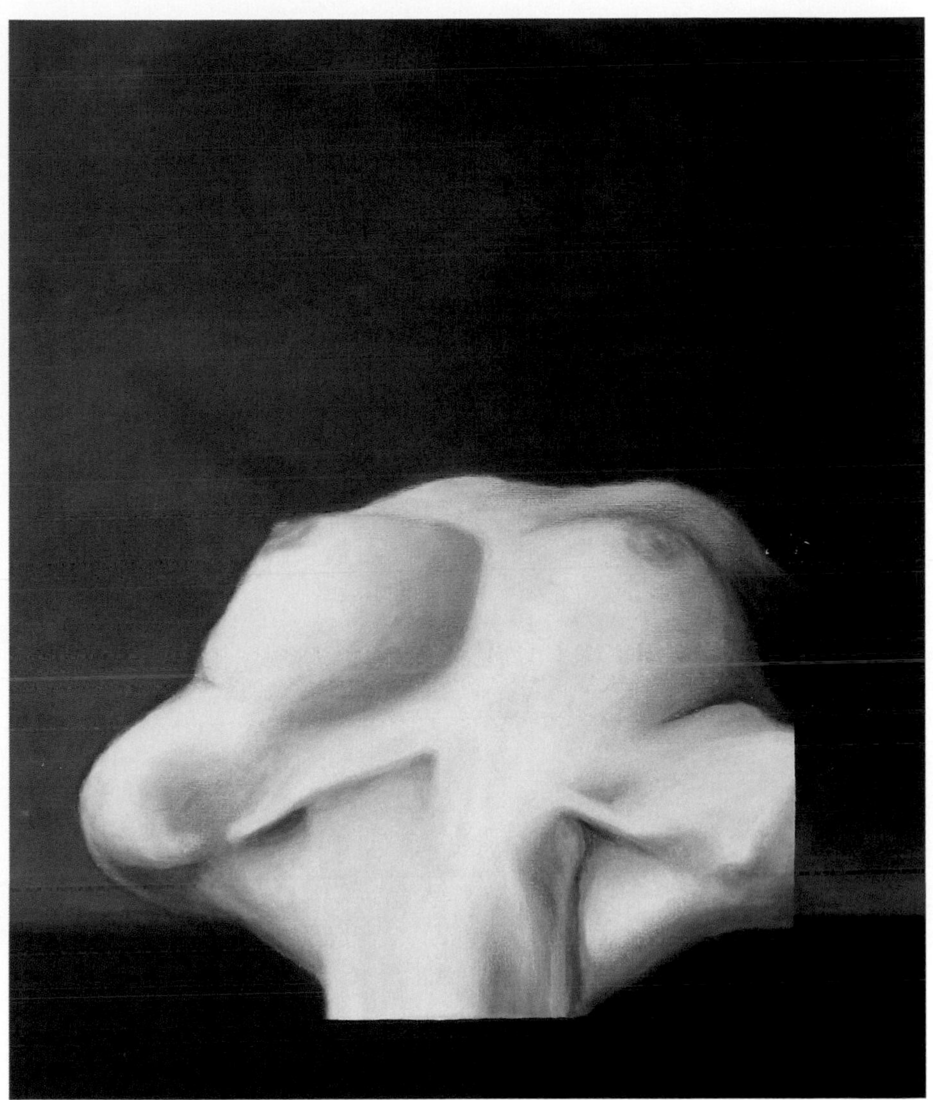

The Crucifixion of Nemesis

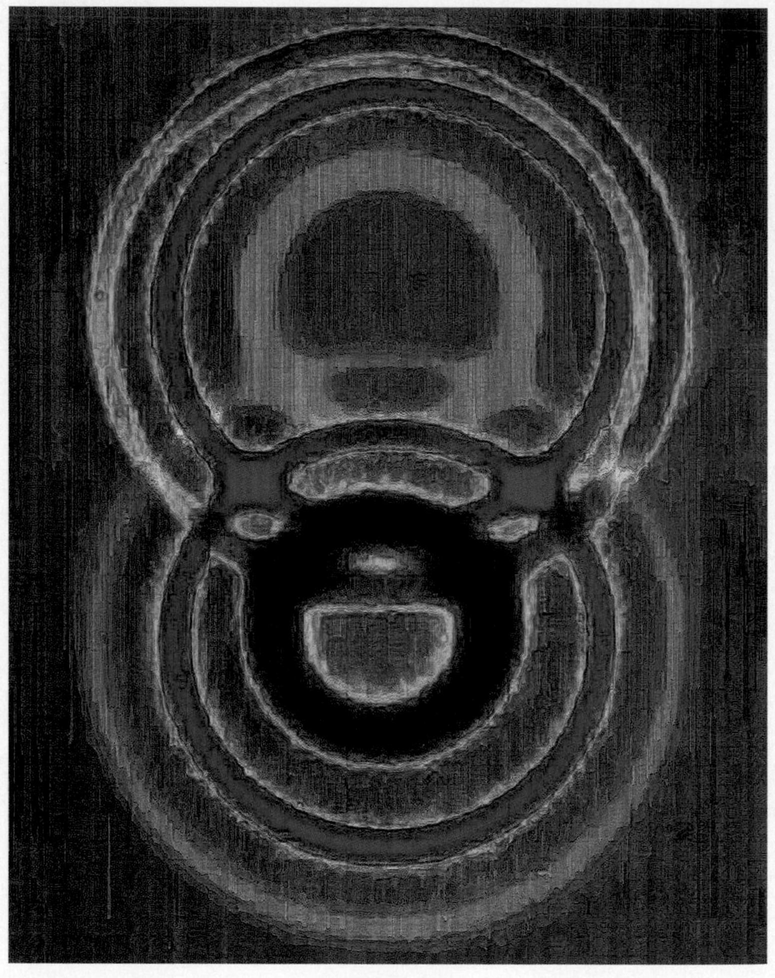

Entanglio

Quioazander

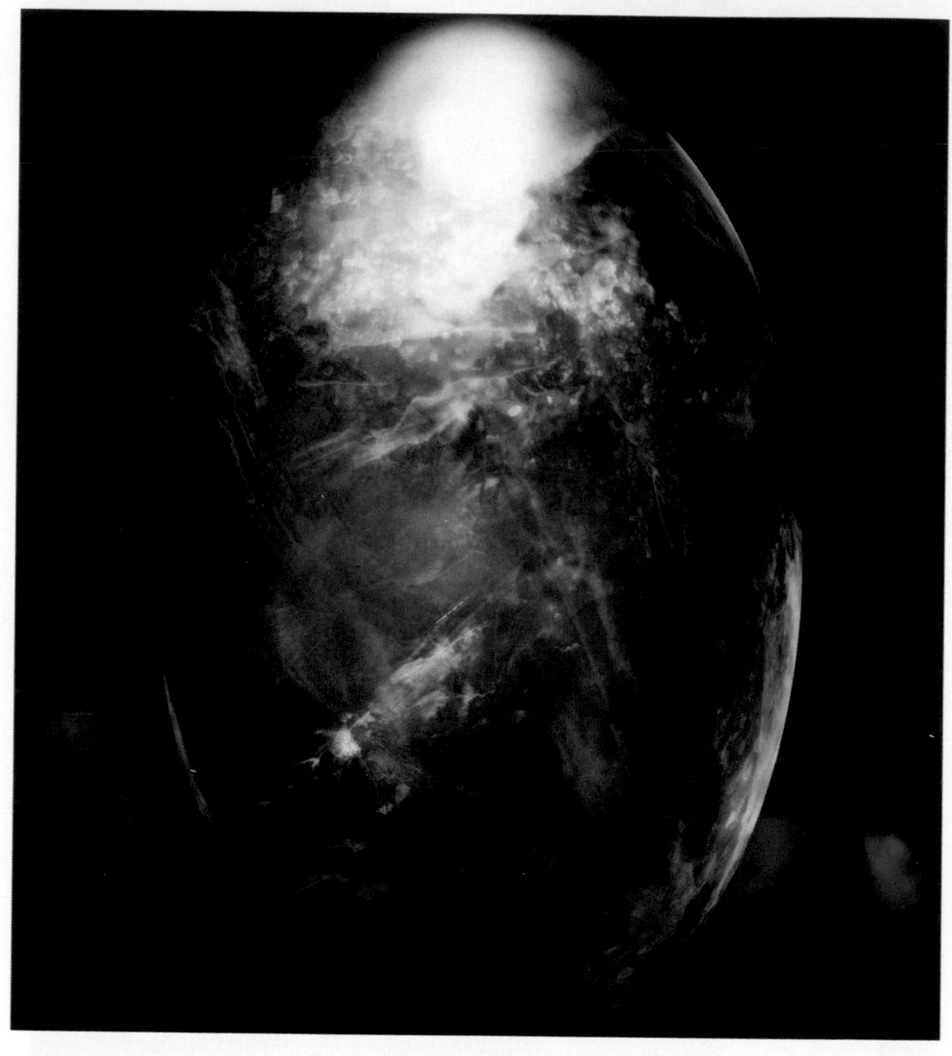

Discovery,

Chapter Figures

Chapter 3, Figure 1

Chapter 3, Figure 2

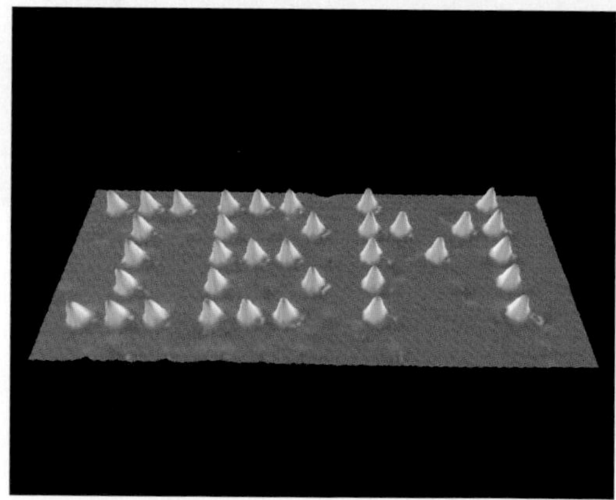

Chapter 3, Figure 3

Chapter 7, Figure 1

Chapter 7, Figure 2

Chapter 7, Figure 3

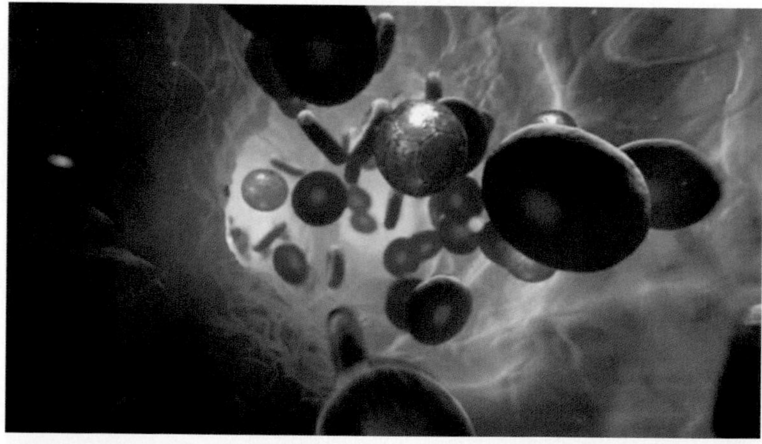

Chapter 7, Figure 4

Chapter 7, Figure 5

Chapter 7, Figure 6

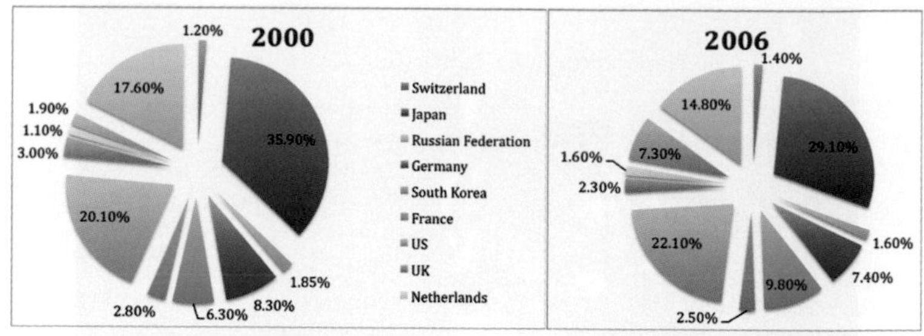

Share of countries in total patents grants

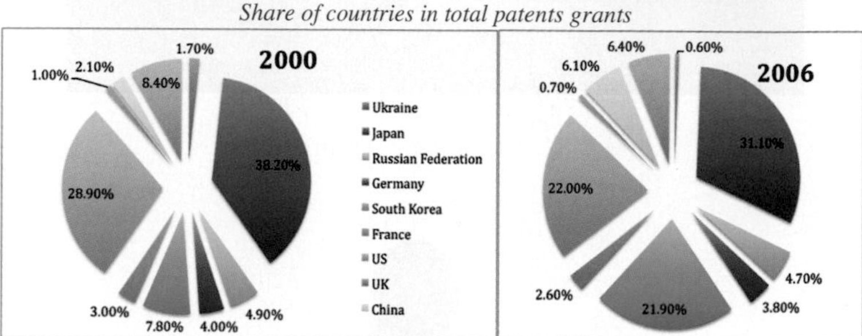

Figure 2

* WIPO's data breaks down the amounts for each European country, a procedure that dilutes Europe's role on patenting tendencies worldwide.

Source: WIPO (2008). *World Patent Report: A Statistical Review 2008*. Switzerland.

Chapter 16, Figure 2

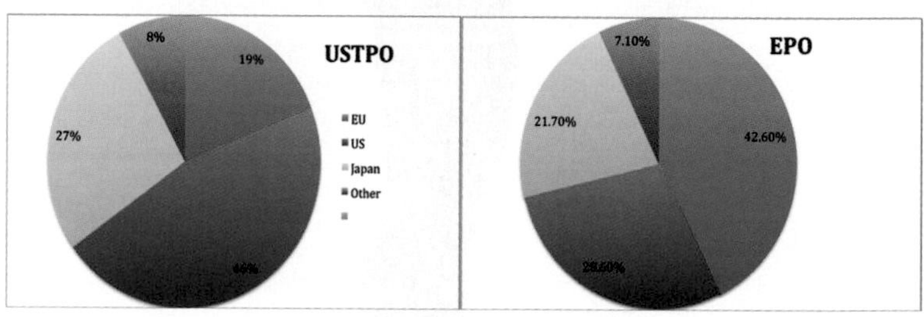

Figure 3

Source: Glänzel *et al* (2003). *Nanotechnology Analysis of an Emerging Domain of Scientific and Technological Endeavour.* Steunpunt O&O Statistieken. Belgium. July: 46.

Chapter 16, Figure 3

Main Nanotechnology Application Areas (2007)

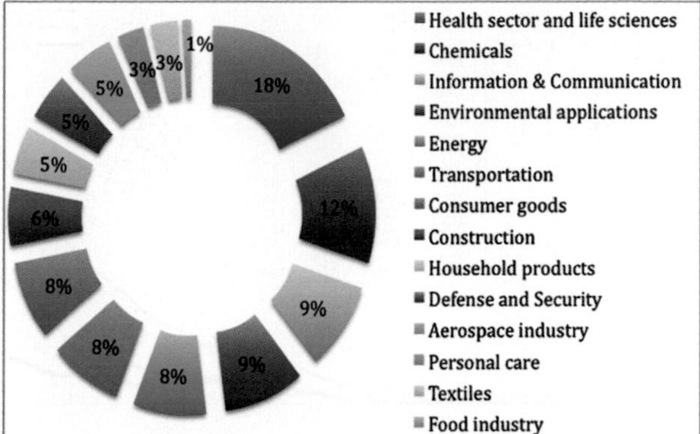

Figure 4

Source: Nanoposts (2007). *Government Policy and Initiatives in Nanotechnology Worldwide 2007*. Canada: 39.

Chapter 16, Figure 4

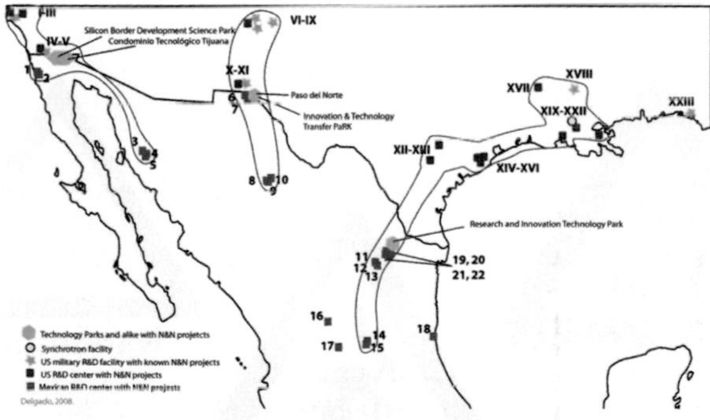

Feasible bi-national N&N corridors

I. University of California – Los Angeles / Integrated Nanosystems Research Facility at Irvine / NASA's Institute for Cell Mimetic Space Exploration; II. Caltech; III. Naval Air Warfare Centre Weapons – Point Mugu / Naval Surface Warfare Centre – Seal Beach / US Air Force Space Missile Centre – El Segundo; IV. University of California – San Diego; V. Space and Naval Warfare Systems Centre – San Diego; VI. Centre for Integrated Nanotechnologies at Alamos National Laboratory; VII. Centre for Integrated Nanotechnologies at Sandia National Laboratory); VIII. Kirtland Airforce Research Laboratory; IX. University of New Mexico; X. New Mexico State University –Las Cruces; XI. White Sands Missile Range – Army Research Laboratory; XII. University of Texas – Austin; XIII. Texas A&M University; XIV. Rice University; XV. Baylor College of Medicine; XVI. University of Houston; XVII. Louisiana Tech University; XVIII. Engineer Research and Development Centre – US Army Corps of Engineers; XIX. Louisiana State University – Baton Rouge; XX. University of New Orleans Research Park; XXI. University of Louisiana – Lafayette; XXII; Centre for Advanced Microstructures & Devices – Baton Rouge; XXIII. Naval Surface Warfare Centre – Panama City, FL / Air Armament Centre.

1. Centro de Nanociencias y Nanotecnología UNAM; 2. Centro de Investigación Científica y de Educación Superior de Ensenada; 3. Universidad de Sonora; 4. Centro de In vestigación en Alimentación y Desarrollo, A.C.; 5. Instituto Tecnológico de Hermosillo; 6. Universidad Autónoma de Ciudad Juárez; 7. Instituto Tecnológico de Ciudad Juárez; 8. Instituto Tecnológico de Chihuahua; 9. Centro de Investigación en Materiales Avanzados, S. C; 10. Universidad Autónoma de Chihuahua; 11. Corporación Mexicanade Investigación en Materiales, S.A. de C.V: 12. Centro de Investigación en Química Aplicada; 13. Instituto Tecnológico de Saltillo; 14. Universidad Autónoma de San Luis Potosí; 15. Instituto Potosino de Investigación Científica y Tecnológica A.C; 16. Universidad Autónoma de Zacatecas; 17. Universidad Politécnica de Aguascalientes; 18. Instituto Tecnológico de Ciudad Madero; 19. Instituto de Innovación y Transferencia de Tecnología de Nuevo León; 20. Universidad Autónoma de Nuevo León; 21. Instituto Tecnológico de Estudios Superiores de Monterrey; 22. Universidad de Monterrey.

Chapter 16, Figure 6

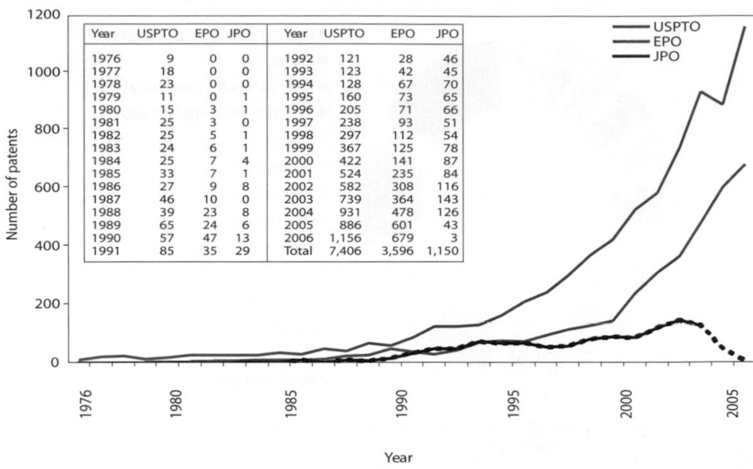

Year	USPTO	EPO	JPO	Year	USPTO	EPO	JPO
1976	9	0	0	1992	121	28	46
1977	18	0	0	1993	123	42	45
1978	23	0	0	1994	128	67	70
1979	11	0	1	1995	160	73	65
1980	15	3	1	1996	205	71	66
1981	25	3	0	1997	238	93	51
1982	25	5	1	1998	297	112	54
1983	24	6	1	1999	367	125	78
1984	25	7	4	2000	422	141	87
1985	33	7	1	2001	524	235	84
1986	27	9	8	2002	582	308	116
1987	46	10	0	2003	739	364	143
1988	39	23	8	2004	931	478	126
1989	65	24	6	2005	886	601	43
1990	57	47	13	2006	1,156	679	3
1991	85	35	29	Total	7,406	3,596	1,150

Chapter 20, Figure 3

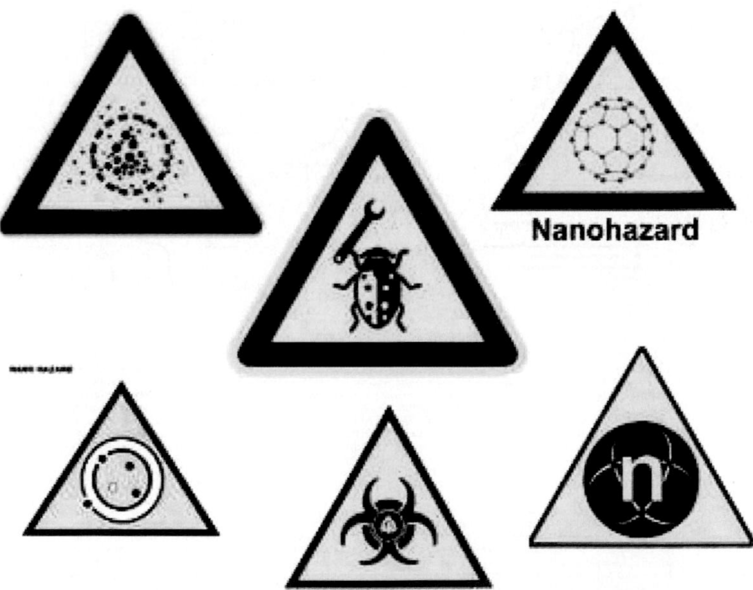

Chapter 20, Figure 6

Chapter 20, Figure 7